SOLID STATE POLYMERIZATION

SOLID STATE POLYMERIZATION

Edited by

CONSTANTINE D. PAPASPYRIDES
STAMATINA N. VOUYIOUKA

School of Chemical Engineering
National Technical University of Athens

WILEY

A JOHN WILEY & SONS, INC., PUBLICATION

Library of Congress Cataloging-in-Publication Data:

Papaspyrides, Constantine D.
  Solid state polymerization / Constantine D. Papaspyrides, Stamatina N. Vouyiouka.
    p. cm.
  Includes index.
  ISBN 978-0-470-08418-2 (cloth)
  1. Polymerization. 2. Solid state chemistry. I. Vouyiouka, Stamatina N. II. Title.
  TP156.P6P37 2009
  668.9′2–dc22

                                                                              2008051572

Printed in the United States of America

10 9 8 7 6 5 4 3 2 1

To my wife Dida
—CDP

To my mother Anastasia
—SNV

CONTENTS

CONTRIBUTORS

U. S. Agarwal, Reliance Technology Center, Reliance Industries Limited, Patalganga, Raigad, Maharashtra, India

W. H. Boon, Reliance Technology Center, Reliance Industries Limited, Patalganga, Raigad, Maharashtra, India

Tokimitsu Ikawa, 2-6-36, San-yo, Akaïwa City 709-0827, Okayama Prefecture, Japan

Saleh A. Jabarin, University of Toledo, Polymer Institute, Toledo, Ohio

Y. A. Liu, SINOPEC/FPCC/Aspen Tech Center of Excellence in Process Systems Engineering, Department of Chemical Engineering, Virginia Polytechnic Institute and State University, Blacksburg, Virginia

V. M. Nadkarni, Reliance Technology Center, Reliance Industries Limited, Patalganga, Raigad, Maharashtra, India

C. D. Papaspyrides, Laboratory of Polymer Technology, School of Chemical Engineering, National Technical University of Athens, Zographou, Athens, Greece

Rudolf Pfaendner, Ciba Lampertheim GmbH, Lampertheim, Germany

F. Pilati, Department of Materials and Environmental Engineering, University of Modena and Reggio Emilia, Modena, Italy

Kevin C. Seavey, SINOPEC/FPCC/Aspen Tech Center of Excellence in Process Systems Engineering, Department of Chemical Engineering, Virginia Polytechnic Institute and State University, Blacksburg, Virginia

M. Toselli, Department of Applied Chemistry and Material Science, University of Bologna, Bologna, Italy

S. N. Vouyiouka, Laboratory of Polymer Technology, School of Chemical Engineering, National Technical University of Athens, Zographou, Athens, Greece

S. A. Wadekar, 204, Mohandeep Society, Almeida Road, Chandanwadi, Panchpakhadi, Thane (West), State-Maharashtra, India

Haibing Zhang, University of Toledo, Polymer Institute, Toledo, Ohio

PREFACE

Solid state polymerization (SSP) constitutes a valuable technique used industrially mainly for condensation polymers such as polyamides and polyesters. It involves heating the starting material in an oxygen-free atmosphere (i.e., under flowing gas or high pressure or in vacuo), at a temperature below the melting point, increasing the molecular weight of the product while the material retains its solid shape. Its advantages over conventional melt-phase operations are the low polymerization temperatures, eliminating decomposition and undesirable by-product formation, the simplicity and low cost of process equipment, and the less expensive catalyst systems required.

SSP technology dates from 1940 and the first relevant patents of Flory (1939) and Monroe (1962) issued in an increase in prepolymer molecular weight via reactions in the solid state. Since then, the industrial scale for polyamides has been expanded to include PA 66, PA 6, and polyesters such as PET, and their use in overall polymerization layouts is often stated as drying or finishing. In parallel to its application, extensive research is on going in universities and in industry to understand the reaction mechanism and to optimize the process, especially with regard to the low reaction rates and sintering problems. The amount of open literature, especially patents, has increased steadily since 1995, and its investigation has spread to the majority of issues of most immediate concern, since it is an indispensable part of polymer production lines.

The answer to the question "Why study SSP in 2009?" is that the value of SSP lies beyond its obvious role as an extension of conventional polymerization techniques. It can also be used as a recycling method, through which the molecular weight of the postconsumer material is increased, thus permitting processing without severe recycled material deterioration. It offers a feasible tool with which to investigate polymer behavior during polymerization or subsequent processing, due to its simplicity regarding technical requirements. For example, the catalytic performance of organic substances can be tested primarily in SSP and extrapolated further with regard to activity in the melt-phase reaction. Even more, the mechanisms prevailing in a solid-phase reaction can be used and interpreted to overcome a series of problems during conventional polymerization and processing.

In our experience, SSP comprises an "exciting" technique for a student or researcher because it combines fundamentals of polymer science, chemistry, physical chemistry, and engineering. In Chapter 1, the complexity of the process is highlighted and how it involves both chemical and physical steps combined with process mechanisms and apparatus is discussed. These SSP steps are found to control the reaction rate separately and/or jointly, depending on the starting material type and process conditions (i.e., temperature, initial stoichiometry and crystallinity, reacting particle size distribution, condensate content in the surroundings, catalysts presence, etc.). The physical chemical aspects of the process are described further in Chapter 2, where the concept of end-group diffusion is indicated as being the primary difference from melt-phase technique. The reacting system morphology shows how it can influence SSP kinetics, and a model is constructed to predict molecular weight achieved. In the following chapters, SSP kinetics are investigated further, also considering other possible rate-controlling steps, such as chemical reactions and by-product removal. Polyesters and polyamides are examined as prevailing SSP polymers. Especially for polyamides, examples are given where SSP serves as a tool to investigate the effect of additives on polymerization rates.

As depicted in SSP kinetics, the slow process rate handled using catalysts and relevant systems is discussed in Chapter 5, covering metallic, nonmetalic, reactive, and inert additives. In Chapter 6 we describe a specialized application of SSP conducted under high pressure on a laboratory scale. It is shown how SSP can serve as a tool to examine the differences in a series of polyamide monomers and polymers in terms of monomer polymerizability, polyamide structure, and degree of orientation. Finally, Chapters 7 and 8 cover engineering aspects regarding process modeling and industrial application. The relevant knowledge can serve as a guide to develop SSP reactors and design pertinent plants.

Based on the book's structure, we have gathered and filtered the literature available on the SSP technique. It is hoped that the reader will find information not only to comprehend the pertinent polymerization technique, but also to initiate future investigations, since it is our belief that SSP processes consist of a continuously developing field.

Finally, we wish to cordially thank the indispensable contribution of each chapter author, with whom we enjoyed an excellent collaboration. Also, our warm thanks to the members of the staff at Wiley for all their help and guidance.

School of Chemical Engineering   CONSTANTINE D. PAPASPYRIDES
National Technical University of Athens   .   STAMATINA N. VOUYIOUKA
Athens, Greece
July 2008

1

FUNDAMENTALS OF SOLID STATE POLYMERIZATION

C. D. PAPASPYRIDES AND S. N. VOUYIOUKA

School of Chemical Engineering, National Technical University of Athens, Athens, Greece

Solid State Polymerization, Edited by Constantine D. Papaspyrides and Stamatina N. Vouyiouka
Copyright © 2009 John Wiley & Sons, Inc.

1.1. INTRODUCTION

1.1.1. Polymers and Plastics

Plastics are indispensable contemporary materials used in nearly all areas of daily life, and their production and fabrication are major worldwide industries. The key word related to plastics is *polymers*, organic substances that are built up by small molecules (monomers) joined together with covalent bonds and forming long carbon or heteroatom chains (macromolecules). There may be hundreds, thousands, ten thousands, or more monomers linked together in a polymer molecule, as noted by the Greek root of the word *polymer*, meaning "many parts." Plastics are highly modified polymeric materials that have been or can readily be formed or molded into useful shapes, and a typical commercial plastic resin may contain one or more polymers in addition to various additives and fillers.

Although humankind has used natural polymers, such as animal skins, silk, cellulose, and natural rubber since prehistoric times, the birth of plastics may be traced back to the mid-nineteenth century. More specifically, the true nature of polymer molecules was elucidated in 1920 by the pioneering work of Herman Staudinger, who proposed and defended the macromolecular structure of polymers, differing from the then- prevailing theory that polymers are colloids. Distinct milestones in plastics history (Fig. 1.1) include development of the vulcanization process by Charles Goodyear, transforming the latex of natural rubber to a useful elastomer for tire use; the invention of celluloid, the first

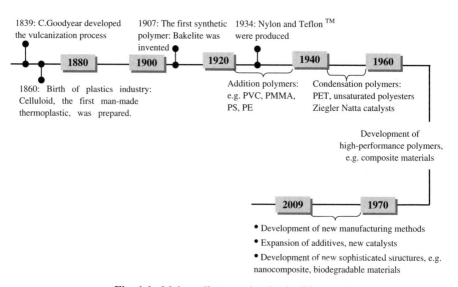

Fig. 1.1. Major milestones in plastics history.

humanmade thermoplastic, created from nitrocellulose and camphor by John Wesley Hyatt; the invention by Leo Baekeland of Bakelite (phenol–formaldehyde resin), which is used due to its excellent thermal and electric insulating properties; the preparation of nylon, an exotic polymer at that time, by Wallace Carothers, who was working at Du Pont during the 1920s and 1930s; and the development of low-pressure catalysts by Ziegler and Natta, which led to the commercialization of polypropylene as a major commodity plastic [1,2]. The last 30 years of plastics development involve progress in polymer chemistry as well as in modification, processing, and fabrication, allowing plastic materials to be shaped at low cost while achieving the desired functionality and characteristics.

The plastics industry is about the same size as that of metals, accounting globally for $120 billion in 2002 for the five major thermoplastics [polyethylenes, polypropylene (PP), polystyrene (PS), poly(vinyl chloride) (PVC), and poly(ethylene terephthalate) (PET)] [3]. Plastics are widely used in commodity products (e.g., textiles, tires, packaging) as well as for engineering materials in the transportation, electrical and electronic, medical, chemical, and biotech industries (Fig. 1.2). Therefore, it is not a false reflection to say that we live in a plastic world, where the characteristics and advantages of plastics render them preferred materials for consumer durables and constructive applications, based on careful end-product design and assessment of the pertinent manufacturing capabilities and costs.

The corollary of all the aforementioned applications is that global plastic consumption was 230,000 kilotons in 2005. In Europe it was estimated at 106 kg/capita, based on 2001 data, with a prediction of a 50% increase by 2010, and in the United States it was 121 kg/capita in 2001 with a 47% increase projected for 2010 [4]. Plastics segmentation in Western Europe for 2004 revealed that packaging is the biggest single sector, at 37% of total plastics demand, followed by building and construction applications at 20%. In addition to automotive uses (7.5%) and E&E applications (7.5%), there is also a large range of other applications (29%), including agriculture, household, medical devices, toys, leisure, and sports goods [5].

1.1.2. Polymerization Processes

Polymers, the core of plastics, are prepared by a process known as *polymerization*, which involves the chemical combination of monomers. The polymerization mechanism is widely used as a criterion for polymer classification. Accordingly, during the development of polymer science, polymers were classified as either condensation or addition, based on structural and compositional differences, a scheme attributed to W. Carothers, the inventor of nylon. Condensation polymers are formed from polyfunctional monomers through various organic chemistry reactions, resulting in joining the repeating units by bonds of one type, such as ester, amide, urethane, sulfide, and ether linkages. In most cases, the condensation reaction is followed by the liberation of small molecules (by-product,

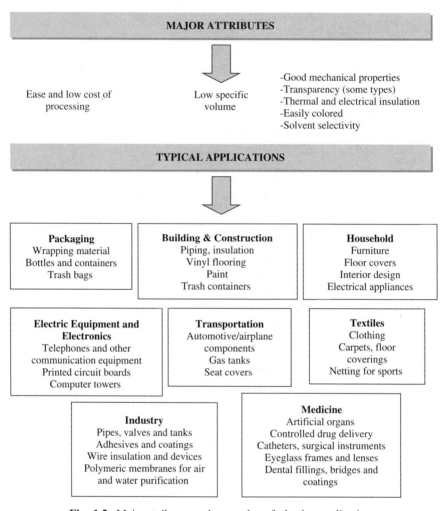

Fig. 1.2. Main attributes and examples of plastics applications.

condensate), such as water in the case of polyamides. On the other hand, addition polymers are formed from monomers without the loss of a small molecule; consequently, the repeating unit has the same composition as the monomer and does not contain functional groups as part of the backbone chain. Typical examples of condensation and addition polymers are depicted in Table 1.1.

Some years later than Carothers, Flory [6] highlighted the variations in the mechanism through which polymer molecules are built up, distinguishing between step- and chain-growth polymers. *Step-growth polymers* are formed by the stepwise reaction between functional groups of the reactants, and the size of polymer molecules increases at a relatively low rate, proceeding slowly from monomer to dimer, trimer, tetramer, pentamer, and so on. As a result,

TABLE 1.1. Typical Condensation and Addition Polymers

Condensation Polymers

Polymer	Characteristic Linkage	Indicative Repeating Units	
Polyamides (PAs)	$\overset{O}{\overset{\|}{-C-N-}}$ $_H$	Poly(tetramethylene oxamide) (PA 42) — [HN(CH$_2$)$_4$NHCOCO]— Poly(tetramethylene adipamide) (PA 46) — [HN(CH$_2$)$_4$NHCO(CH$_2$)$_4$CO]— Poly(hexamethylene adipamide) (PA 66) — [HN(CH$_2$)$_6$NHCO(CH$_2$)$_4$CO]— Poly(hexamethylene sebacamide) (PA 610) — [HN(CH$_2$)$_6$NHCO(CH$_2$)$_8$CO]— Poly(dodecamethylene adipamide) (PA 126) — [HN(CH$_2$)$_{12}$NHCO(CH$_2$)$_4$CO]— Poly(dodecamethylene sebacamide) (PA 1210) — [HN(CH$_2$)$_{12}$NHCO(CH$_2$)$_8$CO]— Polycaproamide (PA 6) — [HN(CH$_2$)$_5$CO]— Polyundecanamide (PA 11) — [HN(CH$_2$)$_{10}$CO]— Polydodecanamide (PA 12) — [HN(CH$_2$)$_{11}$CO]—	Linear PAs (nylons)

Poly(m-phenylene isophthalamide) (Nomex)

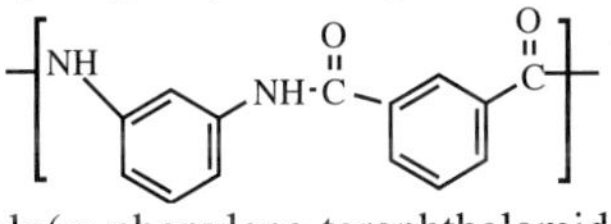

Poly(p-phenylene terephthalamide) (Kevlar)

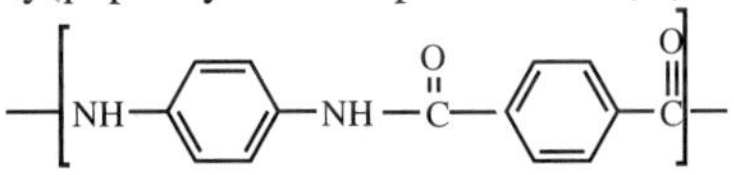

Polyesters (PEs)	$\overset{O}{\overset{\|}{-C-O-}}$	Poly(ethylene terephthalate) (PET) 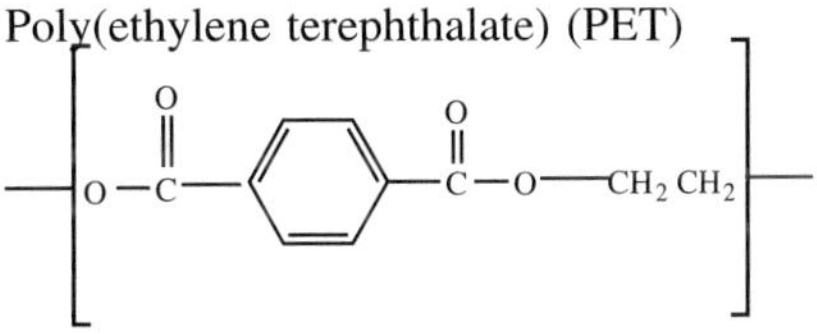

Poly(butylene terephthalate) (PBT)

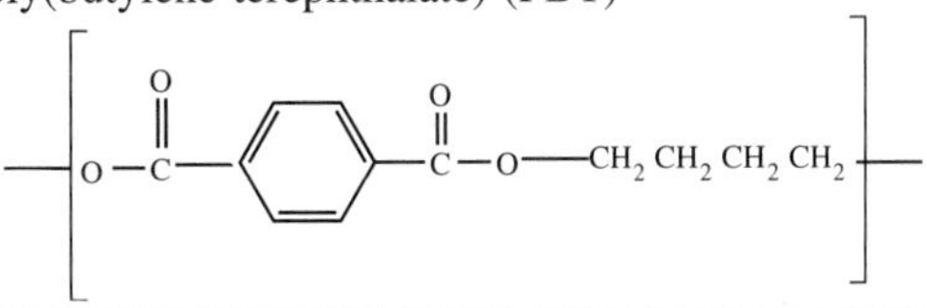

(Continued overleaf)

TABLE 1.1. (*Continued*)

Condensation Polymers		
Polymer	Characteristic Linkage	Indicative Repeating Units

Poly(ethylene naphthalate) (PEN)

Poly(dimethylene cyclohexane terephthalate) (PCT)

Polycarbonate (PCs)	$-O-\overset{O}{\underset{\|}{C}}-O-$	Poly(bisphenol A carbonate) (BPA-PC)
Polyurethane (isocyanate polymers)	$-O-\overset{O}{\underset{\|}{C}}-\overset{}{\underset{H}{N}}-$	Polyurethane (Perlon U or Igamid U)

Addition Polymers	
Polymer	Structure
Polyethylene (PE)	$-\!\!\left[CH_2CH_2\right]\!\!-$
Polypropylene (PP)	
Poly(vinyl chloride) (PVC)	
Poly(vinyl alcohol) (PVOH)	

TABLE 1.1. (*Continued*)

Addition Polymers	
Polymer	Structure
Poly(vinyl acetate) (PVA)	$$\left[CH_2 - \overset{\displaystyle H}{\underset{\displaystyle \underset{\underset{\underset{CH_3}{|}}{\overset{C=O}{|}}}{O}}{\underset{|}{C}} \right]$$
Polystyrene (PS)	$$\left[CH_2 - \overset{\displaystyle H}{\underset{\displaystyle C_6H_5}{\underset{|}{C}}} \right]$$
Poly(methyl methacrylate) (PMMA)	$$\left[CH_2 - \overset{\displaystyle CH_3}{\underset{\displaystyle \underset{\underset{CH_3}{|}}{\overset{C=O}{\underset{O}{|}}}}{\underset{|}{C}}} \right]$$

high-molecular-weight polymer is formed only near the end of polymerization (i.e., at high monomer conversion, typically greater than 98%). *Chain-growth polymers* are prepared in the presence of an initiator, so as to provide reacting species (e.g., free radicals, cations, or anions), which act as reaction centers, and polymerization occurs by successive additions of a large number of monomer molecules in the chain, usually over a short period.

According to Flory's theory [6], step-growth polymerization kinetics can be described as either second or third order, assuming equal reactivity of the functional end groups (i.e., the intrinsic reactivity of all reactive moieties is constant and independent of molecular size). The overall reaction rate is expressed as a rate of decrease in the monomer concentration, and the reaction order depends on whether a catalyst is involved. For a bimolecular stepwise equilibrium reaction $a-A-a/b-B-b$ type (functional groups a and b) [equation (1.1)], the reaction rate is given by (1.2) and (1.3) for catalyzed (second order) and uncatalyzed (third order) polymerization, respectively. It should be noted that certain step-growth polymerizations are self-catalyzed, so third-order kinetics indicate that one of the functional groups exhibits catalytic behavior [e.g., group b in (1.3)], and thus its effect on polymerization must be included in the rate equation. In addition, the reverse (depolymerization) reaction term can be ignored when the condensation by-product is removed continuously as it is formed.

$$na - A - a + nb - B - b \underset{k_r}{\overset{k_f}{\rightleftharpoons}} a - [AB]_n - b + (2n - 1)ab \qquad (1.1)$$

$$r = -\frac{d[a]}{dt} = -\frac{d[b]}{dt} = k_f[a][b][catalyst] - k_r[AB][ab] \xrightarrow{[ab] = 0} -\frac{d[b]}{dt}$$
$$= k_2[a][b] \qquad (1.2)$$

$$r = -\frac{d[a]}{dt} = -\frac{d[b]}{dt} = k_f[a][b][b] - k_r[AB][ab] \xrightarrow{[ab] = 0} -\frac{d[b]}{dt}$$
$$= k_3[a][b]^2 \qquad (1.3)$$

where k_f is the rate constant for polymerization, $k_r(k_f/K_{eq})$ the rate constant for depolymerization, K_{eq} the equilibrium constant, $k_2 = k_f$ [catalyst] the rate constant for catalyzed polymerization (second order), k_3 the rate constant for uncatalyzed polymerization (third order), and [a] and [b] the functional group concentrations (i.e., [a] = 2[a − A − a] and [b] = 2[b − B − b], where [a − A − a] and [b − B − b] are the concentrations of the bifunctional monomers). Integration of (1.2) and (1.3) differs according to whether the functional groups are in equimolar stoichiometry, resulting in relevant kinetic expressions for catalyzed and uncatalyzed step-growth polymerization (Table 1.2).

TABLE 1.2. Kinetic Expressions for Step-Growth Polymerization

	Integrated Kinetic Expressions	Polymerization Conversion and Degree
	[a] = [b] = c	
Catalyzed reaction	$\frac{1}{c_t} - \frac{1}{c_0} = k_2 t,\quad \frac{1}{(1-p_t)} - 1 = \overline{X}_n - 1 = c_0 k_2 t$	$p_t = \frac{c_0 - c_t}{c_0}$
Uncatalyzed reaction	$\frac{1}{c_t^2} - \frac{1}{c_0^2} = 2k_3 t,\quad \frac{1}{(1-p_t)^2} - 1 = \overline{X}_n^2 - 1 = 2c_0^2 k_3 t$	$\overline{X}_n = \frac{1}{1-p_t}$

where k_2 and k_3 are the reaction rate constants for second- and third-order kinetics, c_0 the initial concentration of a or b groups, c_t the concentration of a or b groups at any given time t, $\overline{X}_n$ the number-average degree of polymerization, and p_t the polymerization conversion.

$$[a] \neq [b], [a] < [b], r = \frac{[a]_0}{[b]_0}$$

Catalyzed reaction	$\frac{1}{[b]_0 - [a]_0} \ln \frac{[b]_t [a]_0}{[b]_0 [a]_t} = k_2 t$	$p_t = \frac{[a]_0 - [a]_t}{[a]_0}$
Uncatalyzed reaction	$\frac{1}{([b]_0 - [a]_0)^2} \ln \frac{[b]_t [a]_0}{[a]_t [b]_0} - \frac{1}{[b]_0 - [a]_0}\left(\frac{1}{[b]_t} - \frac{1}{[b]_0}\right) = k_3 t$	$\overline{X}_n = \frac{1+r}{1+r-2rp_t}$

where $[a]_0$ and $[b]_0$ are the initial concentrations of groups a and b, and $[a]_t$ and $[b]_t$ are the concentrations of groups a and b at any given time t.

Chain-growth polymerization requires the presence of initiating species that can be used to attach monomer molecules in the beginning of polymerization. Free-radical, anionic, and cationic chain-growth polymerizations share three common steps: initiation, propagation, and termination, which are described by kinetics [7]. For instance, in the case of free-radical polymerization, initiation may be represented by two steps: the formation of radicals ($R^{\bullet}$) and the reaction of $R^{\bullet}$ with monomer (M). Due to the fact that the formation of free radicals is the slowest step and therefore rate controlling, the rate of initiation (r_i) is described through

$$I \xrightarrow{\text{slow}} nR^{\bullet}$$

$$R^{\bullet} + M \xrightarrow{\text{fast}} RM^{\bullet}$$

$$r_i = \frac{d[R^{\bullet}]}{dt} \quad \Rightarrow \quad -\frac{d[I]}{dt} = k_d[I] \tag{1.4}$$

where k_d is the dissociation rate constant and n is the number of free radicals $R^{\bullet}$ ($n = 1$ or 2) formed during the breakdown of one molecule of the initiator I.

The propagation step can be represented by a single general reaction, and the relevant rate expression is

$$M_i^{\bullet} + M \xrightarrow{k_p} M_{i+1}^{\bullet}$$

$$r_p = -\frac{d[M]}{dt} = k_p[M][M^{\bullet}] \tag{1.5}$$

where k_p is the propagation rate constant and $[M^{\bullet}]$ is the total concentration of all radical species ($[M^{\bullet}] = \sum_{i=1}^{\infty} [M_i^{\bullet}]$).

Termination occurs through combination or disproportionation and the reaction rate is, respectively,

$$r_t = -\frac{d[M^{\bullet}]}{dt} = k_t[M^{\bullet}]^2 \tag{1.6}$$

where k_t is the overall rate constant for termination.

At steady-state conditions, ($d[R^{\bullet}]/dt = -d[M^{\bullet}]/dt$), the rate of polymerization is

$$r_p = -\frac{d[M]}{dt} = k_p[M]\left(\frac{r_i}{k_t}\right)^{1/2} = k_p[M]\left(\frac{k_d[I]}{k_t}\right)^{1/2} \tag{1.7}$$

Apart from polymerization mechanism, the technique applied also plays a fundamental role in polymer industry. There are four commonly used methods for performing polymerization: bulk, solution, suspension, and emulsion (Fig. 1.3). Meanwhile, polymers can be also prepared in the gas or vapor state, in a plasma environment, and in the solid phase.

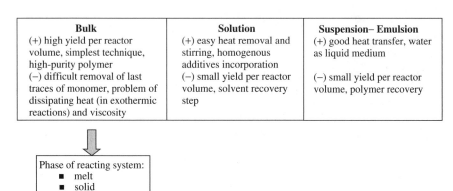

Bulk	Solution	Suspension– Emulsion
(+) high yield per reactor volume, simplest technique, high-purity polymer (−) difficult removal of last traces of monomer, problem of dissipating heat (in exothermic reactions) and viscosity	(+) easy heat removal and stirring, homogenous additives incorporation (−) small yield per reactor volume, solvent recovery step	(+) good heat transfer, water as liquid medium (−) small yield per reactor volume, polymer recovery

Phase of reacting system:
■ melt
■ solid

Fig. 1.3. Polymerization techniques: advantages and drawbacks.

Solid state polymerization (SSP), the book's primary topic, is associated with the bulk (or mass) method and is a widely used technique, especially for condensation polymers. When high degrees of polymerization are required, it is used as an indispensable extension of the melt or solution technique.

1.1.3. Introduction to Solid State Polymerization

Solid state polymerization (SSP) comprises a subcase of bulk polymerization techniques, used for both step- and chain-growth polymers with strong industrial interest in condensation polymers. The fundamental principle of the technique involves heating the starting material in an inert atmosphere or in vacuo, at a temperature below its melting point but permitting the initiation and propagation of typical polymerization reactions.

Dry monomers can be submitted to solid state polymerization as well as solid prepolymers (i.e., low-molecular-weight polymers derived from conventional polymerization techniques). The former process is usually referred as *direct SSP*; meanwhile, in the latter case, post-SSP (SSP finishing) is used to further increase the molecular weight and to improve processability and end-product properties, respectively [8–10]. To the same perspective, SSP is proved to be an efficient recycling technique [11,12], through which the molar mass of the postconsumer material is increased, thus permitting processing without severe recycled material deterioration.

The open literature on SSP dates from 1960 (Fig. 1.4); meanwhile, in-house industrial research and development are also being performed, covering all possible topics of the process: namely, chemistry, chemical physics, and process engineering aspects. Based on the SSP references histogram, it can be seen that the years of high publication activity belong to the period 1960 through 1977. However, since then, SSP still consists of a continuous investigation topic, presenting an increasing number of patents as the years go by, especially after 1995. Additionally, SSP expands to different contemporary peak research issues, such as

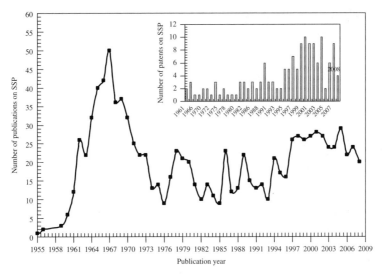

Fig. 1.4. History of the number of publications (patents and journals) on solid state polymerization. (Data from SciFinder Scholar, 2006.)

catalysts and nanocomposites, often serving as a tool to investigate compositional effects and materials behavior as well as to improve end-product performance.

At this point it should be emphasized that SSP industrial application refers primarily to step-growth polymer production, with polyamides (PAs) and polyesters (PEs) being the predominant SSP products. Based on its expansion regarding condensation polymers, the current book focuses on solid state polycondensation; meanwhile, main issues of solid state polyaddition are presented briefly in the following section.

1.2. SOLID STATE POLYMERIZATION OF CHAIN-GROWTH POLYMERS (SOLID STATE POLYADDITION)

Chain-growth polymerization can be carried out in liquids or solutions and also in the solid phase, by irradiating monomers either at the polymerization temperature or at a lower temperature, and subsequently warming the sample to obtain a postradiation reaction. Radiation-induced SSP has been studied since 1960 [13–15], and the relevant advantages are first that a range of active species may be produced so that the chances of successfull induction of polymerization are enhanced. Second, active centers can be formed throughout a large sample of monomer, which is not always possible with other techniques.

Pertinent monomers submitted to radiation-induced SSP involve vinyl or cyclic compounds, such as acrylamide [16–19], vinyl carbazole [16,20], vinylpyrrolidone [20], *N*-vinylpyridinium salts [21,22], isobutene, butadiene [23], phenylacetylene [24], trioxane [25], diketene [26], β-propiolactone [27], and 3,3-bis(chloromethyl)cycloxabutane [28], without also excluding

TABLE 1.3. Cyclic Monomer Structures and Derived Polymer Units

	Monomer Structure	Polymer Repeating Unit
Trioxane ($T_m = 64°C$)		$\left(\!-CH_2\!-\!O\!-\!\right)_n$
Tetraoxane ($T_m = 120°C$)		$\left(\!-CH_2\!-\!O\!-\!\right)_n$
3,3-Bis(chloromethyl) cycloxabutane ($T_m = 18°C$)		
β-Propiolactone ($T_m = -33.4°C$)		
Diketene ($T_m = -6.5°C$)		

copolymerization. The aforementioned cyclic monomer structures and the derived polymer repeating units are depicted in Table 1.3.

An important aspect of radiation-induced SSP is the topotacticity achieved: the extent to which there is three-dimensional correspondence between the product and its host [29]. In this perspective, the first research efforts focused on controlling the initiation stage of the polymerization by monomer orientation within the crystal lattice, so as to increase the polymerization rate and to form crystalline stereospecific polymers. However, this was not found feasible in all cases. In vinyl monomers such as acrylamide, the formation of polymer involves a change in hybridization of the olefinic carbon atoms of the monomer, which in turn requires a change in the geometrical disposition of the substituent groups. In the perfect crystal, the available free volume is insufficient to allow these configurational changes to occur, so polymerization hardly takes place in the perfect regions of the crystal lattice. As a result, it is initiated and propagated at other favorable sites in the lattice (i.e., the defects), resulting, however, in amorphous polymer within the original crystal structure [30,31].

TABLE 1.4. Values of Exponent n for Typical Chain-Growth SSP

Monomer	Exponent n
Vinyl Monomers	
Acrylamide	0.8
Acrylonitrile	1.0
Styrene	1.07
2,4-Dimethyl styrene	0.9
N-Vinyl carbazole	1.0
N-Vinylpyrrolidone	0.6–0.7
Acetylene	1.25
Cyclic Monomers	
β-Propiolactone	1.0
Hexamethylene–cyclotrisiloxane	0.9
Carbonylic Compounds	
Formaldehyde	0.65
Acetaldehyde	1.0

Source: Papaspyrides [36].

On the contrary, in cyclic monomers, where the propagation proceeds through a ring-opening mechanism, the polymer obtained can be well oriented, since polymerization propagates along one of the crystallographic axis of monomer. For example, both trioxane and polyoxymethylene, formed by polymerization of this monomer, have hexagonal crystal structures and in partially polymerized crystals their c-axes are found to be parallel. The polymer formed is fibrous and the fibers are aligned in specific orientations with respect to the crystallographic axes of the parent monomer crystal [32].

Reaction temperature is of great importance for solid state polyaddition, and the polymerization rate presents its maximum value just below the melting point of the monomer [15,33,34]. Indicatively, the maximum SSP rate is cited at $-70°C$ for β-propiolactone and diketene, $10°C$ below its melting point for 3,3-bis(chloromethyl)oxetane ($T_m = 18°C$) and in the range 30 to $60°C$ for trioxane [27,35]. Arrhenius plots often show marked changes in slope; the initial rate increases rapidly with temperature up to the maximum value (very close to T_m), after which it is reduced rapidly as the crystal becomes more disordered and melts [15].

In conjunction with reaction temperature, the radiation dose rate also plays an important role, and the polymerization rate of an irradiated monomer is often expressed as a power (n) of the dose rate [15]. In the literature, there is no universal expression and the exponent n depends on the monomer structure (Table 1.4).

Finally, the *crystal structure* itself (i.e., impurities, modification, crystals) size exercises a strong effect on the relevant SSP process. For instance, acrylonitrile and trioxane polymerize faster as large crystals, in contrast to acrylamide, where small crystals favor the polymerization reaction [37]. Furthermore, if a monomer exists in more that one crystalline modification, each modification will exhibit different reactivity in the solid phase. For example, in the case of tributyl vinyl phosphonium bromide, the less stable form polymerizes faster than the stable form by a factor of about 2. This difference was attributed to the different collisal and steric factors governed by the crystal structures, different packing of the molecules in the two forms, leading to variation in the mobility of the molecules in the lattice and in the imperfections [38].

Overall, solid state polyaddition induced by radiation constitutes a valuable method for investigating the mobility of organic molecules within crystalline and amorphous solids, combined with the nature and distribution of defects and dislocations on their behavior.

1.3. SOLID STATE POLYMERIZATION OF STEP-GROWTH POLYMERS (SOLID STATE POLYCONDENSATION)

In condensation polymers, high molecular weights may be reached through solid state polymerization at temperatures between the glass transition and the onset of melting. The SSP starting materials can be dry monomers (e.g., polyamide salt, amino acid) as well as solid prepolymers. The SSP reactions are considered the same as in the melt method, obeying step-growth chemistry. Moreover, exchange reactions are suggested to provide a mechanism for end-group functionality, especially in the case of post-SSP.

In Figure 1.5, the condensation schemes of four important step-growth polymers, PA 6, PA 66, PET, and BPA-PC, are presented, emphasizing the fact that the pertinent reactions are reversible and may reach equilibrium unless the condensate is sufficiently removed from the reaction zone. In addition to the required chemical affinity of the reactants, physical factors strongly influence SSP processes, due to the restricted species mobility and the reversible character of condensation reactions [39–42]. Mass transfer phenomena interfere with the chemical reaction, referring to the reactive end groups and/or to the condensation by-products (Fig. 1.6). In total, there can be four rate-limiting steps: the intrinsic kinetics of the chemical reaction, the diffusion of functional end groups, the diffusion of the condensate in the solid reacting mass (interior diffusion), and the diffusion of the condensate from the surface of the reacting mass to the surroundings (surface diffusion).

Accordingly, one should consider that during SSP there can be end-group diffusion limitations, due to the restricted mobility, which is not the case in the melt or solution technique. These restrictions become more severe at long reaction times, when the functional species close to each other have already reacted, their concentration and distribution have been reduced locally, and the migration of unreacted chain ends becomes indispensable for the reaction to continue [43,44]. The concept of polymer end-group diffusion during SSP reactions is used in Chapter 2 to develop a model of molecular morphology and chain movement so as to explain critical SSP characteristics, such the decrease of SSP rate versus reaction time.

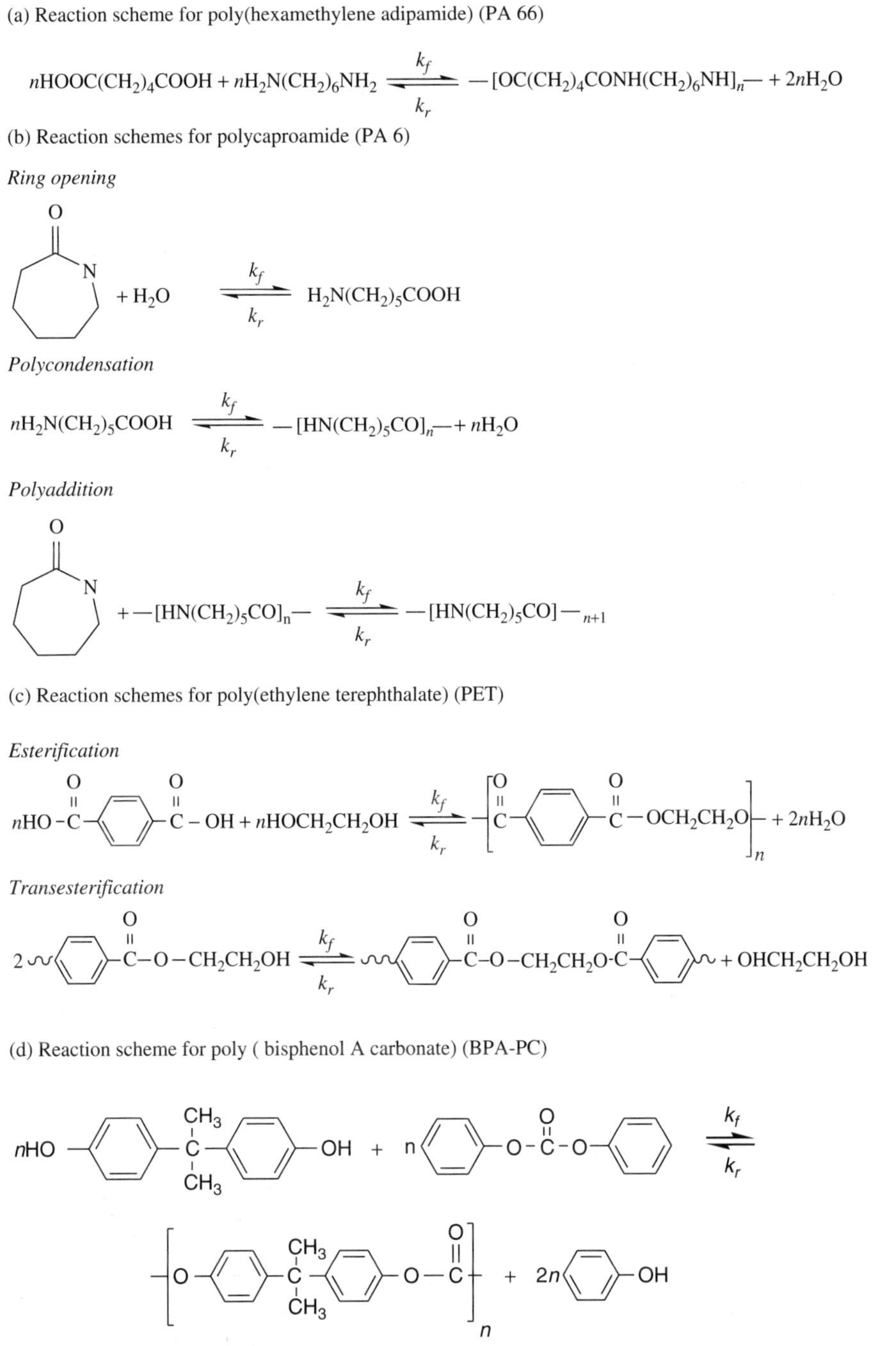

(a) Reaction scheme for poly(hexamethylene adipamide) (PA 66)

$$nHOOC(CH_2)_4COOH + nH_2N(CH_2)_6NH_2 \underset{k_r}{\overset{k_f}{\rightleftharpoons}} -[OC(CH_2)_4CONH(CH_2)_6NH]_n- + 2nH_2O$$

(b) Reaction schemes for polycaproamide (PA 6)

Ring opening

$$+ H_2O \underset{k_r}{\overset{k_f}{\rightleftharpoons}} H_2N(CH_2)_5COOH$$

Polycondensation

$$nH_2N(CH_2)_5COOH \underset{k_r}{\overset{k_f}{\rightleftharpoons}} -[HN(CH_2)_5CO]_n- + nH_2O$$

Polyaddition

$$+ -[HN(CH_2)_5CO]_n- \underset{k_r}{\overset{k_f}{\rightleftharpoons}} -[HN(CH_2)_5CO]-_{n+1}$$

(c) Reaction schemes for poly(ethylene terephthalate) (PET)

Esterification

$$nHO-C-\bigcirc-C-OH + nHOCH_2CH_2OH \underset{k_r}{\overset{k_f}{\rightleftharpoons}} \left[C-\bigcirc-C-OCH_2CH_2O \right]_n + 2nH_2O$$

Transesterification

$$2\sim\bigcirc-C-O-CH_2CH_2OH \underset{k_r}{\overset{k_f}{\rightleftharpoons}} \sim\bigcirc-C-O-CH_2CH_2O-C-\bigcirc\sim + OHCH_2CH_2OH$$

(d) Reaction scheme for poly (bisphenol A carbonate) (BPA-PC)

Fig. 1.5. Reaction schemes for typical condensation polymers.

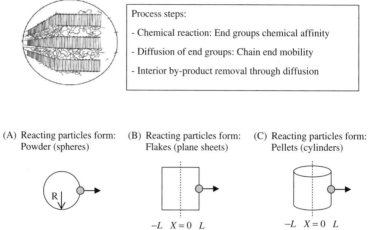

Process steps:

- Chemical reaction: End groups chemical affinity

- Diffusion of end groups: Chain end mobility

- Interior by-product removal through diffusion

(A) Reacting particles form: (B) Reacting particles form: (C) Reacting particles form:
      Powder (spheres)   Flakes (plane sheets)   Pellets (cylinders)

$-L$ $X = 0$ L $-L$ $X = 0$ L

Fig. 1.6. SSP chemical and physical steps occurring in the interior and on the surface of a reacting particle.

Moreover, the condensate removal through diffusion is a prerequisite for high SSP rates and it is distinguished into interior diffusion (i.e., inside the solid reacting particle) and surface diffusion (i.e., from the reacting particle surface to the surrounding atmosphere). Despite this theoretical distinction, interior and surface diffusion are interlinked and influenced by similar parameters, since both stages result in eliminating by-product concentration gradients in the reaction zone, preventing depolymerization and shifting the reaction equilibrium to the right.

Due to the wide application and investigation of solid state polycondensation, a more detailed description of relevant process conditions, mechanisms, and rate parameters follows. The two categories of starting materials, monomers and prepolymers, are examined separately, considering that the main interest in monomer SSP is restricted to laboratory scale; meanwhile, prepolymer SSP is already integrated in production processes.

1.3.1. Monomer Solid State Polymerization (Direct SSP)

There are many monomers capable of undergoing polymerization in the solid (crystalline) state. The monomer crystals react at a temperature lower than the melting point (T_m) of both monomer and polymer (Table 1.5) under inert gas or vacuum or high pressure. In many cases, the direct SSP reactions are topotactic, and the single-monomer crystals can be converted into polycrystalline polymer aggregates, permitting the preparation of highly oriented polymers. Polyamide monomers (salts and amino acids) SSP under high pressure are discussed in Chapter 6.

TABLE 1.5. Direct PA SSP Conditions[a] and Polymerization Conversion Achieved

Polymer	Starting Material	SSP Conditions
PA 46 [45]	Tetramethylenediammonium adipate (PA 46 salt, $T_m = 190°C$)	$T = 151°C$, $t = 7$ h, $p = 0.37$
PA 66 [45]	Hexamethylenediammonium adipate (PA 66 salt, $T_m = 192°C$)	$T = 151°C$, $t = 7$ h, $p = 0.27$
PA 610 [45]	Hexamethylenediammonium sebacate (PA 610 salt, $T_m = 171°C$)	$T = 138°C$, $t = 9.5$ h, $p = 0.12$
PA 126 [46]	Dodecamethylenediammonium adipate (PA 126 salt, $T_m = 151–152°C$)	$T = 138°C$, $t = 14$ h, $p = 0.90$
PA 1210 [45]	Dodecamethylenediammonium sebacate (PA 1210 salt, $T_m = 138°C$)	$T = 126°C$, $t = 8$ h, $p = 0.25$
PA 6 [47]	ε-Aminocaproic acid ($T_m = 210–212°C$)	$T = 170°C$, $t = 12$ h, $p = 0.98$

[a] T, reaction temperature; t, time; p, polymerization conversion.

Despite its laboratory scale, direct SSP presents considerable practical interest, since polymerization occurs from the beginning in the solid state, and consequently, all the problems associated with the high temperatures of melt technology (e.g., energy consumption and polymer degradation) are avoided.

a. Prevailing Mechanisms in Direct Solid State Polymerization　In many studies of direct SSP, polymerization is considered to follow the nucleation and growth model, according to well-known principles of solid state chemistry [48]. As an example, Frayer and Lando [49] proposed that the PA 66 salt polymerization can be considered to involve two stages, initiation and propagation. The initiation or nucleation stage can occur either on the surface of the crystallites or at internal surfaces within the crystallites, where crystal edges, defects, and impurities may act as active centers suitably orienting reacting species and facilitating molecular mobility. Similar is the picture from Macchi et al. [50], who heated single crystals of ε-aminocaproic acid and found that the kinetics of the process were characterized by three steps: an induction period, a subsequent stage in which monomer disappears at a constant rate, and finally, a slow step of polycondensation of polyamide chains after exhaustion of the monomer.

Apart from the crystal lattice initial characteristics, significant contribution in the nucleation stage also seems to have the reacting mass composition. More specifically, during the initial stages of SSP, the volatile hexamethylenediamine (HMD) escapes, as has been observed during the SSP of PA 66 and PA 610 salts [51–53] and also of different aromatic polyamide salts [54]. Several measures have been adopted during PA 66 production to treat HMD loss, such as introduction of the diamine from the beginning of the polycondensation reaction in an amount sufficient to counteract diamine loss (proposed $[NH_2]/[COOH]$ ratios $= 1.1$ to 2) [55], the use of nitrogen gas containing HMD, and finally, the decrease in reaction temperature to minimize HMD loss [56]. In addition, for the same reason, a two-step process has been proposed: first the reaction is carried out under autogenous conditions (pressurized system) to or beyond the point where diamine ends have reacted; subsequently, the system is vented so as to remove the water produced and to drive the condensation reaction toward higher degrees of polymerization [57,58].

In a recent study [51], it was even proved that the HMD loss preceded water formation and caused partial decomposition of the salt structure. This early evolution, combined with the theory of the nucleation-growth model, suggests an SSP mechanism. Considering nucleation, the crystal lattice and its characteristics, such as the size of the crystals, the number and type of lattice defects, and the presence of impurities, may significantly influence polymerization in the solid state. For instance, crystal edges and defects may in some cases inhibit the propagation of polymerization through physical separation of the polymerizing units, while in other cases they may act as active centers, since the orientation of the reacting species at the defective surfaces within the crystallites may differ and promote nucleation of the polymer phase (initiation stage). Impurities may act by creating lattice defects which subsequently affect polymerization, may act as a physical diluent to impede polymerization, or may facilitate molecular mobility and assist the polymerization [48]. Based on these well-known principles of solid state chemistry, the early evolution of HMD may be associated with the nucleation stage: The diamine volatilization results in creating new defective surfaces in the crystal lattice and in increasing the active centers for nucleation of the polymer phase generated, which is grown further following water formation.

Following nucleation, the growth stage often proceeds unexpectedly. In cases of hygroscopic monomers and depending on reaction temperature, a transition has been observed from the solid state to the melt state, dominating in the moderately organized salt structures of long diamines–diacids with high polar site concentrations. This phenomenon has been explained by Papaspyrides et al. [39,45,46,59,60] and correlated with the condensation water accumulation in the reacting mass. In particular, the water produced during the solid state reaction hydrates the polar groups of the reactant, and as the amount of water increases, the crystal structure of the salt is destroyed completely by the formation of highly hydrated regions. More specifically, the reaction begins at the defective sites of the monomer crystalline structure, being the active centers of the reaction [Fig. 1.7(a)]. For active centers up to or very near the grain surface, the water formed can easily be removed to the surrounding heating medium without affecting the reacting mass. On the contrary, in the inner grain, the water cannot be removed easily and hydrates the polar hydrophilic groups of the salt structure. In the case of low reaction rates (i.e., low rates of water formation), an organized accommodation of the by-product within the crystal structure is performed. As the amount of water accumulated increases, a "highly hydrated" area of monomer surrounds the active centers. This "highly hydrated" area has a lower melting point and soon falls in the melt state [Fig. 1.7(b)].

After the formation of these melt areas, the reaction proceeds mainly in the melt state and the rate is increased considerably, while the water accumulation

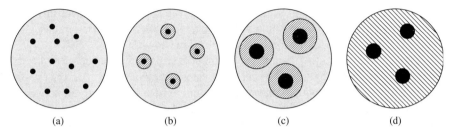

(a)   (b)   (c)   (d)

Fig. 1.7. Solid–melt transition phenomenon (SMT). •, Defects of the monomer crystalline structure; dark areas, polymer nuclei insoluble in water; shaded areas, highly hydrated and eventually, melt area. (From Kampouris and Papaspyrides [59] by permission of Elsevier.)

leads to an increase in the total melt area [Fig. 1.7(c)]. Eventually, these melt areas overlap, which explains the transition of the reaction from the solid to the melt state (SMT phenomenon) observed experimentally [Fig. 1.7(d)]. As the reaction proceeds, the molecular weight increases, the hygroscopicity of the reacting system decreases, and finally, the solid character of the system is restored.

Summarizing the prevailing direct SSP mechanisms, a more precise process definition is that direct SSP often follows the nucleation-growth model, and the monomer is transformed into polymer at a reaction temperature lower than the melting point of both monomer and polymer. However, the relevant reaction rarely takes place in a real solid state, depending on the reaction conditions and the monomer structure.

b. Significant Rate-Controlling Parameters in Direct Solid State Polymerization
The *reaction temperature* is probably the most important factor of direct SSP, due to its interaction with almost all aspects of the process [53]. It can be correlated directly with the intrinsic chemical reaction, the molecular mobility of the monomer, and the number of active sites on its surface, significantly affecting nucleation and growth stage duration as well as by-product diffusivity. The higher the temperature, the shorter the first stage of the SSP reaction, above which the polymerization rate increases more intensively [47,62]. Therefore, the monomer SSP processes are usually characterized by high-temperature coefficients; for example, in the case of amino acids [61] and PA salts [62], the rate doubles with every 2°C increment of the reaction temperature, also establishing a maximum temperature above which quasi-melt transitions are observed. Furthermore, it is observed that when the melting point of the monomer is high, the temperature range of SSP becomes wider, the SSP temperature coefficient decreases, and thus the impact of the reaction temperature on the SSP rate is diminished [61].

By-product diffusion limitations may also occur during a direct SSP process, and this effect is generally more intense at high operating temperatures, where the chemical reaction is no longer the controlling step [63–65]. The effect of by-product diffusion may be concluded when investigating the reacting particle size and the flow rate of the inert gas. The former is somewhat disregarded by most researchers, since it was found that its effect is significant for grain sizes below 20 to 25 mesh [66]. On the other hand, the surface by-product diffusion is influenced primarily by the flow of the inert gas [40], and in many cases the process is found to be diffusion controlled at low flow rates.

Another widely studied parameter of direct SSP is the presence of *catalysts*, used to overcome the slow SSP reaction rate and potential sintering. Acidic, basic, and neutral compounds have been examined for their catalytic action. In particular, regarding the acid-catalyzed polyamidation, the activity of catalysts follows the scheme of the nucleophilic acyl substitution. The proton of an acid (e.g., H_3PO_4, H_3BO_3, H_2SO_4) becomes attached to the carbonyl oxygen, making the carbonyl group even more susceptible to nucleophilic attack by NH_2 [reaction (1.8)]; oxygen can now acquire the π electrons without having to accept a negative charge as is in the uncatalyzed substitution shown in (1.9) [67–69].

$$(1.8)$$

$$(1.9)$$

Zinc chloride ($ZnCl_2$), phosphoric acid (H_3PO_4), magnesium oxide (MgO), boric acid (H_3BO_3), and others have been examined as catalysts during melt polycondensation of amino acids or diamines with dicarboxylic acids [68]. In particular, for the solid state polymerization of ω-amino acids, the catalytic effectiveness was found to be 1% H_3BO_3, 0.2% MgO > 0.5% $(COONH_4)_2$ > 0.5% $(CH_3COO)_2Zn$ > 0.2% Na_2CO_3 > 0.6% CH_3COOH > 0.5% $(NH_4)_2SO_4$ > 1% $SnCl_2$ [70]. For the SSP of PA 66 salt, reported catalyst performance is H_3BO_3 > $(COOH)_2$ > H_3PO_4 > MgO. On the other hand, Na_2CO_3, $NaHSO_4$, and $(SiO_2)_n$ were proved to be inactive [53].

A variety of catalysis mechanisms are described in the literature. Khripkov et al. [71] conjecture regarding the SSP of PA 66 salt that the uncatalyzed polycondensation involves reactions between the end groups of monomers and propagating polymer chains; the initiation and propagation reactions are carried out in the defective parts of small crystals of the salt. In catalyzed processes, the presence of linear oligomers is reported after a short period of SSP reaction; thus, they assume that polymer chain growth is achieved not only with the reaction between the monomer and the propagating polymer chain, but also between the oligomers themselves.

On the other hand, Katsikopoulos and Papaspyrides [72] correlated the effect of catalysts to the proposed mechanism of solid-to-melt transition in the PA salts. It was proposed that the presence of a good catalyst in the reacting structure contributes to an easier removal of the water formed, away from the reacting sites (Fig. 1.7). In other words, hydration seems restricted and diffusion of the water is favored, so that the right-hand reaction is encouraged.

1.3.2. Prepolymer Solid State Polymerization (Post-SSP, Solid State Finishing)

Post-polymerization in the solid state is carried out on low- or medium-molecular-weight semicrystalline or amorphous prepolymers at a temperature below their melting point under inert gas or vacuum [57,64,73–83]. Prepolymers are in the form of pellets, flakes (mean diameter, $\bar{d} > 1.0$mm), or powders ($\bar{d} < 100$ μm) (preextrusion SSP), or even of fibers and thin films (post-extrusion SSP), thus allowing easier by-product removal. Indicative post-SSP conditions and average molecular weights achieved are presented in Table 1.6.

The SSP of prepolymers is widely integrated into industrial production processes and usually comprises the finishing stage. A characteristic example is that of polyamide production, and more specifically the case of PA 66, one of the most important commercial nylons, corresponding to 40% of total PA demand. Its production is a two- or three-step process, depending on the desired molecular weight (Fig. 1.8). First, the aqueous solution of PA 66 salt (70 to 90 wt%) is reacted in an autoclave at temperatures in the range 175 to 200°C while increasing the pressure to minimize loss of the volatile organic compounds (e.g., hexamethylenediamine). Then the temperature is increased further (250 to 270°C) and the pressure is released to bleed off steam and to drive the condensation reaction toward polymerization.

While maintaining the same approximate temperature, the melt reaction mixture is held at a low constant pressure (even vacuum) for a sufficient time, which, however, is restrained due to problems arising from the high melt viscosity, the difficulties in dissipating heat transfer, and the thermal degradation of polymer. When high molecular weight is required, post-polymerization in the solid state may be performed, substantially increasing the degree of polymerization of the polymer while the material retains its solid shape. For example, a typical value of number-average molecular weight ($\overline{M_n}$) derived from polyamide melt techniques is in the range 15,000 to 25,000 g mol^{-1}; meanwhile, resins of $\overline{M_n} > 30,000$ g mol^{-1} are required for injection and blow molding and prepared through solid state finishing [8].

a. Prevailing Mechanisms in Post–Solid State Polymerization Regarding post-SSP, Zimmerman [43,74] has suggested a two-phase model according to which the polymerization proceeds by stepwise reactions in the amorphous regions of the semicrystalline polymer, where the end groups and low-molecular-weight substances (condensate, oligomers) are excluded. It is

TABLE 1.6. Indicative Post–Solid State Polyamidation Conditions and Molecular Weight Achieved

Prepolymer	SSP Conditions	Changes in Mean Molecular Weight
PA 6 [44]	$T = 115-205°C$ $t = 0-24$ h $\bar{d} = 0.2-0.5$ mm $Q_{N_2} = 1.2$ L min^{-1}	$\overline{M_{n0}} = 2500-18,700$ g mol^{-1} $\overline{M_{nt}} = 18,000-37,000$ g mol^{-1} $(t = 24$ h$)$
PA 46 [67]	$T = 190-280°C$ $t = 0-8$ h $\bar{d} = 0.1-0.2$ mm $v_{N_2} = 4$ cm s^{-1}	$\overline{M_{n0}} = 2000$ g mol^{-1} $\overline{M_{nt}} = 8200-46,800$ g mol^{-1} $(t = 8$ h$)$
PA 6 [82]	$T = 190-220°C$ $t = 0-12$ h $\bar{d} = 1.2-1.4$ mm $Q_{N_2} = 0.06$ L min^{-1}	$\overline{M_{n0}} = 17,000$ g mol^{-1} $\overline{M_{nt}} = 24,000-32,000$ g mol^{-1} $(t = 12$ h$)$
PA 46 fibers [81]	$T = 240-260°C$ $t = 0-4$ h $Q_{N_2} = 13$ L min^{-1}	$\overline{M_{v0}} = 41,000$ g mol^{-1} $\overline{M_{vt}} = 60,000-110,000$ g mol^{-1} $(t = 2$ h$)$
PA 66 [73]	$T = 160-200°C$ $t = 0-4$ h $\bar{d} = 1.4-1.7$ mm $Q_{N_2} = 0.26$ L min^{-1}	$\overline{M_{n0}} = 17,300$ g mol^{-1} $\overline{M_{nt}} = 19,300-25,700$ g mol^{-1} $(t = 4$ h$)$
PA 66 [80]	$T = 90-135°C$ $t = 0-10$ h $\bar{d} = 1.8$ mm $Q_{N_2} = 0.7-3.6$ L min^{-1}	$\overline{M_{n0}} = 10,000$ g mol^{-1} $\overline{M_{nt}} = 12,000-20,500$ g mol^{-1} $(t = 10$ h$)$
PA 66 fibers [76,81]	$T = 220-250°C$ $t = 0-2$ h $Q_{N_2} = 14$ L min^{-1}	$\overline{M_{v0}} = 40,000$ g mol^{-1} $\overline{M_{vt}} = 66,000-280,000$ g mol^{-1} $(t = 2$ h$)$

assumed that reactions and equilibrium in the amorphous regions are the same as for a completely amorphous or molten polymer at the same temperature, as shown in Figure1.5. The diffusion of end groups in the amorphous phase is considered to occur either through translation of a low-molecular-weight molecule (oligomers), through motion of terminal segments (segmental diffusion), or through exchange reactions (chemical diffusion) that allow reactive end groups to approach to a distance suitable for reaction. These mechanisms determine different kinetic regimes during post-SSP process, as discussed in Section 3.2.2.

Based on the two-phase model, Duh also proposed that two categories of end groups exist in the amorphous polymer regions: active and inactive. The inactive end groups include chemically dead chain ends and functional groups that are firmly trapped in the crystalline structure and cannot participate in the reaction [79].

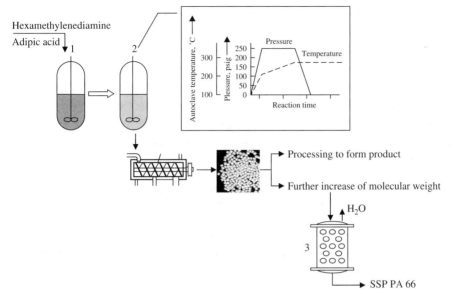

Fig. 1.8. PA 66 production procedure: (1) PA 66 salt aqueous solution preparation, (2) solution-melt polymerization, and (3) solid state polymerization.

To summarize, a more accurate post-SSP definition is that during post-SSP, the prepolymer is heated at a temperature higher than the glass-transition point (T_g) and lower than its melting point (T_m), so as to make the end groups mobile enough to react and to promote condensation reactions in the amorphous regions.

b. Significant Rate-Controlling Parameters in Post–Solid State Polymerization
The literature of prepolymer SSP comprises a significantly higher number of publications than that of direct SSP, due to prepolymer SSP's industrial use as a finishing stage. In most studies the effect of the various process parameters is evaluated and connected to the prevailing chemical and/or physical courses. In particular, *reaction temperature* emerges as the most critical parameter, interfering with the chemical reaction, the mobility of the functional end groups, and the by-product diffusivity [57,64,76–86]. The dependence of the reaction temperature on the SSP rate constant is indicated by the values of the SSP activation energy (E_a), reported to be between 43.9 and 340.7 kJ mol^{-1} in the case of PAs and between 62.7 and 177.6 kJ mol^{-1} for polyesters, being in general higher than those for melt processes [79]. Optimized temperature ranges have been set for a variety of polymers: 20 to 160°C below the final T_m, and the most preferred temperatures are just below T_m [67,68,82,87–91]. Temperature-step processes are also not excluded, and in some cases they are preferred so as to avoid polymer grain agglomeration through gradual increases in the prepolymer softening temperature [87,92–95]. In addition, problems related to oligomer formation and

to initial moisture and impurities in the prepolymer are also reduced in this way [8,79].

An additional SSP parameter studied is the initial *end-group concentration*, which is assumed to affect significantly segmental mobility and diffusion of the reactive chain ends. The lower the concentration of end groups (the higher the value of initial number-average molecular weight, $\overline{M_{n_0}}$), the higher the number-average molecular weight, $\overline{M_n}$, at the end of the SSP reaction [44,67,85]. The high $\overline{M_{n_0}}$ value ensures a more effective confinement of the amorphous phase and therefore a high concentration and homogeneous distribution of reactive chain ends in the reaction zone according to the two-phase model. A similar explanation is given by Duh [96], who indicates that in a lower-$\overline{M_{n_0}}$ prepolymer it is easier for polymer chains to fit into crystal lattices and to form rigid crystals; as a result, a greater number of end groups will be trapped and become inactive. The effect of remelting on the SSP rate is also associated with end-group diffusion; a redistribution of the reactive chain ends is achieved through remelting the prepolymer some time after starting SSP, shortening end-group distances and facilitating their reaction [44,65,85,97–100].

Prepolymer geometry and *gas flow rate* also play important roles, due to the reversible character of most condensation reactions. In particular, particle size strongly influences the overall rate when diffusion of the by-product within the polymer particle (interior diffusion) controls, but this influence gets weaker when the process is controlled by both diffusion and reaction [85,101]. Accordingly, smaller prepolymer particles can lead to an increase in the SSP rate due to the shorter diffusion distance and to the larger particle surface area per unit volume (*S/V* ratio) [40,96]. On the other hand, surface by-product diffusion is influenced principally by the flow of the inert gas: acceleration in the gas flow can increase the mass and heat transfer rates in the gas–solid system and decrease resistance to the diffusion of by-product from the particle surface into the bulk of the gas phase [40, 979,102–104]. It is reported that at a given reaction temperature, increasing the gas flow velocity in the SSP of small-sized PET grains or particles results in changing the limiting step from surface diffusion control to chemical reaction control [105].

However, it should be mentioned that the effect of these two SSP parameters is strongly dependent on such chemical reaction characteristics as the equilibrium constant (K_{eq}), which is proportional to the requirements for by-product removal. Characteristic is the case of polyamides and polyesters, where the equilibrium constant for PAs is hundreds of times larger than that for polyesters, and much higher by-product concentration can thus be tolerated in the first case, having much less severe removal requirements [40]. For example, in polyamides, K_{eq} varies between 100 and 750, depending on the water content in the reacting system [67,106–108]; meanwhile, the relevant values for transesterification and esterification, used widely in PET melt condensation models, are 0.5 to 1 and 1.25, respectively. Correlating these K_{eq} values with SSP data, it was found that in the case of PET SSP, a decrease in the particle diameter from 0.266 cm to 0.14 cm results in reducing the residence time by 56%, whereas the relative decrease in PA 66 SSP is only 3% [40].

Crystallinity (ϕ_w or ϕ_v) is thought to influence the SSP rate due to its inter-action with other controlling critical parameters, such as end-group mobility and by-product diffusion, and its effect is often two-sided theoretically. In particu-lar, based on the two-phase model, SSP reactions in the amorphous phase are anticipated to be favored in well-crystallized semicrystalline polymers, where an increase in ϕ_w leads to higher concentration of end groups rejected in the amor-phous phase and thus to an increase in the reaction rate [85,87,101]. On the other hand, the escape of by-products from the reacting mass may be hindered by a high degree of crystallinity [92] due to diffusion restrictions set by the rigid and well-organized crystal lattice. Notably, the ϕ_w effect is not a one-way determining agent and is strongly correlated with the reaction rate–controlling mechanism. In by-product-diffusion-limited reactions, high crystallinity reduces the SSP rate by imposing a higher degree of resistance to mass transfer, whereas in chemical reaction–controlled process, high crystallinity results in an increased SSP rate because of the effect of concentrating end groups in the amorphous regions [63]. It has been suggested that for optimum behavior, the reacting particles should have sufficiently high crystallinity to prohibit particle agglomeration [40,109]; specifically, Wu et al. [85] proposed a value of 40%. On the other hand, con-tradictory opinions have been expressed concerning the change of crystallinity during SSP. According to Wu et al. [85], crystallinity can be assumed constant in SSP. By contrast, Li et al. [92] observed that the crystal perfection and/or size gradually increase during the SSP of PA 66 and PET. Kim et al. [63] observed that PET crystallinity increases significantly within the first few hours of SSP and then stabilizes, showing only small variations with increased reaction time. Finally, Vouyiouka et al. [110] found a slight decrease of crystallinity during PA 66 post-SSP. This trend was attributed to the increase in molecular weight, which inhibited chain folding in the crystalline phase, clearly for mobility reasons. Syn-ergistic action for the crystallinity decrease observed was also thought to exhibit potential exchange reactions occurring during SSP, which induce morphological changes in the polymer structure through creating loops and bridges and therefore producing structural reorganization of the amorphous and crystalline regions of the polymer.

Finally, the use of *catalysts* currently constitutes a significant area of research into overcoming slow reaction rates, the main drawback in industrial use of SSP. Due to its importance, an entire chapter (Chapter 5) is devoted to this subject. In particular, the addition of easily diffusing mainly acidic compounds (e.g., H_3PO_4, H_3BO_3, H_2SO_4) leads to higher reaction rates, whereas in their absence the reac-tion rate is limited by diffusion of the autocatalyzing acid chain-end groups [44]. In PA SSP, examples of catalysts used are mainly phosphorus compounds such as 2-(2′pyridyl)ethyl phosphonic acid (PEPA) and sodium and manganous hypophosphite [8,109]. It should be mentioned here that the moisture in the pellets is considered to deactivate the catalyst; therefore, evaporation of water from the surface should be encouraged by the use of a low-dew-point inert gas [8]. The use of thermoplastic polyurethane [111] and of sterically hindered hydroxylpheny-lalkylphosphonic ester or monoester [112] as reaction accelerators has been

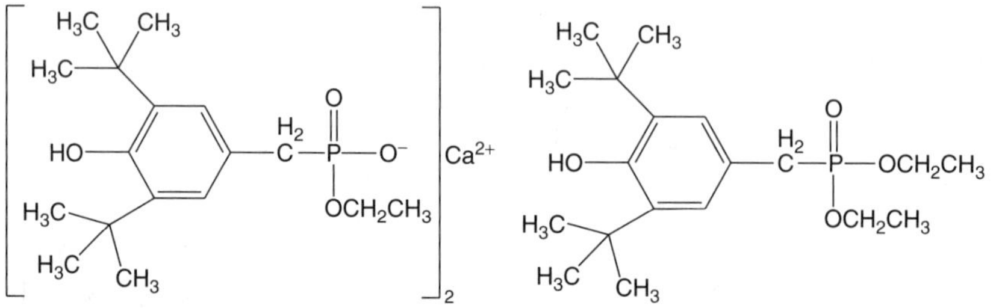

Fig. 1.9. Molecular structures of the phosphonates used as catalysts in PA 66 SSP. (From Pfaender et al. [112], with permission.)

proposed. More recently, hydroxyphenylmethyl phosphonate esters (Fig. 1.9) were found to catalyze the post-polyamidation in the solid phase through increasing the solution relative viscosity of the product up to 57% compared to the uncatalyzed process. This catalytic efficiency was correlated with the structure of the phosphonates: more specifically, to the additive mobility within the solid polymer, implying the possibility of partial incorporation of the catalyst molecule into the polyamide structure as an end group [110]. Finally, recent publications on PET SSP indicate the catalytic effect of nanomaterials such as montmorillonite [113] and silica [114] as a result of increasing the nucleation sites for polymerization.

1.4. SOLID STATE POLYMERIZATION APPARATUS AND ASSEMBLIES

SSP processes are characterized by simple equipment requirements, low reaction temperatures, and the need for an oxygen-free atmosphere, achieved through inert gas flow, vacuum, or high pressure. The choice of one of these three alternatives depends mainly on the application scale and starting material. A vacuum process is usually preferred for small capacity production, since on a larger scale the danger of oxygen rushing in and subsequent oxidation and coloration of the polymer is enhanced [115,116]. High pressure (196 to 490 MPa) can be applied in the case of monomers so as to provide well-oriented polymers, but it gives low polymerization rates, due to the high by-product diffusion resistance [117].

On the other hand, inert gas flow is used widely on an industrial or bench scale to serve three objectives: to remove the condensate, to exclude oxygen from the reactor atmosphere in order to inhibit polymer oxidation, and to heat the reacting mass. The drawback here relates to the energy requirements for heating and drying the gas. The inert gases used most often in SSP processes are nitrogen, carbon dioxide, helium [118,119], superheated steam [67], and supercritical carbon dioxide [102,103,120,121]. In addition, three alternatives exist for the inert surroundings:

1. Heating under continuous inert gas flow (open system), where by-product removal is ensured [44,67,76,92,102,122]; this is used primarily in post-SSP.

2. Heating in an inert gas atmosphere (closed system) under low overpressure, where the loss of monomers and oligomers is hindered [52,123]. This system was used primarily in earlier decades during monomer SSP.

3. A combination of the aforementioned two systems in which first, heating is carried out in an inert atmosphere (closed system) and later, inert gas passes (open system) [56,57]. The latter comprises the advantages of both systems and provides maintenance of the monomers in the reactor and satisfactory removal of the by-product in order to favor the polymerization reaction.

The SSP reactions can be performed in a variety of apparatus (plug and batch flow reactors) depending on the scale: for example, in glass tubes [44,77,92,124], in fluidized- and fixed-bed reactors [78,93,125], in rotating flasks [122], in tumbler dryers [67,76,116,118], in an inert liquid medium [47,48,60,75,126], in vertical reactors with stirring blades [127], and in rotating blenders [128]. The process layout may also comprise inert gas recycling circuit and cleaning apparatus (e.g., bag filters, gas washing, catalytic gas cleaning), so as to remove vaporous reaction by-products, oligomers, and atmospheric oxygen that has penetrated (Fig. 1.10) [126]. Details on PET SSP industrial processes are given in Chapter 8 when we discuss recent developments.

In many cases, heat and mass transfer as well as sintering problems are overcome through mechanical agitation of a solid's reacting particles [93,94,129]. Additionally, microwave energy has been used to increase the SSP rate through

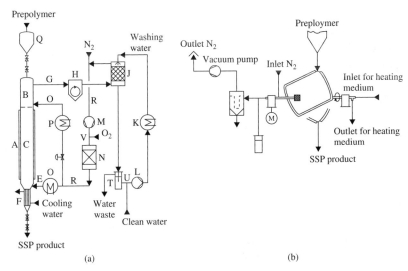

Fig. 1.10. Solid state polymerization layout for (a) continuous and (b) batch processes. A: fixed bed reactor, Q: prepolymer silo, B: reactor first zone (e.g., 90–150°C), C: reactor second zone (e.g., 180–230°C), F: cooling zone, D: inert gas inlet for zone B, E: inert gas inlet for zone C, P, O, K: heat exchangers, R: inert gas piping, N: inert gas catalytic cleaning, V: O_2 supply, G: inert gas outlet, H: cyclone, J: inert gas washing, L: clean water pump.

exciting and heating the condensate in a polymer, thus allowing higher diffusion rates [87,95]. However, despite the promise, this technique has not yet been employed on an industrial scale.

1.5. SOLID STATE APPLICATIONS IN THE POLYMER INDUSTRY

1.5.1. Solid State Polymerization Advantages

Solid state polymerization, especially post-SSP, has important advantages that render its use attractive. First, it is employed as an extension of the melt technique for the production of high-molecular-weight resins. Melt-based techniques are usually not carried out to high conversion, so as to prevent the thermal degradation of the product as well as various problems arising from melt viscosity increase, such as stirring, heat and mass transfer dissipation, and reactor handling [8,126].

SSP polymers often have improved properties, because monomer cyclization and other side reactions are limited or even avoided, due to the low SSP operating temperatures [115]. Only linear chains seem to be formed [47] and SSP products usually show greater heat stability in the molten state than those samples prepared in the melt phase [129,130]; on the other hand, they contain small amounts of monomers and oligomers, so that no stage of monomer removal is needed [131].

Furthermore, the molecular weight increase during SSP may be accompanied by an increase in crystallinity and crystal perfection [76] while drying the polymer, which is important because of the moisture-negative effect on resin processability for yarn manufacture [77]. In addition, there is practically no environmental pollution because a solvent is not used, the process requires simple equipment, and operation can be continuous [92].

1.5.2. Post–Solid State Polymerization Application in Polyamides

To provide an overview of SSP application and current significance in the polymer industry, a case study of SSP commercialization and products is presented. Polyamides are examined with respect to market demands and segmentation for the relevant SSP polymers (i.e., the polyamide fragment, which is produced through SSP). In particular, SSP PAs are high-molecular-weight resins suitable for [132] (1) industrial filament yarns, accounting for 15% of the total PA demand; (2) carpet fibers, counting for 16% of total PA demand; and (3) engineering plastics to prepare molded products and films through injection and extrusion, accounting for 34% of total PA demand.

The sum of the percentages above reflect SSP PA demand: 65% of total polyamide consumption is anticipated to be produced through SSP as a finishing stage. And in particular, based on 2004 data [132], this number corresponds to nearly 4200 kilotons out of 6440 kilotons of total demand.

The SSP PAs can be segmented further by application sector. Engineering SSP polyamides end up in the automotive industry processed through injection molding (29.9%), in other industrial markets (28.5%), in electrical and electronic (E&E; 21.8%), in films processed through extrusion (11%), and for other uses (8.8%) [133]. The SSP PA demand by application is depicted in Table 1.7. The four major sectors for SSP PAs are carpet fiber, industrial, automotive, and E&E, with percentages of 11 to 25%, while film covers 6% of SSP products.

The development areas for SSP products are definitely engineering plastics in automotive (i.e., underhood, interior, exterior) and E&E applications (e.g., cable ties, connectors, plugs, switches). Automotive engineering continues to be the application segment with the highest consumption of engineering PAs, since PAs are replacing metal in a large number of applications. In particular, the relevant growth rate is up to seven times higher than that of the fiber market, due mainly to the strong competition from lower-cost materials, such as PET fibers [133–136].

Considering that PA 6 and PA 66 are the most popular polyamides, accounting for more than 90% of PA uses, one could also estimate the distribution of SSP processes per polyamide type. In particular, based on the demand for PA 6, PA 66, and high-performance polyamides (HPPAs) such as PA 11, PA 12, PA 46, PA 612 with respect to SSP applications [132], figures for the SSP processes per polyamide type can be assessed (Table 1.8), showing that more than the half (57%) of the SSP processes are related to PA 6 prepolymers. There is also strong interest in the specialty polyamides, which have an estimated capacity of 250 kilotons and a consumption of 200 kilotons in 2007, and the annual average growth rate of consumption is expected to exceed 8% in the coming years [137].

Finally, based on the largest polyamide producers, in 1995 North America was the largest regional PA consumer (35% of global consumption), with Asia/Pacific excluding Japan following (22% of global consumption), also reflecting the relevant SSP assets. By 2015 it is estimated that Asia/Pacific excluding Japan will account for 35% of global PA consumption, followed by North America (26% of global consumption) [137].

TABLE 1.7. Segmentation of SSP PA Consumption Based on 2004 Figures

SSP PA Applications	Kilotons	SSP PAs (%)
Fibers		
Carpet fibers	1030	25
Industrial yarns	966	23
Engineering PAs		
Automotive	655	16
Other industrial markets	624	14
E&E	477	11
Film	241	6
Others	193	5
Total	4186	

TABLE 1.8. Distribution of SSP PA Consumption by Polyamide Type and Application Based on 2004 PA Total Demand

SSP PA Demand (wt%) for:	Polyamide Type		
	PA 6	PA 66	HPPA
Carpet and industrial fibers	26	16	
Engineering plastics	25	22	3
Film	6	1	
Total	57	39	4

1.6. CONCLUSIONS

Solid state polymerization (SSP) processes are widely used in the commercial production of polyamide and polyester to increase the degree of polymerization and to improve the quality of the end product. The most important commercial advantages of SSP focus on the use of uncomplicated, inexpensive equipment and on avoiding some of the drawbacks of conventional polymerization processes. In this chapter we provide a theoretical background on the fundamentals of step- and chain- growth polymerization and on using the solid state technique. We also define SSP, presenting its main characteristics, such as prevailing mechanisms, process parameters, apparatus required, and assemblies and applications.

REFERENCES

1. Fried J. *Polymer Science and Technology*. Prentice Hall, Upper Saddle River, NJ, 1995, pp. 1–4.
2. Strong A, *Plastics: Materials and Processing*. Prentice Hall, Upper Saddle River, NJ, 2000, pp. 1–22.
3. London Metal Exchange. Plastics industry overview. http://www.lme.co.uk/plastics_industryoverview_pp.asp.
4. *Plastics News*, Sept. 29, 2003, http://www.plasticsnews.com.
5. Plastics Europe, Association of Plastics Manufacturers. Plastics major applications. http://www.plasticseurope.org.
6. Flory P. *Principles of Polymer Chemistry*. Cornell University Press, Ithaca, NY, 1975, pp. 75–83.
7. Young R, Lovell P. *Introduction to Polymers*. Chapman & Hall, London, 1991, pp. 43–51.
8. Dujari R, Cramer G, Marks D. Method for solid phase polymerization (E.I. du Pont de Nemours & Company). WIPO Patent WO 98/23666, 1998.
9. Flory P. Polymerization process (E.I. du Pont de Nemours & Company). U.S. Patent 2,172,374, 1939.
10. Monroe G. Solid phase polymerisation of polyamides (E.I. du Pont de Nemours & Company). U.S. Patent 3,031,433, 1962.

11. Cruz S, Zanin M. PET recycling: evaluation of the solid state polymerization process. *J. Appl. Polym. Sci.* 2006;99:2117–2123.

12. Karayannidis G, Kokkalas D, Bikiaris D. Solid-state polycondensation of poly(ethylene terephthalate) recycled from postconsumer soft-drink bottles: I. *J. Appl. Polym. Sci.* 1993;50(12):2135–2142.

13. Charlesby A. Solid-state polymerization induced by radiation. *Rep. Prog. Phys.* 1965;28:464–518.

14. Eastmond G. Solid-state polymerization. *Prog. Polym. Sci.* 1970;2:1–46.

15. Hayashi K, Okamura S. Kinetics of radiation-induced solid-state polymerization of cyclic monomers. *Makromol. Chem.* 1961;47:230–236.

16. Restaino A, Mesrobian R, Morawetz H, Ballantine D, Dienes G, Metz D. γ-Ray initiated polymerization of crystalline monomers. *J. Am. Chem. Soc.* 1956;78:2939–2943.

17. Adler G. Amorphous and unoriented character of polymers formed in radiation-induced solid-state polymerization. *J. Chem. Phys.* 1959;31:848–849.

18. Jager P, Waight E. Solid-state polymerization of methacrylamide and N-aryl-methacrylamides. *J. Polym. Sci. A.* 1963;1:1909.

19. Ueda H. Electron-spin resonance (ESR) studies of irradiated single crystals of methacrylamide. *J. Polym. Sci. A.* 1964; 2(5) 2207–2216.

20. Chapiro A, Hardy G. Radiochemical polymerization of N-vinylcarbazole in liquid and solid phase. *J. Chim. Phys. Phys. Chim. Biol.* 1962;59:993–998.

21. Duling I, Price C. Polymerization of N-vinylpyridinium salts. *J. Am. Chem. Soc.* 1962;84:578–583.

22. Fujioka S, Hayashi K, Okamura S. Radiation-induced polymerization of 2-vinylpyridinium salts. *Nippon Hoshasen Kobunshi Kenkyu Kyokai Nempo.* 1962;4:199–207.

23. Magat M. Ionic polymerizations initiated by ionizing irradiation. *Makromol. Chem.* 1960;35:159–173.

24. Barkalov I, Goldanskii V, Enikolopyan N, Terekhova S, Trofimova G. Radiation-induced, solid-phase polymerization: I. Polymerization of acrylonitrile. *J. Polym. Sci. C.* 1964; 4:897–908.

25. Okamura S, Hayashi K, Kitanishi Y, Watanabe H, Nishii M. Crystalline polymers by radiation-induced polymerization of ring compounds. *Nippon Aisotopu Kaigi Hobunshu.* 1961;4:450–456.

26. Okamura S, Hayashi K, Nishii M. Radiation-induced polymerizations of vinyl monomers by a pre irradiation method. *Kogyo Kagaku Zasshi.* 1961;64:487–488.

27. Okamura S, Hayashi K, Nishii M. Polymer crystals obtained by radiation polymerization of trioxane in the solid state. *J. Polym. Sci.* 1962;60:526–529.

28. Watanabe H, Hayashi K, Okamura S. The mechanism of radiation-induced solid-state polymerization of 3,3-bis(chloromethyl)oxetane. *Nippon Hoshasen Kobunshi Kenkyu Kyokai Nempo.* 1962;4:119–126.

29. Dent Glasser L, Glasser F, Taylor H. Topotactic reactions in inorganic oxy-compounds. *Q. Rev. (London).* 1962;16:343–360.

30. Adler G, Ballantine D, Baysal B. The mechanism of free radical polymerization in the solid state. *J. Polym. Sci.* 1960;48:195–207.

31. Baysal B, Adler G, Ballantine D, Colombo P. Solid state polymerization of acrylamide initiated by γ-radiation. *J. Polym. Sci.* 1960;44:117–127.

32. Hayashi K, Kitanishi Y, Nishii M, Okamura S. Crystalline polymers prepared by radiation-induced solid state polymerization. *Makromol. Chem.* 1961;47:237–241.

33. Hsia C, Catherine S. Solid-state polymerization of acenaphthylene induced by ionizing radiation. *J. Polym. Sci.* 1962;62: 174.

34. Bensasson R, Dworkin A, Marx R. Radiopolymerization of solid acrylonitrile: the effect of the conditions of crystallization and phase transition. *J. Polym. Sci. C.* 1964; 4:881–895.

35. Hayashi K, Ochi H, Okamura S. Radiation-induced post-polymerization of trioxan in the solid state. *J. Polym. Sci. A* 1964; 2(6) 2929–2946.

36. Papaspyrides C. Solid state polymerization of nylon 12,6 salt. Ph.D. dissertation, National Technical University of Athens, Greece, 1982.

37. Fadner T, Morawetz H. Polymerization in the crystalline state: I. Acrylamide. *J. Polym. Sci.* 1960;45:475–501.

38. Chen C, Grabar D. Effect of polymorphism in solid-state polymerization. *J. Polym. Sci. C* 1964;4:849–868.

39. Papaspyrides C. Solid state polyamidation. In: *The Polymeric Materials Encyclopedia*, Salamone JC, ed. CRC Press, Boca Raton, FL, 1996, pp. 7819–7831.

40. Yao K, Ray W. Modeling and analysis of new processes for polyester and nylon production. *AIChE J.* 2001; 47(2) 401–412.

41. Achilias D. A review of modelling of diffusion controlled polymerization reactions. *Macromol. Theory Simul.* 2007;16:319–347.

42. Lucas B, Seavey K, Liu Y. Steady-state and dynamic modelling for new product design for the solid state polymerization of poly(ethylene terephthalate). *Ind. Eng. Chem. Res.* 2007; 46:190–202.

43. Zimmerman J, Kohan M. Nylon: selected topics. *J. Polym. Sci. A* 2001;39:2565–2570.

44. Gaymans R, Amirtharaj J, Kamp H. Nylon 6 polymerization in the solid state. *J. Appl. Polym. Sci.* 1982;27:2513–2526.

45. Papaspyrides C. Solid state polyamidation of aliphatic diamine–aliphatic diacid salts: a generalized mechanism for the effect of polycondensation water in reaction behaviour. *Polymer*. 1990; 31(3) 490–495.

46. Papaspyrides C, Kampouris E. Solid-state polyamidation of dodecamethylenediammonium adipate. *Polymer*. 1984;25:791–796.

47. Yamazaki T, Kaji K, Kitamaru R. Polymerization kinetics on the thermo induced solid state polycondensation of ε-aminocaproic acid and nylon 66 salt. *Bull. Kyoto Univ. Educ. Ser. B.* 1983;63:53–63.

48. Grabar D, Hsia C. Catherine S. Morphological aspects of polymerization in the solid state. *J. Polym. Sci. C* 1963;3:105–107.

49. Frayer P, Lando J. Polymerization of crystalline hexamethylenediammonium adipate. *Mol. Cryst. Liq. Cryst.* 1969; A-1:465–483.

50. Macchi E, Morosoff N, Morawetz H. Polymerization in the crystalline state: X. Solid-state conversion of 6-aminocaproic acid to oriented nylon 6. *J. Polym. Sci. A-1* 1965;6:2033–2049.

51. Papaspyrides C, Vouyiouka S, Bletsos I. New aspects on the solid state polyamidation of PA 6,6 salt. *Polymer*. 2006;47:1020–1027.

52. Oya S, Tomioka M, Araki T. Studies on polyamides prepared in the solid state: I. Polymerisation mechanism. *Kobunshi Kagaku*. 1966; 23(254) 415–421.

53. Khripkov E, Kharitonov V, Kudryavtsev G. Some features of the polycondensation of hexamethylene diammonum adipinate. *Khim. Volokna*. 1970;6:63–65.

54. Volokhina A, Kudryavstev G, Raeva M, Bogdanov M, Kalmykova V, Mandrosova F, Okromchedidze N. Polycondensation reactions in the solid phase: V. Polycondensation of the diamine salts of terephthalic and hexahydroterephthalic acids in the solid state. *Khim. Volokna*. 1964;6:30–33.

55. Silverman B, Raleigh N, Stewart L. Process for producing ultrahigh molecular weight polyamides (Monsanto Company). U.S. Patent 3,562,206, 1971.

56. Wlloth F. Solid state preparation of polyamides (Vereinigte Glasstoff-Fabriken A.G.). U.S. Patent 3,379,696, 1968.

57. Papaspyrides C, Vouyiouka S, Bletsos I. Preparation of polyhexamethyleneadipamide prepolymer by a low temperature process. *J. Appl. Polym. Sci.* 2004;92:301–306.

58. Tynan G, Papaspyrides C, Bletsos I. Preparation of low-water-content, diamine–dicarboxylic acid monomer salts (E.I. du Pont de Nemours & Company). U.S. Patent 5,941,634, 1998.

59. Kampouris E, Papaspyrides C. Solid state polyamidation of nylon salts: possible mechanism for the transition solid-melt. *Polymer*. 1985;26:413–417.

60. Papaspyrides C. Solid-state polyamidation of nylon salts. *Polymer*. 1988;29:114–117.

61. Volokhina A, Kudryavtsev G, Skuratov S, Bonetskaya A. The polyamidation process in the solid state. *J. Polym. Sci.* 1961;53:289–294.

62. Zeng H, Feng L. Study of solid-state polycondensation of nylon 66 salt. *Gaofenzi Tongxun*. 1983; 5(5) 321–327.

63. Kim T, Lofgren E, Jabarin S. Solid-state polymerization of poly(ethylene terephthalate): I. Experimental study of the reaction kinetics and properties. *J. Appl. Polym. Sci.* 2003;89:197–212.

64. Chang T. Kinetics of thermally induced solid state polycondensation of poly(ethylene terephthalate). *Polym. Eng. Sci.* 1970; 10(6) 364–368.

65. Chen S, Chen F. Kinetics of polyesterification: III. Solid-state polymerization of polyethylene terephthalate. *J. Polym. Sci. A.* 1987;25:533–549.

66. Pilati F. Solid-state polymerization. In *Comprehensive Polymer Science*, Vol. 5. Pergamon Press, New York, 1989, pp. 201–216.

67. Gaymans R, Schuijer J. Polyamidation in the solid phase. In: *Polymerization Reactors and Processes*, Henderson JN, Bouton CT, eds. ACS Symp. Ser. American Chemical Society, Washington, DC, 1979.

68. Korshak V, Frunze T. *Synthetic Heterochain Polyamides*. IPST, Jerusaleum, Israel, 1964, p. 120.

69. Hiemenz P. *Polymer Chemistry: The Basic Concepts*. Marcel Dekker, New York, 1984, p. 306.

70. Volokhina A, Kudryavtsev G, Skuratov S, Bonetskaya A. The polyamidation process in the solid state. *J. Polym. Sci.* 1961;53:289–294.

71. Khripkov E, Lavrov B, Kharitinov V, Kudryavtsev G. Some problems in solid-phase polycondensation of hexamethylenediammonium adipate. *Vysokomol. Soedin. B.* 1976; 18(2) 82–85.

72. Katsikopoulos P, Papaspyrides C. Solid-state polyamidation of hexamethylene-diammonium adipate: II. The influence of acid catalysts. *J. Polym. Sci. A.* 1994;32:451–456.

73. Vouyiouka S, Papaspyrides C, Weber J, Marks D. Polyamide solid state polymerization: evaluation of pertinent kinetic models. *J. Appl. Polym. Sci.* 2005: 97; 671–681.

74. Zimmerman J. Equilibria in solid phase polyamidation. *J. Polym. Lett.* 1964;2:955–958.

75. Kampouris E. New solid state polyamidation process. *Polymer.* 1976; 17(5) 409–412.

76. Srinivasan R, Desai P, Abhiraman A, Knorr R. Solid-state polymerisation vis-à-vis fiber formation of step-growth polymers: I. Results from a study of nylon 66. *J. Appl. Polym. Sci.* 1994;53:1731–1743.

77. Fujimoto A, Mori T, Hiruta S. Polymerization of nylon-6,6 in solid state. *Nippon Kagaku Kaishi.* 1988; 3:337–342.

78. Mallon F, Ray W. Modeling of solid state polycondensation: II. Reactor design issues. *J. Appl. Polym. Sci.* 1998;69:1775–1788.

79. Vouyiouka S, Karakatsani E, Papaspyrides C. Solid state polymerization. *Prog. Polym. Sci.* 2005; 30(1) 10–37.

80. Griskey R, Lee B. Thermally induced solid-state polymerisation in nylon 66. *J. Appl. Polym. Sci.* 1966;10:105–111.

81. Srinivasan R, Almonacil C, Narayan S, Desai P, Abhiraman A. Mechanism, kinetics and potential morphological consequences of solid-state polymerization. *Macromolecules.* 1998;31:6813–6821.

82. Xie J. Kinetics of the solid-state polymerization of nylon-6. *J. Appl. Polym. Sci.* 2002;84:616–621.

83. Agarwal A, Mhaisgawali V. Post-extrusion solid state polymerization of fully drawn polyester yarns. *J. Appl. Polym. Sci.* 2006;102:5113–5122.

84. Wang X, Deng D. A comprehensive model for solid-state polycondensation of poly(ethylene terephthalate): combining kinetics with crystallization and diffusion of acetaldehyde. *J. Appl. Polym. Sci.* 2002;83:3133–3144.

85. Wu D, Chen F, Li R, Shi Y. Reaction kinetics and simulations for solid-state poly-merisation of poly(ethylene terephthalate). *Macromolecules.* 1997;30:6737–6742.

86. Kuran W, Debek C, Wielgosz Z, Kuczyńska L, Sobczak M. Application of a solid-state postpolycondensation method for synthesis of high molecular weight polycarbonates. *J. Appl. Polym. Sci.* 2000;77:2165–2171.

87. Mallon F, Ray W. Modeling of solid-state polycondensation: I. Particle models. *J. Appl. Polym. Sci.* 1998;69:1233–1250.

88. Goodner M, DeSimone J, Kiserow D, Roberts G. An equilibrium model for diffusion-limited solid-state polycondensation. *Ind. Eng. Chem. Res.* 2000;39:2797–2806.

89. Gostoli C, Pilati F, Sarti G, Di Giacomo B. Chemical kinetics and diffusion in poly(butylene terephthalate) solid-state polycondensation: experiments and theory. *J. Appl. Polym. Sci.* 1984;29:2873–2887.

90. Goodner M, Gross S, DeSimone J, Roberts G, Kiserow D. Modeling and experimental studies of the solid state polymerization of polycarbonate facilitated by supercritical carbon dioxide. *Polym. Prepr. (Am. Chem. Soc., Div. Polym. Chem.)* 1999; 40(1) 97–98.

91. Kumar A, Gupta S. Simulation and design of nylon 6 reactors. *J. Macromol. Sci. Rev. Macromol. Chem. Phys.* 1986; 26(2) 183–247.

92. Li L, Huang N, Liu Z, Tang Z, Yung W. Simulation of solid-state polycondensation of nylon-66. *Polym. Adv. Technol.* 2000;11:242–249.

93. Beaton D. Continuous, solid-phase polymerization of polyamide granules (E.I. du Pont de Nemours & Company). U.S. Patent 3,821,171, 1974.

94. Chang S, Sheu M, Chen S. Solid-state polymerization of poly(ethylene terephthalate). *J. Appl. Polym. Sci.* 1983;28:3289–3300.

95. Mallon F, Ray W. Enhancement of solid-state polymerization with microwave energy. *J. Appl. Polym. Sci.* 1998;69:1203–1212.

96. Duh B. Reaction kinetics for solid-state polymerisation of poly(ethylene terephthalate). *J. Appl. Polym. Sci.* 2001;81:1748–1761.

97. Duh B. Effects of the carboxyl concentration on the solid-state polymerization of poly(ethylene terephthalate). *J. Appl. Polym. Sci.* 2002;83:1288–1304.

98. Kulkarni M, Gupta S. Molecular model for solid-state polymerization of nylon 6: II. An improved model. *J. Appl. Polym. Sci.* 1994;53:85–103.

99. Li L, Huang N, Tang Z, Hagen R. Reaction kinetics and simulation for the solid-state polycondensation of nylon 6. *Macromol. Theory Simul.* 2001;10:507–517.

100. Kaushik A, Gupta S. A molecular model for solid-state polymerisation of nylon 6. *J. Appl. Polym. Sci.* 1992;45:507–520.

101. Ravindranath K, Mashelkar R. Modeling of poly(ethylene terephthalate) reactors: IX. Solid state polycondensation process. *J. Appl. Polym. Sci.* 1990;39:1325–1345.

102. Shi C, DeSimone J, Kiserow D, Roberts G. Reaction kinetics of the solid-state polymerization of poly(bisphenol A carbonate) facilitated by supercritical carbon dioxide. *Macromolecules.* 2001;34:7744–7750.

103. Shi C, Gross S, DeSimone J, Kiserow D, Roberts G. Reaction kinetics of the solid state polymerisation of poly(bisphenol A carbonate). *Macromolecules.* 2001;34:2062–2064.

104. Gao Q, Nan-Xun H, Zhi-Lian T, Gerking L. Modelling of solid state polycondensation of poly(ethylene terephthalate). *Chem. Eng. Sci.* 1997; 52(3) 371–376.

105. Huang B, Walsh J. Solid-phase polymerization mechanism of poly(ethylene terephthalate) affected by gas flow velocity and particle size. *Polymer.* 1998; 39(26) 6991–6999.

106. Mallon F, Ray H. A comprehensive model for nylon melt equilibria and kinetics. *J. Appl. Polym. Sci.* 1998;69:1213–1231.

107. Steppan D, Doherty M, Malone M. A kinetic and equilibrium model for nylon 6,6 polymerization. *J. Appl. Polym. Sci.* 1987;33:2333–2344.

108. Ogata N. Studies on polycondensation reactions of nylon salt: I. The equilibrium in the system of polyhexamethylene adipamide and water. *Makromol. Chem.* 1960;42:52–65.

109. Blanchard E, Cohen J, Iwasyk J, Marks D, Stouffer J, Aslop A, Lin C. Process for preparing polyamides (E.I. du Pont de Nemours & Company). WIPO Patent WO 99/10408, 1999.

110. Vouyiouka S, Papaspyrides C, Pfaendner R. Catalyzed solid state polyamidation. *Macromol. Mater. Eng.* 2006;291:1503–1512.

111. Brink A, Owens J. Thermoplastic polyurethane additives for enhancing solid state polymerisation rates (Eastman Chemical Company). WIPO Patent WO 99/11711, 1999.

112. Pfaendner R, Hoffman K, Herbst H. Increasing the molecular weight of polycondensates (Ciba-Geigy AG). WIPO Patent WO 96/11978, 1996.

113. Huimin Y, Keqing H, Muhuo Y. The rate acceleration in solid-state polycondensation of PET by nanomaterials. *J. Appl. Polym. Sci.* 2004;94:971–976.

114. Achilias D, Bikiaris D, Karavelidis V, Karayannidis. Effect of silica nanoparticles on solid state polymerization of poly(ethylene terephthalate). *Eur. Polym. J.* 2008;44(10):3096–3107.

115. Heinz H, Schulte H, Buysch H. Process for the manufacture of high molecular weight polyamides (Bayer AG). European Patent EP 410,230/91 A2, 1991.

116. Hosomi H, Kitamura K. Ultra-high-molecular-weight polyhexamethyleneadipamides (Asahi Kasei Kogyo K.K.). Japanese Patent JP 1–284525, 1989.

117. Ikawa T. Nylons (by high pressure solid-state polycondensation). In: *The Polymeric Materials Encyclopedia*, Vol. 6. Salamone JC, ed. CRC Press, Boca Raton, FL, 1996, pp. 4689–4694.

118. Muller K, Eugel F, Gude A. Process for the preparation of polyamide powders processing high viscosities (Chemische Werke Huls Aktiengesellschaft). U.S. Patent 3,476,711, 1969.

119. Mallon F, Beers K, Ives A, Ray W. The effect of the type of purge gas on the solid-state polymerisation of polyethylene terephthalate. *J. Appl. Polym. Sci.* 1998;69:1789–1791.

120. Gross S, Roberts G, Kiserow D, DeSimone J. Crystallization and solid state polymerization of poly(bisphenol A carbonate) facilitated by supercritical CO_2. *Macromolecules*. 2000;33:40–45.

121. Gross S, Flowers D, Roberts G, Kiserow D, DeSimone J. Solid-state polymerisation of polycarbonates using supercritical CO_2. *Macromolecules*. 1999;32:3167–3169.

122. Roerdink E, Warnier J. Preparation and properties of high molar mass nylon-4,6: a new development in nylon polymers. *Polymer*. 1985;26:1582–1588.

123. Gaymans R, Van Utteren T, Van Den Berg J, Schuyer J. Preparation and some properties of nylon 46. *J. Polym. Sci. Polym. Chem. Ed.* 1977;15:537–545.

124. Fortunato B, Pilati F, Manaresi P. Solid state polycondensation of poly(butylene terephthalate). *Polymer*. 1981;22:655–657.

125. Gaymans R, Venkatraman V, Schuijer J. Preparation and some properties of nylon-4,2. *J. Polym. Sci. Polym. Chem. Ed.* 1984;22:1373–1382.

126. Weger F, Hagen R. Method and apparatus for the production of polyamides (Karl Fischer Industrieanlagen GmbH). U.S. Patent 5,773,555, 1998.

127. Shimizu K, Ise S. Polyhexamethyleneadipamide with restricted three-dimensional formation and process for the manufacture (Asahi Chemical Industry Ltd.). Japanese Patent JP 4–93323, 1992.

128. Sheetz H. Solid phase polymerisation of nylon (DSM N.V.). U.S. Patent 5,461,141, 1995.

129. Bruck S. New polyoxamidation catalysts. *Ind. Eng. Chem. Prod. Res. Dev.* 1963; 2(2) 119–121.

130. Morawetz H. Polymerization in the solid state. *J. Polym. Sci. C.* 1966; 12:79–88.

131. Mizerovskii L, Kuznetsov A, Bazarov Y, Bykov A. Equilibrium in the system polycaproamide–caprolactam–water below the melting point of the polymer. *Polym. Sci. U.S.S.R.* 1982; 24(6) 1310–1326.

132. Charaf F. Polyamide chain revisited. Presented at the 6th World Congress, Polyamide 2005: The Polyamide Chain, Duesseldorf, Germany, *Congress Proceedings*, 2005.

133. Growth in PA demand set to weaken. Plast. Additives Compound. Sept.–Oct. 2007.

134. De Bievre B. Advantages of PA 6 in automotive applications. Presented at the First World Congress, Polyamide 2000: The Polyamide Chain, Session I/2, Zurich, Switzerland, *Congress Proceedings*, 2000.

135. Modern Plastics Editorial Staff, *World Encyclopedia* article, Jan. 1, 2006.

136. Ramos R. An overview of current PA 6 and PA 66 South American markets and future outlook. Presented at the 4th World Congress, Polyamide 2003: The Polyamide Chain, Session 1/4, Zurich, Switzerland, *Congress Proceedings*, 2003.

137. Scheidl K. Presented at the 9th World Congress, Polyamide 2008: The Polyamide Chain, Zurich, Switzerland, 2008.

2

SOLID STATE POLYMERIZATION CHEMISTRY AND MECHANISMS: UNEQUAL REACTIVITY OF END GROUPS

HAIBING ZHANG AND SALEH A. JABARIN

College of Engineering, University of Toledo, Toledo, Ohio

Solid State Polymerization, Edited by Constantine D. Papaspyrides and Stamatina N. Vouyiouka
Copyright © 2009 John Wiley & Sons, Inc.

2.1. INTRODUCTION

Solid state polymerization (SSP) is an industrially important process used in the production of high-molecular-weight polyesters and polyamides [1–6]. The main aspects of SSP were presented in Chapter 1, and the process kinetics of the two most important categories of polycondensation polymers, polyesters and polyamides, are the focus of Chapters 3 and 4.

Investigation of the SSP mechanism, kinetics, and modeling generally follows three approaches. The first, the experimental approach, results in many empirical equations describing SSP reaction kinetics [6–15]. Accordingly, studies are performed on the changes in intrinsic viscosity (IV) or molecular weight as a function of time and temperature as well as of the flowing gas rate and reacting particle size, in order to extract the apparent SSP rate constants and corresponding activation energies. Regarding the second approach, modeling is developed based on the SSP reaction mechanisms. Previous work [8,10,11,16] considered the main polymerization reactions in developing the SSP kinetics. In the case of polyesters, these reactions include ester-interchange [8,10,11,16] and esterification reactions and also some side reactions [12,14,17–19]. Advanced comprehensive models are also developed, taking into consideration the chemical reactions as well as the diffusion of by-products, such as ethylene glycol (EG) [12,17,19,20]. The simulation models are able to predict variations in molecular weight during SSP as a function of reaction time and temperature, of the changes in the concentrations of end groups, and of the levels of EG, terephthalic acid (TA), acetaldehyde (AA), and vinyl ester end groups [17,20]. In addition, the models provide an optimum molar ratio of the end-group concentration necessary to achieve maximum SSP rates [20].

The two approaches described above are examined in Chapters 3, 4, and 7. In the present chapter we emphasize a third approach, based on the concept of polymer end-group diffusion during the SSP process [13,14,17,18,21,22]. A summary and brief description of the various relevant SSP models has been considered elsewhere [20]. The end-group diffusion concept has been extended to include active and inactive polymer end groups [22]. Inactive end groups are those that are not able to participate in the reactions for a variety of reasons, such as their confinement between crystalline structures. Based on this consideration [22], a semiempirical rate expression was proposed and found to fit experimental data of the molecular weight attained during the SSP of poly (ethylene terephthalate) (PET) under various polymerization conditions. In addition, it has the ability to predict the ultimate IV or molecular weight achievable by SSP.

2.2. SPECIAL CHARACTERISTICS OF SOLID STATE POLYMERIZATION

The large amount of experimental work on the SSP of PET indicates the existence of certain specific characteristics, documented initially by Bamford and Wayne [23] and recently by Duh [21]:

- At long SSP reaction times, the intrinsic viscosity (IV) or the molecular weight of the polymer tends to level off at an ultimate value.
- The higher the SSP temperature, the higher the ultimate IV.
- For each initial IV of precursor or prepolymer, there exists an ultimate IV after SSP.
- The ultimate IV after SSP increases with increasing prepolymer IV.
- The SSP rate constant increases with increasing prepolymer IV.

In addition to the aforementioned experimental observations, it has been documented by Moore et al. [24] that after 8 hours of SSP, an amount of the PET powder was removed from the SSP reactor, melted, quenched, ground, crystallized at 130°C for 30 minutes, and then solid state–polymerized under identical environment and temperature conditions. The remelted and recrystallized sample attained a higher SSP rate and higher ultimate IV than those of the original prepolymer, regardless of the SSP time.

Relevant kinetic models for the SSP of PET should be able to explain the specific characteristics outlined above. The kinetic model developed by Duh [21,22] offers an explanation of the foregoing characteristics of the SSP of PET. Duh's model is based on the assumptions that there are two types of functional groups: active and inactive end groups. The inactive end groups include a few chemically dead end groups, the majority of which are "firmly trapped in or severely restrained by the crystalline structure" [21]. However, the concentration of end groups trapped in the crystalline structure must be very small, as indicated by the following experiments. Results of etching PET by hydrolysis [25] show that the crystalline region was not hydrolyzed, even at 210°C, indicating that water could not penetrate the crystalline region. Furthermore, Zhang et al.'s research results [26] showed that deuterated samples of PET, with equivalent end-group concentrations but different crystallinities, exhibit very similar end-group absorption in their Fourier transfer infrared (FTIR) spectra and that all the end groups are accessible to the heavy water. These results indicate that the carboxyl and hydroxyl end groups of variously crystallized PET samples are within amorphous regions, since they are accessible to deuteration.

In this chapter, an alternative approach is proposed to describe the SSP of PET, based on a model of molecular morphology and movement of polymer chains. This model can explain successfully many of the characteristics of the SSP of PET.

2.3. CLASSICAL KINETIC EQUATIONS IN SOLID STATE POLYMERIZATION

Although the chemical kinetics of PET SSP have been the subject of many studies, there is no general agreement on the relevant expression describing the chemical kinetic equation, as has been pointed out [1]. In general, the kinetic models developed for PET are based on the classical second-order or power-law kinetics. The limitation of the classical second-order kinetic application to the SSP of PET has been summarized by other authors [1,21,22], and some of the relevant concerns related to the development of kinetic models are described below.

Reactive end groups are not all available for all other reactive end groups, because of the restriction of the polymer chains and the hindered mobility of polymer chain ends due to the crystalline phase. It is reasonable to assume that polymer chain ends can only move in a limited region. DeGennes [27–29] also pointed out that the classical second-order kinetic equation $r_A = k_r$ [A] [B] cannot be used in the kinetic analysis of polymer melt reactions because of the restriction of the polymer chains on the reactants. In the SSP process, there are polymer chain restrictions on the movement of end groups as well as restrictions due to the crystalline regions of the polymer.

Most existing empirical rate equations, which describe the IV and molecular-weight increase during SSP, have a common deficiency, which has been discussed by Duh [22]. These expressions do not limit the IV or molecular-weight increase; when SSP times goes to infinity, IV and molecular weight also increase infinitely. Duh's [22] rate equation does, however, fit the experimental data very well at both short and long SSP times. It also includes characteristics of SSP leveling off after a long SSP time.

2.4. MODEL OF MOLECULAR MORPHOLOGY AND CHAIN-END MOVEMENT

2.4.1. Definition of Chain-End Length in the Amorphous Phase

The molecular morphology of a PET prepolymer during SSP can be divided into crystalline and amorphous regions. All end groups are considered to be located in the amorphous regions where SSP reactions take place. For a single macro-molecule, some chain segments are packed in the crystalline regions (lamellae), and others, in the amorphous regions. The portions within the amorphous phase can take the form of tie chains, taut tie chains, loose loops, sharp folds (adjacent reentry), and loose ends. Previous research results [30], obtained with small-angle neutron scattering, show that there is little adjacent reentry in melt-crystallized polymers, and the probability that cold-crystallized polymer will have adjacent reentry is even lower because of its restricted polymer chain mobility. The poly-mer chain in a crystalline region is perpendicular to the lamellar surface. These results mean that the length of each segment in a crystalline region is the lamella thickness (l_c).

For PET with weight fraction crystallinity (ϕ_w), one polymer molecule is taken out randomly and stretched in a straight line, as shown in Figure 2.1. The figure illustrates a single polymer molecule passing through the amorphous phase to connect two separate crystalline lamellae. In addition, it shows a loop within the amorphous phase, with two attachments in a single crystalline lamella. Two chain ends are also shown in the amorphous region. For this work, we have

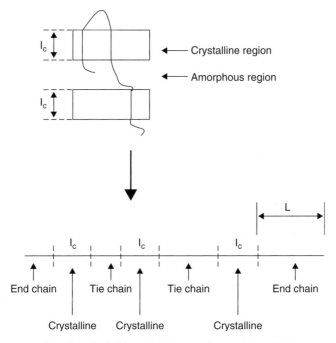

Fig. 2.1. Definition of L in a semicrystalline PET.

included all polymer chain segments residing in the amorphous phase, without chain ends, as tie molecules (n_t). The contour length (N) of this macromolecule can be calculated as

$$N = \frac{\overline{M_n}}{M_0} l_r \qquad (2.1)$$

where $\overline{M_n}$ is the number-average molecular weight of PET (g mol^{-1}), M_0 the molecular weight of the repeating unit of PET (192 g mol^{-1}), and l_r the length of a repeating unit of PET (1.08 nm) [31]. The number of lamellae (n_l) in the polymer molecule can be calculated as

$$n_l = \frac{\phi_w N}{l_c} \qquad (2.2)$$

The number of tie chains (n_t) in a single macromolecule will be

$$n_t = n_l - 1$$

Since the crystalline regions are randomly distributed in the polymer molecule, it is reasonable to assume that there is no difference between the length of the tie chain (L_t) and that of the chain end (L) (i.e., $L_t = L$).

The length of a chain end will be calculated as

$$2L = (1 - \phi_w)N - n_t L_t$$
$$L = \frac{(1 - \phi_w)N}{1 + \phi_w N / l_c} \qquad (2.3)$$

L in equation (2.3) is only an average number. The average L is on the order of 5 nm if a simple calculation is based on IV $= 0.60$ dL g^{-1}, $\phi_w = 0.50$, and $l_c = 5$ nm. Because the crystallization is random, statistically L will exhibit a Gaussian distribution for the same molecular weight, as shown in Figure 2.2(a). The cumulative distribution is shown in Figure 2.2(b). For a polymer with a certain molecular-weight distribution (MWD), the distribution of L will be similar in form to its MWD. For the convenience of qualitative analysis, a Gaussian

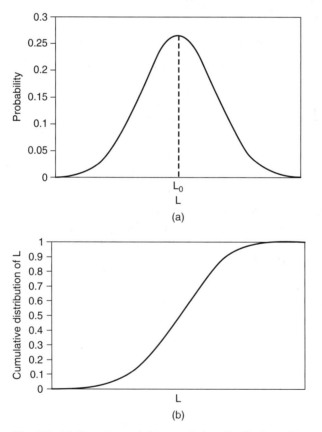

Fig. 2.2. (a) Gaussian and (b) cumulative distribution of L.

distribution is used in this chapter:

$$P(L) = \frac{1}{\sigma\sqrt{2\pi}} e^{-(L-L_0)^2/2\sigma^2} \tag{2.4}$$

where P is the probability, L the length of chain end, L_0 the average length of the chain end, σ^2 the variance, and π a constant (3.14).

2.4.2. How End Groups Move During Solid State Polymerization

The tube model of Edwards and Doi [32–34] and the reptation model of DeGennes [35,36] have been developed to describe the movement of a polymer molecule in the melt and in elastomers. According to the reptation model, in a given instant, every polymer chain is confined within a tube because it cannot intersect the neighboring chains, due to entanglements. The chain thus moves inside the tube like a snake, as shown in Figure 2.3. The diameter (δ) of a tube in the polymer melt is on the order of 5 nm [37]. The displacement s (nm) of the chain within the tube at a time t (s) can be described by the equation [27]

$$s^2 = 2D_{\text{tube}}t \tag{2.5}$$

where D_{tube} (nm^2 s^{-1}) is the diffusion coefficient, which can be related to the mobility, (μ_{tube} (nm^2 s^{-1} J^{-1}) by the Einstein equation,

$$D_{\text{tube}} = kT\mu_{\text{tube}} \tag{2.6}$$

The mobility (μ_{tube}) is inversely proportional to the chain length N [27]:

$$\mu_{\text{tube}} = \frac{\mu_0}{N} \tag{2.7}$$

where T is the temperature (K), k the Boltzmann constant (1.38×10^{-23} J K^{-1}), and μ_0 a constant (nm^3 s^{-1} J^{-1}). The equations above are applicable in the

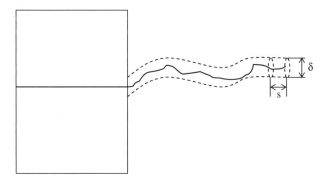

Fig. 2.3. Model of end-group movement in a semicrystalline polymer.

melt. The situation during SSP is different, because the chains are fixed in the crystalline regions. Consequently, we have $\mu_{\text{tube}} \to 0$.

In solid state polymerization, the reaction time is very long ($t \to +\infty$). This leads to

$$s^2 = 2kT\mu_{\text{tube}}t \qquad (2.8)$$

For $\mu_{\text{tube}}t = 0 \cdot \infty$, there are three possible results:

1. Zero is impossible, because there is Brownian motion.
2. Infinity is impossible, because the chain is fixed in the crystalline region.
3. Constant $= \beta$ is the only possibility, and

$$s^2 = 2k\beta T \qquad (2.9)$$

Based on (2.9), the displacement is a function of temperature only. The end group moves within a segment of tube only. For simplicity of analysis, we assume that the end group moves within a sphere. The radius (R_0) of the sphere can be calculated as follows, assuming that the volume of the tube is similar to the volume of the sphere:

$$\frac{4}{3}\pi R_0^3 \approx \pi \left(\frac{\delta}{2}\right)^2 s$$

$$R_0 = \sqrt[3]{\frac{3}{16}\delta^2 s}$$

$$= 0.57\delta^{2/3}(k\beta T)^{1/6} \qquad (2.10)$$

From the equation, R_0 is found to be a function of reaction temperature, $R_0(T)$. The physical meaning of R_0 is as the radius of a relative sphere space in which end groups move. It can be calculated roughly by (2.9). Then β can also be determined roughly. The value of β depends on polymer configuration, chemical bond angle and length, and other factors. The physical meaning of β is as the effect of polymer type on the value of R_0.

2.4.3. How Chain-End Length Affects the Movement of End Groups

In reality, not all end groups have a radius R_0, as shown in Figure 2.4. The chain-end length is a distribution of lengths. There are some long chain ends and some short chain ends. The chain-end length (L) has an effect on the radius of the sphere within which end groups move, as shown in Figure 2.4. When L is longer than $2R_0$, $R = R_0(T)$, and when L is shorter than $2R_0$, $R = L/2$. The relationship between R and L is shown in Figure 2.5(a). The average of R ($\overline{R}$) is half of the shadowed area in Figure 2.5(b). The equation $\overline{R} = \overline{R}(R_0, L)$ can be attained only if the exact distribution of L and the exact value of R_0 are measured.

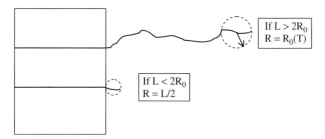

Fig. 2.4. Effect of the chain-end length on the movement of end groups.

2.5. REACTIVITY OF END GROUPS

2.5.1. Principles of Equal Reactivity of End Groups in Melt Polymerization

We quote Flory [38]: "The combined results of kinetic studies on condensation polymerization reactions and on the degradation of various polymers by reactions which bring about chain scission demonstrate quite clearly that the chemical reactivity of a functional group does not ordinarily depend on the size of the molecule to which it is attached. Exceptions occur only when the chain is so short as to allow the specific effect of one end group on the reactivity of the other to be appreciable."

2.5.2. Principles of Unequal Reactivity of End Groups in Solid State Polymerization

In solid state polymerization, the situation is much different than in melt polymerization, since crystalline regions restrict the mobility of polymer chains, and consequently, the reactivities of end groups are anticipated to be unequal and dependent on the value of R. During SSP, end groups are evenly distributed in the amorphous region and the pertinent concentration can be calculated as

$$C = \frac{n_g \rho}{1 - \phi_v} \tag{2.11}$$

where n_g is the number of end groups in 1.0 g of PET, ρ the PET density, and ϕ_v the volume fraction crystallinity.

When considering a chain end with radius of movement (R), the end groups (n_a) available to react with it are located in the sphere of its movement. The number of these end groups (n_a) can be calculated as

$$n_a = \frac{4}{3}\pi R^3 C = \frac{4}{3}\pi R^3 \frac{n_g \rho}{1 - \phi_v} \tag{2.12}$$

The reactivity of an end group depends on n_a, and in turn, n_a is proportional to R: High R values result in increased chain-end reactivity. In addition, the value of R

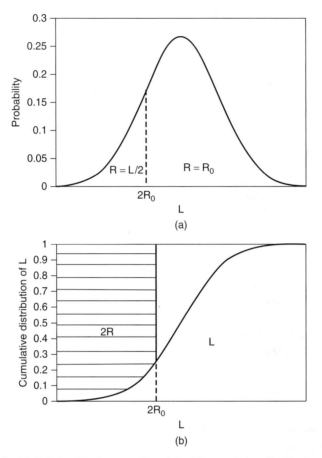

Fig. 2.5. (a) Relationship between R and L; (b) cumulative distribution of R.

depends on the L value: For end groups with L higher than $2R_0[R = R_0(T)]$, the reactivities are the same and depend primarily on temperature. For end groups with L lower than $2R_0$ ($R = L/2$), the reactivities depend on the chain-end length (R or L); more specifically, the greater the value of R or L, the higher the reactivity of the end group.

2.5.3. Sources of Low Reactivity of End Groups

The reasons for low reactivity of end groups include:

1. *Crystallization and morphology*, as discussed previously.
2. *Hydrolytic and glycolysis degradation*. The reverse reactions of esterification and transesterification are hydrolysis and glycolysis, which not only decrease the molecular weight attained but also reduce the reactivity of end groups through shortening the chain-end length (L). As shown

in Figure 2.6, after the reverse reactions of esterification or transesterification, the number-average molecular weight remains constant, but end
groups with low R (or low reactivity) are created. In other words, the
reversible reaction in a polymer differs from the reversible reaction of
small molecules, where the number of molecules and their reactivity do
not change after forward and backward reactions.

3. *Excess of carboxyl end groups* (COOH). Although there are many types of
   end groups in a PET sample, the majority are hydroxyl and carboxyl end
   groups and to a lesser extent, vinyl end groups. Hydroxyl end groups react
   not only with hydroxyl end groups through transesterification, but also with
   carboxyl end groups through esterification. This means that hydroxyl end
   groups are active to all end groups in PET. On the other hand, carboxyl
   end groups can react with hydroxyl end groups only by esterification. If
   there is an excess of carboxyl end groups in PET, some of these groups
   may have no chance to meet hydroxyl end groups. Although they have
   access to other carboxyl end groups (the spaces of end-group movement
   overlap with each other), they cannot react with each other, as shown in
   Figure 2.7. The number of inactive carboxyl end groups is proportional to
   the COOH/OH ratio.

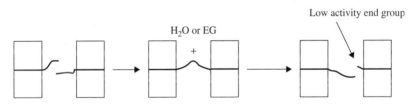

Fig. 2.6. Effect of degradation on reactivity of an end group.

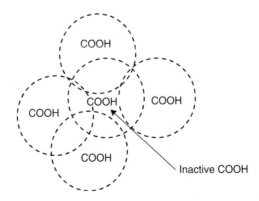

Fig. 2.7. Scheme for inactive carboxylic groups (COOH).

2.6. WHY INTRINSIC VISCOSITY LEVELS OFF DURING SOLID STATE POLYMERIZATION

2.6.1. Definition of Residual Average Radius and Residual End-Group Concentration

Every end group can move only in a limited region, which is a sphere with radius R, as shown in Figure 2.8(a), and there is no diffusion of end groups, due to the difference in end-group concentration. After a very long SSP time, some end groups cannot react with each other, even if SSP is continued for much longer. The reason is that these end groups cannot meet each other, as shown in Figure 2.8(b). That is also why the IV levels off after a very long SSP time. These end groups are defined as residual end groups, and their concentration in the amorphous region is defined as residual end-group concentration (C_r). The average R of these end groups is defined as residual R ($\overline{R}_r$).

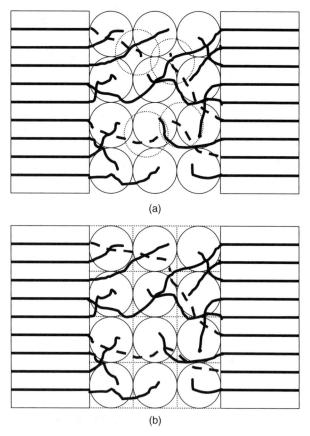

(a)

(b)

Fig. 2.8. Movement of end groups during SSP: (a) at initial stage; (b) after IV levels off (after SSP).

2.6.2. Relationship Between $\overline{R}_r$ and $\overline{R}$

At the initial stage of SSP, the average radius $(\overline{R})$ of the sphere within which an end group moves can be expressed as

$$\overline{R} = \frac{\sum_{t=0} R}{n} \tag{2.13}$$

The number of end groups with different reactivity changes after IV levels off, as shown in Figure 2.9. Initially, there are n_2 moles of end groups with $L < 2R_0$ and their R values are equal to $L/2$. Since the value of L is a distribution, as shown in Figure 2.5(a), these n_2 moles of end groups have different reactivity. The average of R is used for analysis. The average value of R should be R_0/a, where $a > 1$. For convenience of qualitative analysis, $R_0/2$ is used in this chapter. If further analysis is conducted, it will be clear that there is no difference between $R_0/2$ and R_0/a. Interested readers can contact the authors for further analysis.

$$\overline{R} = \frac{n_1 R_0 + n_2(R_0/2)}{n} \tag{2.14}$$

If $x = n_1/n$, then

$$\overline{R} = x R_0 + (1 - x)\frac{R_0}{2} \tag{2.15}$$

After a long period of SSP, the average radius $(\overline{R}_r)$ of a sphere in which an end group moves can be expressed by the equation

$$\overline{R}_r = \frac{\sum_{t=\infty} R}{n'}$$

There are n'_2 moles of end groups with $L < 2R_0$, and their R values are equal to $L/2$. Similarly, it can be assumed that $R = R_0/2$, resulting in the average residual radius:

$$\overline{R}_r = \frac{n'_1 R_0 + n'_2(R_0/2)}{n'} \tag{2.16}$$

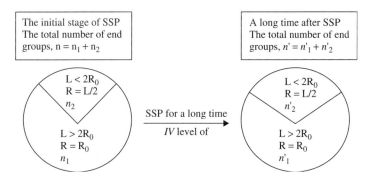

Fig. 2.9. Change of end groups after a long SSP time.

If $x' = n'_1/n'$, then

$$\overline{R}_r = x'R_0 + (1 - x')\frac{R_0}{2} \tag{2.17}$$

The possibility for an end group to react during SSP will depend on the number of end groups (n_a) available, and obviously, the larger the movement volume, the more end groups are available to it, and consequently, the higher the possibility that the end group can react. The residual number of end groups is therefore considered disproportional to n_a, also defining a constant λ.

$$n'_1 = \frac{\lambda}{n_a}n_1 \tag{2.18}$$

Combining (2.18) with (2.12), the following two equations can describe the relationship between n_1, n_2 and n'_1, n'_2:

$$n'_1 = \frac{\lambda}{(4/3)\pi R_0^3[n_g\rho/(1 - \phi_v)]}n_1 \tag{2.19}$$

$$n'_2 = \frac{\lambda}{(4/3)\pi(R_0/2)^3[n_g\rho/(1 - \phi_v)]}n_2 \tag{2.20}$$

where λ is a constant. Then

$$x' = \frac{n'_1}{n'_1 + n'_2} = \frac{\dfrac{\lambda}{(4/3)\pi R_0^3[n_g\rho/(1 - \phi_v)]}n_1}{\dfrac{\lambda}{(4/3)\pi R_0^3[n_g\rho/(1 - \phi_v)]}n_1 + \dfrac{\lambda}{(4/3)\pi(R_0/2)^3[n_g\rho/(1 - \phi_v)]}n_2}$$

$$= \frac{n_1}{n_1 + 8n_2} = \frac{xn}{xn + 8(1 - x)n}$$

$$x' = \frac{x}{8 - 7x} \tag{2.21}$$

$$\overline{R}_r = \frac{7R_0 - 6\overline{R}}{15R_0 - 14\overline{R}}R_0 \tag{2.22}$$

During SSP, the end groups with larger values of R will have higher probabilities of reacting than will end groups with smaller R values. For this reason, the distribution of R and L will change after SSP, as shown in Figures 2.10 and 2.11. The proportions of smaller R or L values will increase after SSP.

2.6.3. Relationship Between C_r, $\overline{R}_r$, and Ultimate IV

The relationship between the number-average molecular weight and IV (dL g^{-1}) [39] is given by

$$IV = 7.50 \times 10^{-4}\overline{M_n}^{0.68} \tag{2.23}$$

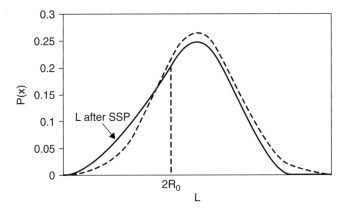

Fig. 2.10. Change in R and L distribution.

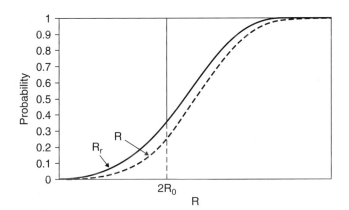

Fig. 2.11. Change in R and L cumulative distribution.

Assuming that the end groups are evenly distributed throughout the amorphous phase, the residual end-group concentration in the amorphous regions (C_r) can be calculated as

$$C_r = \frac{2\rho}{(1 - \varphi_v)\overline{M_n}} \qquad \text{mol cm}^{-3} \qquad (2.24)$$

where ϕ_v is the volume fraction crystallinity.

The volume (V) occupied by each end group can be calculated as

$$V = \frac{10^{21}}{C_r \times 6.02 \times 10^{23}} \qquad \text{nm}^3 \qquad (2.25)$$

The volume is a cube, as shown in Fig. 2.8(b), and the length of each side (S) is the diameter of the sphere within which the end group moves:

$$2\overline{R}_r = S = V^{1/3} \quad \text{nm} \tag{2.26}$$

$$C_r = \frac{10^{21}}{8 \times 6.02 \times 10^{23} \times \overline{R}_r^3} = \frac{2.08 \times 10^{-4}}{\overline{R}_r^3} \quad \text{mol cm}^{-3} \tag{2.27}$$

Combining (2.23) and (2.24) yields

$$\text{IV}_u = 7.5 \times 10^{-4} \left(\frac{9.63 \times 10^3 \rho \overline{R}_r^3}{1 - \phi_v} \right)^{0.68} \tag{2.28}$$

Since $\phi_v = (\rho - 1.333)/(1.45 - 1.333)$[31],

$$\text{IV}_u = 9.2 \times 10^{-2} \left(\frac{\rho \overline{R}_r^3}{1.45 - \rho} \right)^{0.68} \tag{2.29}$$

If a simple calculation is based on $\rho \approx 1.4$ g/cm^3, $\text{IV}_u = 1.0$ to 4.0 dL g^{-1}, $\overline{R}_r$ will be on the order of 1 to 2 nm. The calculation indicates that $\overline{R}_r$ and R_0 are much shorter than the average chain-end length (L_0), which is on the order of 5 nm.

Equations (2.3), (2.22), and (2.29) can be used to explain specific characteristics of the SSP of PET. In Sections 2.6.4 and 2.6.5 we explain the effects of SSP temperature and the initial IV on the ultimate IV after the SSP process has been completed. Other parameters, such as the effects of comonomer, remelting, moisture content, particle size, and the ratios of carboxyl to hydroxyl end-group concentrations, have been analyzed in detail elsewhere [40].

2.6.4. Explanation of Temperature Effect on Solid State Polymerization

Temperature during SSP has two effects. It causes an increase in the level of crystallinity as well as a decrease in L. If temperature is increased within the SSP temperature range from 200 to 230°C, lamella thicknesses also increase. After spherulites impinge, however, no significant number of additional lamellae form with increasing temperature. The increase in crystallinity results primarily from increasing lamella thicknesses [41]. In (2.3), describing L, lamella number, $n_l = \phi_w N / l_c$ is constant; therefore, L is a function only of ϕ_w. An increase in SSP temperature will increase ϕ_w, and consequently, decrease L; however, the effect of temperature on ϕ_w is very small within the temperature range 200 to 230°C. A 10°C increase in SSP temperature results in only a 1 to 2% increase in ϕ_w [42]. This also results in only a 1 to 2% reduction of L.

Increasing SSP temperature can also increase R_0, as shown by (2.10). The effect of increasing R_0 on $\overline{R}$ is much greater than the effect of a decrease in L. When SSP temperature changes $(T_2 > T_1)$, $\overline{R}(T_2) > \overline{R}(T_1)$ or $d\overline{R}/dT > 0$:

$$
IV_u = 9.2 \times 10^{-2} \left(\frac{\rho \overline{R}_r^{-3}}{1.45 - \rho} \right)^{0.68}
$$

$$
\frac{dIV_u}{dT} = \frac{dIV_u}{d\overline{R}_r} \frac{d\overline{R}_r}{d\overline{R}} \frac{d\overline{R}}{dT} + \frac{dIV_u}{d\rho} \frac{d\rho}{dT}
$$

where

$$
\frac{d\overline{R}_r}{d\overline{R}} = \frac{8R_0^2}{(15R_0 - 14\overline{R})^2} > 0 \qquad \frac{dIV_u}{d\overline{R}_r} = 0.19 \left(\frac{\rho}{1.45 - \rho} \right)^{0.68} \overline{R}_r^{-1.04} > 0
$$

$$
\frac{dIV_u}{d\rho} = 0.063 \overline{R}_r^{-2.04} \left(\frac{\rho}{1.45 - \rho} \right)^{-0.32} \frac{1.45}{(1.45 - \rho)^2} > 0 \qquad \frac{d\rho}{dT} > 0
$$

As a result, $\Rightarrow d\,IV_u/dT > 0$. Using calculations based on previous experiments, we can determine that the effect of density (ρ) is very small. This means that most of the IV increase is due to $d\overline{R}/dT > 0$.

2.6.5. Explanation of Initial IV Effect on Solid State Polymerization

It is known that at a given crystallization temperature, a higher-molecular-weight PET has lower crystallinity than an equivalent lower-molecular-weight PET [42]. There is enough evidence to say that lamella thicknesses are constant when IV is higher than the critical entanglement molecular weight $(\overline{M}_c = 3400$ g mol^{-1} or IV $= 0.19$ dL g^{-1}) [42–44]. The derivative of L with respect to IV in (2.3) leads to

$$
\frac{dL}{dIV} = \frac{dL}{dN} \frac{dN}{dIV} + \frac{dL}{d\phi_w} \frac{d\phi_w}{dIV}
$$

where

$$
\frac{dL}{dN} = \frac{1 - \phi_w}{(1 + \phi_w N/l_c)^2} > 0 \qquad \frac{dN}{dIV} > 0
$$

$$
\frac{dL}{d\phi_w} = \frac{-(1 + 1/l_c)N}{(1 + \phi_w N/l_c)^2} < 0 \qquad \frac{d\phi_w}{dIV} < 0
$$

As a result, $\Rightarrow dL/d\,IV > 0$. This means that L increases with increasing N, and that L increases with increasing molecular weight. Therefore, when initial

$IV_2 > IV_1$, $\overline{R}(IV_2) > \overline{R}(IV_1)$ or $d\overline{R}/dIV > 0$,

$$IV_u = 9.2 \times 10^{-2} \left(\frac{\rho \overline{R}_r^3}{1.45 - \rho} \right)^{0.68}$$

$$\frac{dIV_u}{dIV} = \frac{dIV_u}{d\overline{R}_r} \frac{d\overline{R}_r}{d\overline{R}} \frac{d\overline{R}}{dIV} + \frac{dIV_u}{d\rho} \frac{d\rho}{dIV}$$

where

$$\frac{dIV_u}{d\overline{R}_r} = 0.19 \left(\frac{\rho}{1.45 - \rho} \right)^{0.68} \overline{R}_r^{1.04} > 0 \quad \text{and} \quad \frac{d\overline{R}_r}{d\overline{R}} = \frac{8R_0^2}{(15R_0 - 14\overline{R})^2} > 0$$

The expression $(dIV_u/d\overline{R}_r)(d\overline{R}_r/d\overline{R})(d\overline{R}/dIV)$ increases IV_u.

$$\frac{dIV_u}{d\rho} = 0.063\overline{R}_r^{2.04} \left(\frac{\rho}{1.45 - \rho} \right)^{-0.32} \frac{1.45}{(1.45 - \rho)^2} > 0$$

$$\frac{d\rho}{dIV} < 0$$

The expression $(dIV_u/d\rho)(d\rho/dIV)$ decreases IV_u.

The experimental results show that higher initial precursor IV gives higher product ultimate IV. After calculation it is found that the effect of density on IV_u is very small; therefore, it can be concluded that the reason for the effect of initial IV is

$$\frac{dIV_u}{d\overline{R}_r} \frac{d\overline{R}_r}{d\overline{R}} \frac{d\overline{R}}{dIV} > 0 \quad \text{or} \quad \frac{d\overline{R}}{dIV} > 0$$

2.7. SOLID STATE POLYMERIZATION KINETICS

2.7.1. Kinetic Equation of Ideal Solid State Polymerization

Ideal SSP is herewith defined as a process wherein the radii of spheres within which all end groups move are the same and equal to R_0. According to this definition, the radii do not change during SSP ($R_r = R_0$), there is no degradation, and all end groups are of the same chemical species. The assumptions above are valid when the initial IV is very high, all end groups are hydroxyls, and the particle size is very small. The higher the IV or molecular weight, the greater the L. The distribution of L will move to the right side in Figure 2.5(a). Consequently, more end groups have a movement diameter of $2R_0$. A hydroxyl end group can react to all other end groups, but a carboxyl end group can only react to hydroxyl. A small particle size avoids degradation and easily removes condensates.

In an ideal SSP process, the end groups are evenly distributed. If the radii of spheres within which the end groups move are R_0 and the distance between two

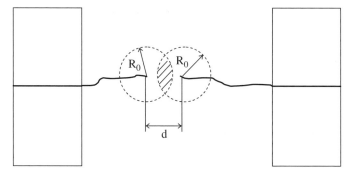

Fig. 2.12. Overlapping of two spheres of end groups.

nearby end groups is d, the overlapping volume between two spheres is indicated by the shaded area in Figure 2.12 and is described by

$$v = \frac{4\pi R_0^3}{3} - \pi R_0^2 d + \frac{\pi d^3}{12} \qquad 0 < d < 2R_0 \qquad (2.30)$$

A given end group moves evenly within its own sphere of radius R_0. This means that the possibility that the end group appears in any point of its own sphere is the same. Only when the end group appears in the overlapping volume can it react with other end groups. The probability that the end group will appear in the overlapping volume can be calculated as

$$p = \frac{4\pi R_0^3/3 - \pi R_0^2 d + \pi d^3/12}{4\pi R_0^3/3} \qquad (2.31)$$

The rate at which one end group reacts with another end group depends on the overlapping volume or the probability. The kinetic equation of ideal SSP is presented by

$$\frac{dC}{dt} = -k_r(pC)^2 \qquad (2.32)$$

The units of C, t, and k_r are, respectively, mol cm^{-3}, h, and cm^3 mol^{-1} h^{-1}. Equation (2.32) is reasonable, as can be seen by observing its boundary conditions. If $d \geq 2R_0$, $p = 0$ and $dC/dt = 0$, which means that end groups cannot meet to react with each other, since there is no overlapping volume. If $R_0 \to +\infty$, $p \to 1$ and the rate is $dC/dt = -k_r C^2$, which represents the kinetics of small molecules, where the reacting species mobility is high and the respective movement volume can be considered infinite ($R_0 \to +\infty$).

It is ideal SSP, as there is only amorphous area. If we think that every end group occupies one cube, the distance between neighbor end groups will be the side length of the cube, as shown in Figures 2.8(b) and 2.12. If there are n (mol) end groups, the volume they occupy will be

$$V = (n \cdot 6.02 \times 10^{23})d^3$$

$$C = \frac{n}{V} = \frac{n}{(n \cdot 6.02 \times 10^{23})d^3} = \frac{1}{6.02 \times 10^{23}d^3} \quad (2.33)$$

where d units are centimeter or nanometers.

$$C_r = \frac{1}{6.02 \times 10^{23} \times (2R_0)^3} \quad (2.34)$$

$$\frac{dC}{dt} = -k_r \left[\frac{(4\pi R_0^3/3) - \pi R_0^2 d + (\pi d^3/12)}{4\pi R_0^3/3} C \right]^2 \quad (2.35)$$

Dividing (2.34) by (2.33) gives

$$\frac{d}{R_0} = 2 \left(\frac{C_r}{C} \right)^{1/3} \quad (2.36)$$

Substituting (2.36) into (2.35) gives

$$\frac{dC}{dt} = -k_r \left(C - \frac{3}{2} C_r^{1/3} C^{2/3} + \frac{1}{2} C_r \right)^2 \quad (2.37)$$

$$\frac{dC}{dt} = -k_r \left[C - \frac{C_r^{1/3}}{2} (3C^{2/3} - C_r^{2/3}) \right]^2 \quad (2.38)$$

In (2.38), if the reaction is of small molecules, then $C_r = 0$, $\rightarrow dC/dt = -k_r C^2$. When the reaction proceeds for a long time, $C \rightarrow C_r$, and we have $dC/dt \rightarrow 0$, which means that the reaction stops.

A method that can be used to verify the IV increase during SSP utilizes (2.38) as follows:

1. Calculate the molecular weight from IV using

$$\overline{M_n} = 3.92 \times 10^4 \text{IV}^{1.47} \quad (2.39)$$

2. Calculate the concentration (C) of end groups in the amorphous phase according to

$$C = \frac{2\rho}{(1 - \phi_v)M_n} \quad (2.40)$$

where ρ is the density of PET and ϕ_v is the volume fraction crystallinity.

3. Calculate $\Delta C/\Delta t$ at a given SSP time.

4. Consider $\Delta C/\Delta t \approx F = dC/dt$.
5. Plot $\Delta C/\Delta t$ versus C using the software of a Sigma plot.
6. Choose the user-defined regression and enter it into (2.38), whose dC/dt is also F or $\Delta C/\Delta t$.
7. Choose suitable initial values for k_r and C_r.
8. Simulate the experimental data and obtain k_r and C_r by the simulation.
9. The ultimate IV can be calculated from C_r by (2.39) and (2.40).

The procedure above was applied to the data given in Figure 2.13 and results are tabulated in Tables 2.1 to 2.3 [40]. The corresponding simulations of the data are given in Figures 2.14 to 2.16 utilizing the following equation:

$$F = \frac{\Delta C}{\Delta t} \approx \frac{dC}{dt} = -k_r \left[C - \frac{C_r^{1/3}}{2} (3C^{2/3} - C_r^{2/3}) \right]^2 \qquad (2.41)$$

The values of k_r and C_r are obtained by regression. Ultimate $\overline{M_n}$ values are calculated from C_r by (2.40). Ultimate IV values are then calculated from $\overline{M_n}$ using (2.39). The results are shown in Table 2.4. The end groups move in a larger space at a higher temperature. The higher the temperature, the higher the reaction kinetic constant k_r, and the higher the ultimate IV.

2.7.2. Empirical Kinetic Equation of Real Solid State Polymerization

The aforementioned rate expressions can also be used to simulate the IV increase during SSP, and the characteristics of IV leveling off after a long time can be

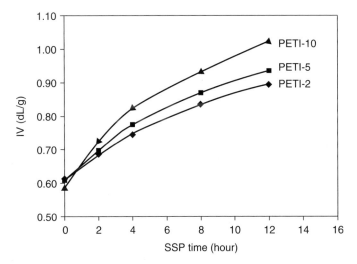

Fig. 2.13. IV increase during SSP ($T = 220°C$) for PETI-2, PETI-5, and PETI-10.

TABLE 2.1. Calculation of $\Delta C/\Delta t$ for PETI-2

SSP Time (h)	IV (dL g^{-1})	$\overline{M_n}$ (g mol^{-1})	C (mol cm^{-3})	$\Delta C/\Delta t$ (average)	$\Delta C/\Delta t$ (at C)
0	0.610	19,046	0.000343	—	—
1	0.650	20,911	0.000313	−3.1E-05	−2.8E-05
2	0.690	22,830	0.000286	−2.6E-05	−2.2E-05
3	0.720	24,305	0.000269	−1.7E-05	−1.5E-05
4	0.745	25,556	0.000256	−1.3E-05	−1.3E-05
5	0.770	26,827	0.000244	−1.2E-05	−1.1E-05
6	0.790	27,858	0.000235	−9.0E-06	−9.8E-06
7	0.815	29,164	0.000224	−1.1E-05	−9.2E-06
8	0.835	30,223	0.000216	−7.8E-06	−6.7E-06
9	0.850	31,024	0.000211	−5.6E-06	−5.5E-06
10	0.865	31,833	0.000205	−5.3E-06	−5.2E-06
11	0.880	32,648	0.000200	−5.1E-06	−5.0E-06
12	0.895	33,470	0.000195	−4.9E-06	—

TABLE 2.2. Calculation of $\Delta C/\Delta t$ for PETI-5

SSP Time (h)	IV (dL g^{-1})	$\overline{M_n}$ (g mol^{-1})	C (mol cm^{-3})	$\Delta C/\Delta t$ (average)	$\Delta C/\Delta t$ (at C)
0	0.610	19,046	0.000343	—	—
1	0.655	21,148	0.000309	−3.4E-05	−3.1E-05
2	0.700	23,319	0.000280	−2.9E-05	−2.5E-05
3	0.740	25,304	0.000258	−2.2E-05	−1.9E-05
4	0.775	27,084	0.000241	−1.7E-05	−1.4E-05
5	0.800	28,378	0.000230	−1.1E-05	−1.1E-05
6	0.825	29,692	0.000220	−1.0E-05	−9.8E-06
7	0.850	31,024	0.000211	−9.5E-06	−8.3E-06
8	0.870	32,104	0.000204	−7.1E-06	−6.9E-06
9	0.890	33,195	0.000197	−6.7E-06	−5.7E-06
10	0.905	34,021	0.000192	−4.8E-06	−4.7E-06
11	0.920	34,853	0.000188	−4.6E-06	−4.5E-06
12	0.935	35,692	0.000183	−4.4E-06	—

obtained by theoretical derivative. However, the rate equation cannot be integrated by a common method. Therefore, several empirical kinetic equations of SSP are introduced to describe the fact that IV levels off after a long time. They are used conveniently, although they are empirical.

Duh [21] introduced the kinetic equation (2.42) to simulate the SSP process, especially for the a long period of time. In this expression, C_i represents the concentration of inactive end groups which are trapped in the crystalline region. The semiempirical kinetic equation is powerful in its simplicity and it is also

TABLE 2.3. Calculation of $\Delta C/\Delta t$ for PETI-10

SSP Time (h)	IV (dL g^{-1})	$\overline{M}_n$ (g mol^{-1})	C (mol cm^{-3})	$\Delta C/\Delta t$ (average)	$\Delta C/\Delta t$ (at C)
0	0.585	17,909	0.000365	—	—
1	0.660	21,386	0.000306	−5.9E-05	−4.8E-05
2	0.720	24,305	0.000269	−3.7E-05	−3.1E-05
3	0.770	26,827	0.000244	−2.5E-05	−2.4E-05
4	0.825	29,692	0.000220	−2.4E-05	−1.8E-05
5	0.860	31,563	0.000207	−1.3E-05	−1.0E-05
6	0.880	32,648	0.000200	−6.9E-06	−8.3E-06
7	0.910	34,298	0.000191	−9.6E-06	−7.8E-06
8	0.930	35,412	0.000185	−6.0E-06	−6.5E-06
9	0.955	36,821	0.000177	−7.1E-06	−6.2E-06
10	0.975	37,960	0.000172	−5.3E-06	−5.8E-06
11	1.000	39,400	0.000166	−6.3E-06	−6.1E-06
12	1.025	40,857	0.000160	−5.9E-06	—

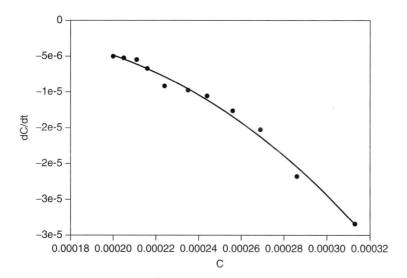

Fig. 2.14. Regression of theoretical equation (2.41) for PETI-2 for SSP at 220°C.

similar to (2.38), which we obtained by theoretical derivation. More specifically, the variable expression of $(C_r^{1/3}/2)(3C^{2/3} - C_r^{2/3})$ in (2.38) is considered as a constant C_i in Duh's equation:

$$\frac{dC}{dt} = -2k_r(C - C_i)^2 \tag{2.42}$$

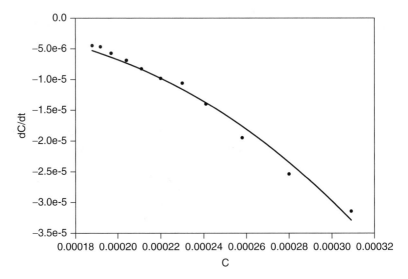

Fig. 2.15. Regression of theoretical equation (2.41) for PETI-5 for SSP at 220°C.

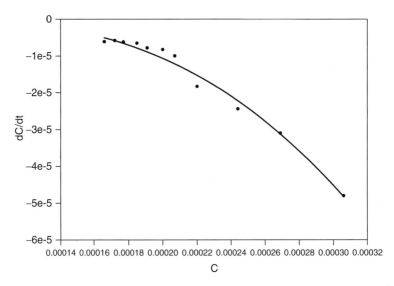

Fig. 2.16. Regression of theoretical equation (2.41) for PETI-10 for SSP at 220°C.

TABLE 2.4. Results of k_r, C_r, and Ultimate IV for SSP at 220° C

	k_r (cm^3 mol^{-1} h^{-1})	C_r (mol cm^{-3})	Ultimate IV (dL g^{-1})
PETI-2	5000	5.6910e-5	2.06
PETI-5	4401	4.5929e-5	2.39
PETI-10	6146	4.3188e-5	2.50

We also propose the following additional convenient empirical kinetic equation:

$$IV = IV_0 + (IV_u - IV_0)e^{1/k_r t^{0.5}} \qquad (2.43)$$

where k_r is the kinetic constant, t the time, and IV_0 represents the initial IV (dL g^{-1}). If $\ln(IV - IV_0)$ is plotted as a function of $1/t^{0.5}$, IV_u can be obtained from the intercept and k_r from the slope of the resulting line:

$$\ln(IV - IV_0) = \ln(IV_u - IV_0) - \frac{1}{k_r t^{0.5}} \qquad (2.44)$$

The experimental data given in Figure 2.13 were analyzed according to (2.44), and the resulting plot is shown in Figure 2.17. Ultimate IV values, obtained by applying kinetic equations (2.38), (2.42), and (2.43) to these data, are summarized in Table 2.5. The ultimate IV values calculated using the two empirical equations, (2.42) by Duh and (2.43) by Zhang and Jabarin, are very similar for all three samples studied. The theoretical equation (2.38) developed in this

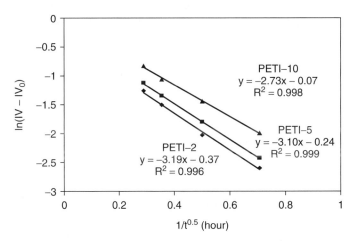

Fig. 2.17. Linear regressions of equation (2.44) for PETI-2, PETI-5, and PETI-10 at SSP $T = 220°C$.

TABLE 2.5. Ultimate IV Results Obtained Using Three Kinetic Equations for SSP at 220° C

	Ultimate IV (dL g^{-1})		
	Eq. (2.38)	Eq. (2.42)	Eq. (2.43)
PETI-2	2.06	1.28	1.30
PETI-5	2.39	1.38	1.40
PETI-10	2.50	1.54	1.52

chapter, however, gives somewhat higher ultimate IV values. The potential reason is that (2.38) comes from ideal SSP. Ideal SSP assumes that all end groups are the same (hydroxyl) and move in the radius of R_0, and that no degradation occurs. These ideal conditions may lead to higher ultimate IV values than those obtained using commercial PET resins, since commercial resins can experience thermal and hydrolytic degradation during melting and processing. Commercial resins also contain varieties of end groups, such as carboxyl and vinyl ester end groups, in addition to hydroxyl end groups.

2.8. CONCLUSIONS

In this chapter a solid state polymerization approach based on the concept of polymer end-group diffusion during the relevant processes has been emphasized. The SSP reaction is described though molecular morphology and movement of polymer chains, focusing on the restrictions set by the solid state nature of SSP and defining critical process parameters such as residual end-group concentration and average movement radius. Through the relevant mechanism, a kinetics model is constructed that explains the intrinsic viscosity leveling off at long SSP times and the effects of reaction temperature and initial IV on the SSP rate.

REFERENCES

1. Vouyiouka S, Karakatsani E, Papaspyrides C. Solid state polymerization. *Prog. Polym. Sci.* 2005;30:10–37.
2. Culbert B, Christel A. Continuous solid-state polycondensation of polyesters. In: *Modern Polyesters*, Scheirs J, Long TE, eds. Wiley, Chichester, UK, 2003, Chap. 4.
3. Goltner W. Solid-state polycondensation of polyester resins. In: *Fundamentals and Industrial Production in Modern Polyesters*, Scheirs J, Long TE, eds. Wiley, Chichester, UK, 2003, Chap. 5.
4. Jabarin SA. Polyethylene terephthalate, chemistry and preparation. In: *Polymeric Materials Encyclopedia*, Salamone JC, ed. CRC Press, Boca Raton, FL, 1996.
5. Pilati F. Solid-state polymerization. *Comprehensive Polym. Sci.* 1989;5:201–216.
6. Ravindranath K, Mashelkar RA. Polyethylene terephthalate: 1. Chemistry, Thermodynamics and Transport Properties. *Chem. Eng. Sci.* 1986;41(9):2197–2214.
7. Huang B, Walsh J. Solid-phase polymerization mechanism of poly(ethylene terephthalate) affected by gas flow velocity and particle size. *Polymer*. 1998;39:6991–6999.
8. Gao Q, Huang N, Gerking L. Modelling of solid state polycondensation of poly(ethylene terephthalate). *Chem. Eng. Sci.* 1997;52:371–376.
9. Ravindranath K, Mashelkar R. Modeling of poly(ethylene terephthalate) reactors. IX. Solid state polycondensation process. *J. Appl. Polym. Sci.* 1990;39:1325–1345.
10. Chen F, Griskey R, Beyer G. Thermally induced solid state polycondensation of nylon 66, nylon 6–10 and polyethylene terephthalate. *AIChE J.* 1969;15:680–685.
11. Chang T. Kinetics of thermally induced solid state polycondensation of poly(ethylene terephthalate). *Polym. Eng. Sci.* 1970;10:364–368.

12. Mallon F, Ray W. Modeling of solid-state polycondensation. I. Particle models. *J. Appl. Polym. Sci.* 1998;69:1233–1250.

13. Chen S, Chen F. Kinetics of polyesterification: III. Solid state polymerization of polyethylene terephthalate. *J. Polym. Sci. A.* 1987;25:533–549.

14. Wu D, Chen F, Li R. Reaction kinetics and simulations for solid-state polymerization of poly(ethylene terephthalate). *Macromolecules*. 1997;30:6737–6742.

15. Jabarin S, Lofgren E. Solid state polymerization of poly(ethylene terephthalate): kinetic and property parameters. *J. Appl. Polym. Sci.* 1986;32:5315–5335.

16. Karayannidis G, Sideridou I, Zamboulis, D, Bikiaris G, Lazaridis N, Wilmes A. Solid-state polycondensation of poly(ethylene terephthalate) films. *Angew. Makromol. Chem.* 1991;192:155–168.

17. Kang C. Modeling of solid-state polymerization of poly(ethylene terephthalate). *J. Appl. Polym. Sci.* 1998;68:837–846.

18. Devotta I, Mashelkar R. Modelling of polyethylene terephthalate reactors: X. Comprehensive model for solid-state polycondensation process. *Chem. Eng. Sci.* 1993;10:1859–1867.

19. Yoon K, Kwon M, Jeon M, Park O. Diffusion of ethylene glycol in solid state poly(ethylene terephthalate). *Polymer*. 1993;25:219–226.

20. Kim T, Jabarin S. Solid-state polymerization of poly(ethylene terephthalate). II. Modeling study of the reaction kinetics and properties. *J. Appl. Polym. Sci.* 2003;89:213–227.

21. Duh B. Reaction kinetics for solid-state polymerization of poly(ethylene terephthalate). *J. Appl. Polym. Sci.* 2001;81:1748–1761.

22. Duh B. Effects of the carboxyl concentration on the solid-state polymerization of poly(ethylene terephthalate). *J. Appl. Polym. Sci.* 2002;83,1288–1304.

23. Bamford C, Wayne R. Polymerization in the solid phase: a polycondensation reaction. *Polymer*. 1969;10:661–681.

24. Moore I, Rule M, Wicher T. Process for preparing high molecular weight polyesters (Eastman Kodak Company). U.S. Patent 4,446,303, 1984.

25. Miyagi A, Wunderlich B. Etching of crystalline poly(ethylene terephthalate) by hydrolysis. *J. Polym. Sci. Polym. Phys.* 1972;10:2073–2083.

26. Zhang H, Rankin A, Ward I. Determination of the end-group concentration and molecular weight of poly(ethylene naphthalene-2,6-dicarboxylate) using infra-red spectroscopy. *Polymer*. 1996;37:1079–1085.

27. DeGennes PG. *Introduction to Polymer Dynamics*. Cambridge University Press, Cambridge, UK, 1990.

28. DeGennes P. Kinetics of diffusion-controlled processes in dense polymer systems. I. Nonentangled regimes. *J. Chem. Phys.* 1982;76:3316–3321.

29. DeGennes P. Kinetics of diffusion-controlled processes in dense polymer systems: II. Effects of entanglements. *J. Chem. Phys.* 1982;76:3322–3326.

30. Sperling L. *Introduction to Physical Polymer Science*. Wiley, New York, 1992, pp. 244–252.

31. Daubeney RP, Bunn CW, Brown CJ. The crystal structure of polyethylene terephthalate. *Proc. R. Soc.* 1954;226A:531–542.

32. Edwards S. The statistical mechanics of polymerized material. *Proc. Phys. Soc.* 1967;92:9–16.

33. Doi M, Edwards S. *The Theory of Polymer Dynamics*. Clarendon Press, Oxford, UK, 1986.

34. Doi M. *Introduction of Polymer Physics*. Clarendon Press, Oxford, UK, 1996.

35. DeGennes P. Reptation of a polymer chain in the presence of fixed obstacles. *J. Chem. Phys.* 1971;55:572–579.

36. DeGennes P. *Scaling Concepts in Polymer Physics*. Cornell University Press, Ithaca, NY, 1979.

37. DeGennes P. *Phys. Today.* June 1983, p. 33.

38. Flory P. *Principles of Polymer Chemistry*. Cornell University Press, Ithaca, NY, 1953, pp. 69–105.

39. Moore L. Comparison of solution properties of poly(ethylene terephthalate) and poly(1,4-cyclohexylenedimethylene terephthalate). *Am. Chem. Soc. Div. Polym. Chem. Prepr.* 1960;1(1):234–243.

40. Zhang H. Solid state polymerization of PET. Ph.D. dissertation, University of Toledo, Polymer Institute, 2004.

41. Schultz J. *Polymer Crystallization*. Oxford University Press, New York, 2001, pp. 57–92.

42. Kim T, Lofgren E, Jabarin S. Solid-state polymerization of poly(ethylene terephthalate): I. Experimental study of the reaction kinetics and properties. *J. Appl. Polym. Sci.* 2003;89:197–212.

43. Fetters L, Lohse D, Richter D, Witten T, Zirkel A. Connection between polymer molecular weight, density, chain dimensions, and melt viscoelastic properties. *Macromolecules*. 1994;27:4639–4647.

44. Fetters L, Luhse D, Milner S, Graessley W. Packing length influence in linear polymer melts on the entanglement, critical, and reptation molecular weights. *Macromolecules*. 1999;32:6847–6851.

3

KINETIC ASPECTS OF POLYESTER SOLID STATE POLYMERIZATION

F. PILATI

University of Modena and Reggio Emilia, Modena, Italy

M. TOSELLI

University of Bologna, Bologna, Italy

Solid State Polymerization, Edited by Constantine D. Papaspyrides and Stamatina N. Vouyiouka
Copyright © 2009 John Wiley & Sons, Inc.

3.1. INTRODUCTION

Solid-state polymerization (SSP) is a process used widely in industry to obtain high-molecular-weight polyesters for uses such as bottles, frozen-food trays, and tire cords, which it would be difficult or impossible to obtain using a melt technique. The reason for the latter derives from the relatively high rate of side chain–scission reaction, which at the high temperature required to maintain polyesters in the molten state with relatively low viscosity, counterbalances or becomes faster than chain-growth reactions. The SSP of solid prepolymers of relatively low intrinsic viscosity (IV) is carried out at significantly lower temperatures than that of melt polymerization (well above the glass-transition temperature, T_g, but below the melting temperature, T_m) and takes advantage of the high activation energy typical of side reactions to reduce the side chain–scission rate more than the polycondensation rate. SSP can also be useful, as it allows us to obtain polyesters of either improved quality [e.g., in the case of poly(ethylene terephthalate) (PET)] or at a higher yield [such in the case of poly(L-lactic acid) (PLLA)]. Again, the reason is that SSP allows overcoming problems related to side reactions, which can lead to the formation of acetaldehyde (AA), carboxyl end groups, and diethylene glycol moieties in the case of PET, and of high amounts of cyclic oligomers in the case of PLLA.

Most reactions during melt polymerization of polyesters [1–10] can, in principle, also occur during SSP, and a discussion of the possible role of these reactions is reported first to provide the reader with some criteria to select the reactions that have to be considered when developing a kinetic model. The reactions involved in molecular-weight (MW) growth are equilibrium reactions, typically with a low value of the equilibrium constant ($K_{eq} = 0.1$ to 1), and low-molecular-weight by-products have to be removed to shift the equilibrium toward high-MW polyesters. The SSP is therefore a process that involves chemical reactions as well as physical phenomena, such as the diffusion of low-MW by-products within solid particles to the particle surface and then into the gas phase. The overall rate of the SSP process can be controlled by one or more of these phenomena, depending on operating conditions. Reaction kinetics and transport phenomena may therefore play a key role in the control and optimization of SSP processes. Kinetic models can be developed to describe the experimental results of the SSP of a given polyester under certain operating conditions, and to investigate and predict the effects of variables (such as temperature, particle sizes, type and flow rate of the inert gas stream or vacuum employed to remove by-products) without performing time-consuming and expensive experiments.

Many different kinetic models have been proposed in the scientific literature; they include very simple empirical as well as more complex models developed by taking into account several reactions and the diffusion of various by-products. A model based on end-group diffusion and the relevant probability for reaction has been analyzed in Chapter 2. Whereas simple empirical models are suitable to describe SSP only under well-defined conditions, more complex models should in principle be able to describe SSP under different operating conditions. Of course, the several reactions involved in SSP and the related kinetic parameters are different for various polyesters, so that models suitable to describe SSP for a given polyester are not necessarily valid for another polyester. For this reason, in the last part of this chapter we summarize the information available for the SSP of the most important polyesters and discuss the main features that should be included in a kinetic model for each type of polyester.

The design of SSP reactors can also play a relevant role in the optimization of a SSP industrial process [11–14]. However, the discussion relative to this point is beyond the scope of this chapter, where only the SSP occurring in a single particle is discussed. Even though many papers, some books, and good reviews have dealt with the SSP of polyesters [15–18], there are still a lot of open questions in the scientific literature, and none of the kinetic models proposed is completely satisfactory. The aim of this chapter is to discuss the main kinetic aspects of the SSP of polyesters and to present a critical review of the kinetic models that have been proposed in the scientific literature to describe and, hopefully, to optimize SSP processes.

3.2. PHENOMENA INVOLVED IN SOLID STATE POLYMERIZATION OF POLYESTERS

Before starting with the kinetic models proposed in the literature, it is convenient to consider and to discuss in some detail the main chemical reactions and the physical phenomena involved in SSP.

3.2.1. Possible Reactions in Solid State Polymerization of Polyesters

It is generally accepted that reactions during SSP take place in the amorphous regions where, due to a temperature well above T_g, the chain mobility is high enough to allow reactions to take place. All the reactions occurring during the polymerization of polyesters in the molten state can in principle also occur during SSP, and we discuss them below, emphasizing their specific features with respect to the SSP processes [1,3–5,17,19,20].

a. Alcoholysis or Transesterification Reactions The reaction of an alkyl hydroxyl group with an ester group is generally called an *alcoholysis* or *transesterification reaction*. When the terminal ester group is involved, the reaction leads to the formation of a glycol molecule along with a polyester chain having a higher MW:

$$\text{\textasciitilde\textasciitilde\textasciitilde R} - \overset{\overset{\text{O}}{\|}}{\text{C}} - \text{O} - \text{R}' - \text{OH} + \text{HO} - \text{R}' - \text{O} - \overset{\overset{\text{O}}{\|}}{\text{C}} - \text{R\textasciitilde\textasciitilde\textasciitilde} \underset{k_{-1a}}{\overset{k_{1a}}{\rightleftharpoons}} \text{\textasciitilde\textasciitilde\textasciitilde R} - \overset{\overset{\text{O}}{\|}}{\text{C}} - \text{O} - \text{R}' - \text{O} - \overset{\overset{\text{O}}{\|}}{\text{C}} - \text{R\textasciitilde\textasciitilde\textasciitilde} + \text{HO} - \text{R}' - \text{OH}$$

$$(3.1a)$$

In most cases, this is the reaction contributing most to the molecular-weight increase during the SSP of polyesters.

When the nucleophilic attack of a terminal hydroxyl group occurs on the third-to-last carbonyl group with respect to the chain end, oligomers bearing either two hydroxyl groups or one hydroxyl and one carboxyl group are formed according to the following reaction (where the attacked carbonyl group is underlined):

$$\text{\sim\sim R-}\underset{O}{\overset{O}{\underset{\|}{C}}}\text{-O-R'-O-}\underset{O}{\overset{O}{C}}\text{-R-}\overset{O}{C}\text{-OX} \quad \underset{k_{-1b}}{\overset{k_{1b}}{\rightleftharpoons}}$$

(reaction scheme 3.1b, with products on right)

$$+ \text{ HO-R'-O-C-R} \sim\sim\text{(=O)} \qquad \sim\sim\text{R-C(=O)-O-R'-O-C-R}\sim\sim \qquad + \text{ HO-R'-O-C-R-C-OX}$$

X = H or R'−OH

(3.1b)

Similar products are formed when a glycol molecule attacks a carbonyl group (either the next to last or the third-to-last ester group:

$$\text{\sim\sim R-}\underset{}{\overset{O}{C}}\text{-O-R'-O-}\overset{O}{C}\text{-R-}\overset{O}{C}\text{-OX} \quad \underset{k_{-1c}}{\overset{k_{1c}}{\rightleftharpoons}}$$

$$+ \text{ HO-R'-OH} \qquad \qquad \sim\sim\text{R-C(=O)-O-R'-OH} \qquad + \text{ HO-R'-O-C-R-C-OX}$$

X = H or R'−OH

(3.1c)

Even though the formation of these oligomers is usually neglected in SSP modeling, it has been demonstrated that the oligomers can be removed under some operating conditions (very small particles, high temperature, high vacuum) and can give a not-negligible contribution to the molecular weight increase, as evidenced in some papers [19,21,22]. In the absence of specific kinetic data it can be assumed that the kinetic constant is the same as in reaction (3.1a), as the mechanism of all these reactions is expected to be very similar.

When the transesterification reaction occurs as an intermolecular reaction of a terminal hydroxyl group with an ester group in the backbone of another polyester chain,

$$\text{\sim\sim R-}\overset{O}{C}\text{-O-R'-O-}\overset{O}{C}\text{-R}\sim\sim \quad \underset{k_{-1d}}{\overset{k_{1d}}{\rightleftharpoons}}$$

$$+ \text{ HO-R'-O-C-R}\sim\sim\text{(=O)} \qquad \sim\sim\text{R-C(=O)-O-R'-O-C-R}\sim\sim \qquad + \text{ HO-R'-O-C-R}\sim\sim$$

(3.1d)

also referred to as an ester interchange reaction, it leads to changes in the MWs of the original chains, but no volatile products are formed, and as a consequence, it cannot contribute to the MW increase during SSP. However, reaction (3.1d) can contribute to rearranging the molecular-weight distribution (MWD), and as discussed later, it may play a significant role, with respect to the mobility of reactive end groups within the amorphous domains (chemical diffusion), in the last stage of SSP.

A similar transesterification reaction can also occur as the intramolecular reaction of a terminal hydroxyl group with an ester group along the same chain. Cyclic molecules are formed in this case according to the reaction

$$\text{www-O}\!\left(\!\!\begin{array}{cc} O & O \\ \| & \| \\ C-R-C-O-R'-O \end{array}\!\!\right)_{\!\!n}\!\!\begin{array}{cc} O & O \\ \| & \| \\ C-R-C-O-R'-OH \end{array} \underset{k_{-1e}}{\overset{k_{1e}}{\rightleftharpoons}} \begin{array}{l} \text{www-O}\!\left(\!\!\begin{array}{cc} O & O \\ \| & \| \\ C-R-C-O-R'-O \end{array}\!\!\right)_{\!\!n-m}\!\!H \\[2em] + \left(\!\!\begin{array}{cc} O & O \\ \| & \| \\ C-R-C-O-R'-O \end{array}\!\!\right)_{\!\!m+1} \end{array}$$

$$(3.1e)$$

The relative number of cyclic molecules containing $m + 1$ repeating units depends on the type of polyester [1,23–26]; when m is small $(0, 1, 2, \ldots)$ the cyclic oligomers can diffuse through the polyester particles and can be removed from particle surfaces. Even though it has been demonstrated that this reaction can occur in almost all polyesters, its role in the final MW is negligible, and this reaction is typically ignored in kinetic models. However, for some polyesters, it may happen that the removal of these cyclic oligomers can create problems in an SSP plant due to pipe obstruction and/or generation of "dust," or can lead to reduction in the polyester yield (as, e.g., for PLLA). In these cases it should be included in the kinetic model in order to optimize this aspect of the SSP process.

The rate of all these transesterification reactions is typically very low in the absence of suitable catalysts; protic acids (inorganic or organic) could be used to increase the reaction rate; however, too strong protic acids typically increase also, and often still more, the rate of side reactions, and therefore metal-based transesterification catalysts are commonly used [1].

b. Esterification Reactions   The reaction of one hydroxyl group with one carboxyl group is generally called an *esterification* (or *direct esterification*) *reaction* and leads to the formation of one water molecule along with a polyester molecule of higher MW:

$$\text{www-R}\!-\!\overset{\displaystyle O}{\overset{\|}{C}}\!-\!OH + HO\!-\!R'\!-\!O\!-\!\overset{\displaystyle O}{\overset{\|}{C}}\!-\!R\text{-www} \underset{k_{-2a}}{\overset{k_{2a}}{\rightleftharpoons}} \text{wwwR}\!-\!\overset{\displaystyle O}{\overset{\|}{C}}\!-\!O\!-\!R'\!-\!O\!-\!\overset{\displaystyle O}{\overset{\|}{C}}\!-\!R\text{www} + H_2O$$

$$(3.2a)$$

When the concentration of carboxyl end groups in the polyester is relatively high, (3.2a) can make a significant contribution to the MW increase in SSP.

A particular case of esterification is the reaction of a carboxyl terminal group with a hydroxyl group of a glycol:

$$\text{\textasciitilde} R-\overset{\overset{O}{\|}}{C}-OH \;+\; HO-R'-OH \;\underset{k_{-2b}}{\overset{k_{2b}}{\rightleftharpoons}}\; \text{\textasciitilde} R-\overset{\overset{O}{\|}}{C}-O-R'-OH \;+\; H_2O$$

(3.2b)

This reaction does not contribute to a significant MW increase, but it leads to changes in the carboxyl/hydroxyl end-group molar ratio (c_{COOH}/c_{OH}) and should be included in kinetic models when the water diffusion is the controlling step (for large particles and pellets). In principle, when the esterification reaction occurs as an intramolecular reaction of one hydroxyl and one carboxyl end group of the same macromolecule, a cyclic macromolecule would be formed. However, due to the low probability of this reaction, it is generally ignored.

The rate of reaction (3.2a) can also be relatively high in the absence of added catalyst, because it has been demonstrated that the acidic carboxyl groups are able to catalyze this reaction to some extent [1,20]. The metal-based catalysts typically used to catalyze transesterification reactions can also catalyze esterification reactions [1,20,27–29], and in particular, titanium- and tin-based catalysts have been found to be most efficient when different metal-based catalysts had been compared [30–35].

c. Acidolysis Reactions   The reaction of one carboxyl group with one ester group is generally called an acidolysis reaction. When the ester group involved is the next to last, close to the chain end, it leads to the formation of a low-molecular-weight molecule bearing one or two carboxyl groups along with a polyester molecule of higher molecular weight:

$$\text{\textasciitilde} R-\overset{\overset{O}{\|}}{C}-O-R'-O-\overset{\overset{O}{\|}}{C}-R-\overset{\overset{O}{\|}}{C}-OX \;\underset{k_{-3a}}{\overset{k_{3a}}{\rightleftharpoons}}\; \text{\textasciitilde} R-\overset{\overset{O}{\|}}{C}-O-R'-O \;+\; HO-\overset{\overset{O}{\|}}{C}-R-\overset{\overset{O}{\|}}{C}-OX$$

$$+\; \text{\textasciitilde} R-\overset{\overset{}{}}{C}-OH \qquad\qquad\qquad X = H \text{ or } R'-OH$$

(3.3a)

Often, reaction (3.3a) is ignored in kinetic models either because the carboxyl-group concentration is low or because its rate and/or the diffusion of the resulting species is believed to be low. Indeed, this reaction has been little studied, but as discussed later, it can occur, as demonstrated by the analysis of the solid matter condensed on the cold wall of the SSP reactor [19,21,22], and it has to be considered in order to explain the SSP in more detail.

As for alcoholysis, when the acidolysis reaction involves an intermolecular reaction of a terminal carboxyl group with a backbone ester group of another chain,

$$\begin{array}{c} \text{wwR}-\overset{\overset{\displaystyle O}{\|}}{C}-O-R'-O-\overset{\overset{\displaystyle O}{\|}}{C}-R\text{ww} \\ + \text{ wwR}-\underset{\underset{\displaystyle O}{\|}}{C}-OH \end{array} \quad \underset{k_{-3b}}{\overset{k_{3b}}{\rightleftharpoons}} \quad \begin{array}{c} \text{wwR}-\overset{\overset{\displaystyle O}{\|}}{C}-O-R'-O + HO-\overset{\overset{\displaystyle O}{\|}}{C}-R\text{ww} \\ \text{wwR}-\underset{\underset{\displaystyle O}{\|}}{C} \end{array}$$

$$(3.3b)$$

there is no effect on the average MW, but like reaction (3.1d), it could contribute to rearrangement of the chain lengths and can play a role with respect to the "chemical diffusion" of reactive end groups within the amorphous domains, in the last stage of the process.

Similarly, when the acidolysis reaction involves an intramolecular reaction of a terminal carboxyl group with an ester group along the same chain, cyclic molecules could be formed. There is evidence that this reaction occurs at a significant rate at the high temperatures of polymerization [8,36]. Little is known about the rate of acidolysis reactions, even though it has been reported that many different substances are effective catalysts for this reaction [1].

d. Esterolysis Reactions   In principle, the reaction of an ester group with another ester group can be called an *esterolysis reaction*. When the ester groups involved are the last and second-to-last groups, close to the chain ends, it leads to the formation of a low-molecular-weight ester molecule along with a polyester molecule of higher MW:

$$\begin{array}{c} \text{wwR}-\overset{O}{\overset{\|}{C}}-O-R'-O-\overset{O}{\overset{\|}{C}}-R-\overset{O}{\overset{\|}{C}}-OX \\ + \text{ wwR}-\underset{\underset{O}{\|}}{C}-O-R'-OH \end{array} \quad \underset{k_{-4a}}{\overset{k_{4a}}{\rightleftharpoons}} \quad \begin{array}{c} \text{wwR}-\overset{O}{\overset{\|}{C}}-O-R' \\ O \\ \text{wwR}-C \\ \| \\ O \end{array} + \begin{array}{c} \overset{O}{\overset{\|}{C}}-R-\overset{O}{\overset{\|}{C}}-OX \\ O \\ R'-OH \end{array} \quad X = H \text{ or } R'-OH$$

$$(3.4a)$$

As the products deriving from this reaction are similar to those that could result from one or more of the reactions described above, there is no clear evidence that this reaction can occur, and almost no kinetic data are available in the literature. Intermolecular esterolysis reactions:

$$\begin{array}{c} \text{wwR}-\overset{O}{\overset{\|}{C}}-O-R'-O-\overset{O}{\overset{\|}{C}}-R\text{ww} \\ + \text{ wwR}-C-O-R'-O-C-R\text{ww} \\ \| \quad\quad | \\ O \quad\quad O \end{array} \quad \underset{k_{-4b}}{\overset{k_{4b}}{\rightleftharpoons}} \quad \begin{array}{c} \text{wwR}-\overset{O}{\overset{\|}{C}}-O-R' \\ O \\ \text{wwR}-C \\ \| \\ O \end{array} + \begin{array}{c} \overset{O}{\overset{\|}{C}}-R\text{ww} \\ O \\ R'-O-C-R\text{ww} \\ \| \\ O \end{array}$$

$$(3.4b)$$

would contribute to the rearrangement of the chain lengths and to the mobility of reactive end groups within the amorphous phase (chemical diffusion).

An intramolecular esterolysis reaction would lead to the formation of cyclic molecules that could contribute to the overall number of solid oligomers that are formed and removed during SSP.

The rate of these reactions is expected to be very low in the absence of a catalyst. However, nothing is known about the role that acids and metal-based catalysts can play. For all these reasons, esterolysis reactions are typically ignored, and their effects, if they exist, are generally included in the kinetic parameters of other reactions.

e. Side Reactions   The side reactions occurring during the polymerization of various polyesters may be different in importance relative either to their rates with respect to those of chain-growth reactions or their effects on the molecular structures and properties of the polyesters. Due to the high activation energy, their rates are much lower during SSP than during polymerization in the molten state; nevertheless, they can play a significant role in SSP, and several kinetic models that include these reactions have been reported [22,37–40]. A general discussion of side reactions occurring in polyesters must consider two different types of side reactions: those that involve an ester-bond scission in the polyester chain backbone, and those that involve only the end groups, with a change in the chemical nature and/or with the formation of undesired by-products or spurious moieties in the polymer chains.

Chain-Scission Reactions   At high temperatures, polyesters with hydrogen atoms in the β positions with respect to the oxygen of the ester linkage can undergo a well-known ester-bond scission reaction that involves the formation of a six-membered cyclic transition state [41,42]. It leads to the formation of two new end groups, one carboxyl and one vinyl ester group, described schematically as

$$\text{ww-R}-\overset{\overset{\text{O}}{\|}}{\text{C}}-\text{O}-\text{R}'-\text{O}-\overset{\overset{\text{O}}{\|}}{\text{C}}-\text{R ww} \xrightarrow{k_5} \text{ww-R}-\overset{\overset{\text{O}}{\|}}{\text{C}}-\text{O}-\text{R}''-\text{CH}=\text{CH}_2 + \text{HO}-\overset{\overset{\text{O}}{\|}}{\text{C}}-\text{R ww}$$

$$(3.5)$$

where R'' represents, for example, nothing for PET and —CH_2CH_2— for poly(butylene terephthalate) (PBT). The main effect of this reaction is a decrease in MW; however, the changes induced in the chemical nature of the reactive end groups can also be important, as we discuss later. Even though there is some evidence that the catalyst used to increase the polymerization rate may also affect the scission rate for some polyesters [42], it is common to consider this reaction unaffected by catalysts.

Side Reactions That Lead to Changes in End Groups   Some types of side reaction are not effective with respect to changes in MW but can play a role in SSP because of a change in the chemical nature (and reactivity) of the end groups. In general, this type of side reaction can be described schematically as

$$
\text{\textasciitilde}R-\overset{\overset{O}{\|}}{C}-OX \; (+ \; B) \; \xrightarrow{k_6} \; \text{\textasciitilde}R-\overset{\overset{O}{\|}}{C}-OY \; + \; C \tag{3.6}
$$

For instance, it is well known that during the polymerization of PBT, there is the formation of tetrahydrofuran (THF) (C) due to a side reaction that transforms 4-hydroxybutyl end groups (X) into carboxyl groups (Y = H), leaving MW unchanged. As both carboxyl and hydroxyl groups are involved in other reactions, this reaction can affect the overall SSP rate significantly. Another reaction of this type leads to the formation of acetaldehyde (AA) during the polymerization of PET [1]. Depending on the specific polyester, reaction (3.6) can be considered either affected or unaffected by the catalyst. The role of this reaction in SSP is discussed later for each polyester. While reactions (3.1) to (3.4) have to be considered equilibrium reactions, reactions (3.5) and (3.6) are typically not reversible.

f. Other Reactions   In principle, other reactions can lead to the formation of polyesters [1,6,8] in addition to those discussed above, however, they are of little interest for SSP. Only one paper is reported in the scientific literature dealing with the SSP of prepolymers containing acyl chloride reactive groups [43].

Chain extension is quite a common approach to enhancing the MW of polyesters in the molten state, and the reactions of chain extenders with molten polyesters are well known and widely studied [44–47]. A chain-extension reaction can be described schematically as

$$
\text{\textasciitilde}R-\overset{\overset{O}{\|}}{C}-OX \xrightarrow[k_{7a}]{Y\text{\textasciitilde}Y} \text{\textasciitilde}R-\overset{\overset{O}{\|}}{C}-OX-Y\text{\textasciitilde}Y \xrightarrow[k_{7b}]{\overset{\overset{O}{\|}}{XO-C-R\text{\textasciitilde}}} \text{\textasciitilde}R-\overset{\overset{O}{\|}}{C}-OX-Y\text{\textasciitilde}Y-XO-\overset{\overset{O}{\|}}{C}-R\text{\textasciitilde}
$$

$$
X = H \text{ or } R'-OH
$$

$$
\tag{3.7}
$$

In principle, this approach can also be used to increase the MW of polyester prepolymers in the solid state in order to reduce the time required for the SSP process. Instead of withdrawing low-MW by-products, a chain extender Y—Y, able to react with reactive end groups of the polyester, should diffuse from the gas stream into the polyester particles. Chain extenders bearing reactive groups (Y), such as isocyanates, anhydride, or acyl lactams, can be used, and according to the chain extender employed, either hydroxyl or carboxyl end groups, or both [X end groups in reaction (3.7)] can react. Of course, a spurious moiety, chemically different from the monomeric unit, will be included in the final polyester. Branching could also be induced in the final polyester if a chain extender with suitable functionality (>2) was used. From reaction (3.7) it is easy to understand that the relative rates of diffusion and reaction of Y—Y is a critical point: If the diffusion is too fast, the amount of chain extender that enters the polyester particles would be too high and a limited effect (or no effect at all) could derive. In fact, when all the X end groups would be consumed in the first reaction, a polyester with the same MW and Y end groups, unable to react further, would be formed.

At the time of preparation of this chapter, some experiments in solid state chain extension have been carried out by the authors without publication of the results. To the best of the authors' knowledge, no papers using this approach have been reported in the literature.

3.2.2. Chain Mobility and Diffusion of Low-Molecular-Weight By-Products

It is generally accepted that the reactions in the SSP of semicrystalline pre-polymers occur only within the amorphous regions included between crystalline domains (either lamella, crystallites, or spherulites). It is also generally believed that most of the end groups of the polymer chains, as well as low-MW oligomers and the catalyst, are included in these regions [11,38,40,48–50]. Of course, for a reaction between two reactive groups to take place, these have to move and meet at a suitable distance that allows the formation of an "activated complex" in the transition state. In the case of polymer chains, there are two main mechanisms that can bring two terminal reactive groups close each other: translation of a whole molecule, and mobility of the terminal segments (segmental motion). The latter is possible even when translation is hindered or impossible. A third contribution, chemical diffusion, is also possible, as we discuss later.

If during SSP all possible contributions to the mobility of the reactive groups were the same as those occurring in the homogeneous molten state of a bulk polymerization reaction, the same kinetic equations and parameters could be used for SSP. In that case it was enough to take into account the lower temperature typical of SSP and the difference in the reactive-group concentration due to segregation of the end groups in the amorphous regions.

In fact, there is evidence that the reaction rate decreases more than expected as SSP proceeds, and a different kinetic model has to be used in order to describe, in a predictive manner, the series of the events that happen during SSP (at least in the last stage of the SSP process, when high MWs are reached). The reasons are generally ascribed to the lower mobility of the reactive groups. This is mainly because of the higher viscosity in the amorphous domains due to lower temperature, to constraints induced by the crystalline domains, and because most of the end groups belong to macromolecules that have other segments immobilized in the crystalline domains. Of course, for these macromolecules, translational mobility is impossible. The change in chain mobility can differ at different times during SSP, and in the following discussion it is convenient to consider three different regimes, as indicated in Figure 3.1, where the number of end-group pairs that are at a given distance is plotted against the pair distance. Of course, the boundary limits between different regimes in Figure 3.1 are arbitrary, and the curve profile is expected depend on the average shape and dimensions of the amorphous domains and on the number of reactive groups in the prepolymer.

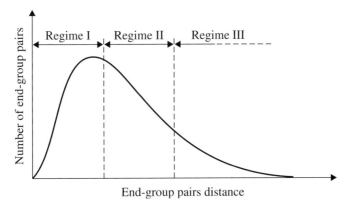

Fig. 3.1. End-group pairs distance distribution in amorphous domains.

In the first stage of SSP, when the MW is relatively low, the amorphous domains contain low-MW molecules such as glycols, water, dicarboxyl acids, and low-MW oligomers (both linear and cyclic), along with chain segments of macromolecules connecting different crystalline domains and terminal segments of polymer chains, which are partially constrained within the crystalline domains, typically bearing hydroxyl and carboxyl groups.

It must be remembered that according to the most probable MWD, typical of polyesters prepared by bulk polycondensation, the molar fraction of macromolecules with different molecular weights is a monotonically decreasing function of the degree of polymerization, which means that the low-MW species are numerically the most abundant [51]. Even though it is sometimes stated that the chain mobility in the amorphous region is strongly affected by boundary crystalline constraints, there is evidence from electron spin resonance (ESR) spectroscopic data [52] that at the temperatures typically used in SSP, the mobility of the terminal segments is high, very similar to that observed for the same polymer chains in a diluted solution. As a consequence, it is reasonable to think that the progress of SSP involves reactions that can be described by the same kinetic constants derived from studies of bulk polymerizations in the molten state (of course, extrapolated to lower temperatures) or those derived from reactions between "model" molecules. The most significant difference with the melt polymerization may derive from the diffusion rate of the low-MW species out of the granules, lower with respect to that observed in the melt for two main reasons: the lower temperature and the tortuosity factor, deriving from the presence of the crystalline domains.

In the first stage of SSP, in regime I, many end groups are very close to each other and the low-MW oligomers are able to translate within the amorphous region. Two reactive groups can meet each other quite easily (the frequency of collision of reactive groups can be assumed high, as in the molten state) and the reaction proceeds relatively fast. The reaction rate is controlled by the rate of formation of the activated complex rather than by chain mobility.

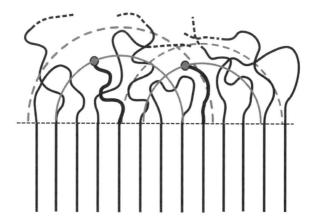

Fig. 3.2. Limited mobility of end groups.

It is important to emphasize that the chain segments bearing end groups in the amorphous regions, belonging to polymer chains that are constrained in crystalline domains, cannot translate but can move freely within a limited space defined by the segment length, as depicted in Figure 3.2. The segment can move and meet other reactive groups within the half-sphere having the maximum radius corresponding to the length of the segment in the extended conformation (dashed line) or, more probably, within the half-sphere described by the radius r corresponding to the average end-to-end distance of the segment in the medium (solid line). In regime I, when the concentration of reactive groups is relatively high, the limited mobility of the segment bearing the end groups does not reduce the overall reaction rate significantly because of the high probability of collision due to the high concentration and translational mobility of the low-MW species.

As SSP proceeds, many of the reactive groups disappear and the concentration of residual reactive groups (end groups of polymer chains and oligomers) in the amorphous regions decreases. In this second stage of the SSP process, regime II (Fig. 3.1), the reaction rate becomes more and more dependent on the segmental diffusion, and the kinetic constants are expected to decrease progressively as far as the closest end-group pairs have reacted. The number of end groups that are able to meet others by segmental mobility becomes smaller and smaller as the SSP progresses, and finally, no end groups remain able to meet each other by simple segmental mobility. The rate of the SSP process is expected to approach close to zero when the distance between the reactive groups is so far that no reactions can occur as a consequence of segmental mobility. However, it must be emphasized that the chain segments bounded to the crystalline domains can actually have higher mobility because of possible intersegment exchange reactions. Few of these can allow a given reactive terminal group to move out of the regions depicted in Figure 3.2, according to the mechanism described in Figure 3.3. Indeed, a few reaction steps may allow a reactive group to move all

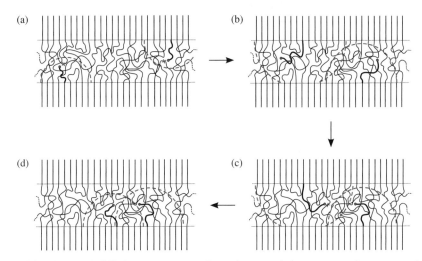

Fig. 3.3. Chemical diffusion. Few steps of reactions can bring two reactive groups close to each other.

over an amorphous domain, and reactions (3.1d), (3.3b), and (3.4b) can play an important role in this respect. The chemical diffusion due to exchange reactions may, in principle, allow reactive groups to reach a distance suitable for reaction in regime III of Figure 3.1.

In a paper by McAlea et al. [53], it was concluded that ester interchange is rapid in the melt and becomes significant in the solid state at temperatures above 225°C. Similar results and an activation energy of 152 kJ mol^{-1} were reported in a paper by Kugler et al. [54]. According to these figures, and assuming that the kinetic constant can be extrapolated to the solid state, the exchange rate (the number of millimole exchange reactions occurring per hour and per kilogram of PET) can easily be calculated (about 1350 10^{-5} mmol kg^{-1} h^{-1} at 230°C). It is interesting to compare this figure with the chain-growth rate due to reaction (3.1a). Taking an average value of $k_1 = 1 \times 10^{-3}$ kg mmol^{-1} h^{-1} at 230°C, we find that in the last stage of the SSP reaction, when the hydroxyl concentration is low (e.g., 50 mmol kg^{-1}), the exchange rate is about 100 times higher than the chain-growth rate. This simple calculation seems to support the conclusions in a recent paper [55], where it is suggested that chemical diffusion is the most important phenomenon allowing reactive groups to meet. As these reactions can occur at a significant rate only at high temperature and in the presence of efficient catalysts, the type and concentration of the catalyst may be quite important in defining the overall reaction rate in regime III.

This picture of the phenomena controlling the SSP rate seems to be supported by considerable experimental evidence, and in particular, it was shown that an increased rate of SSP can be achieved by stopping SSP, melting the polymer, and after solidification, restarting SSP [21,38,56–59]. A rational explanation for these experimental results, as well as an explanation of the effects of the initial intrinsic

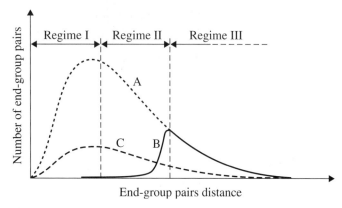

Fig. 3.4. Distribution of end-group pair distances: initial distribution (curve A), after SSP (curve B) and after polymer remelting (curve C).

viscosity (IV) and the temperature [57,60], can be given by assuming that one sample submitted to SSP presents a given distribution of distances between pairs of reactive end groups (see Fig. 3.1 and curve A in Fig. 3.4). Only a fraction of the overall pairs of reactive groups can react at a fast rate similar to that occurring in the molten state (those included in regime I under curve A in Fig. 3.4). Another fraction of reactive groups will react, but at a slower rate, when the segmental mobility becomes the rate-controlling step. For instance, at a given time, the pair distance distribution could be represented by curve B in Figure 3.4, and if at that time the SSP is stopped, the sample melted, solidified, and then resubmitted to SSP, a new statistical distribution of the distances between reactive end groups is expected (curve C in Fig. 3.4). It appears that this new distribution will again present a fraction of the reactive-group pairs in regime I, able to undergo fast reactions.

As most of the reactions involved are equilibrium reactions, efficient removal of the low-MW products is of primary importance in reducing the SSP rate. Fickian diffusion is usually assumed to describe diffusion of low-MW by-products within the particles. Other than from temperature, the diffusion coefficient can depend on type and on amount of crystallinity. The rate of crystallization is typically very fast at the temperature used in SSP (a few minutes to reach the maximum crystallinity even for the slowly crystallizing PET), so that the degree of crystallinity and the diffusion coefficients can be considered as time independent [38,39,61–63].

Every factor that is able to increase the diffusion rate of low-MW by-products would increase the overall rate of diffusion-controlled SSP. Factors such as the reduction of particle dimensions, increase in temperature, or high vacuum or gas flow rate for efficient removal of low-MW products from the surface of the granules are typically used to increase the SSP rate.

A rational and exhaustive explanation of the SSP should therefore be taken into account, and included in a kinetic model should be all possible chemical reactions, including the effects of possible constraints to the mobility of the reactive end groups, the diffusion of low-MW species to the surface of the granules, and the diffusion of these species from the surface of the granules to the surrounding gas medium.

3.2.3. Kinetic and Diffusion Equations

To describe the progress of SSP, a set of differential equations including either kinetic or diffusion equations or both has to be solved either numerically or analytically. The criteria that should be used to choose among the various possible reactions and diffusing species that have to be considered in a kinetic model depend on the information desired. They are discussed later; in this section we discuss the types of kinetic and diffusion equations that can be used for the various reactions described above.

Before discussing the kinetic equations that can be used for different reactions, some general comments might be useful. For all the reactions, it is generally accepted that the reactivity of a functional group does not depend on the chain length of the macromolecules involved [51], and as a consequence, a single kinetic constant for each type of reaction can be used. The kinetic constants of the various reactions are taken most conveniently as those corresponding to the reactive groups rather than to singular molecular species. This means that when a low-MW species such as a glycol contains two reactive groups, the kinetic constant has to be multiplied by 2. When SSP is concerned with A–A + B–B polyesters, the repeating units contain two ester groups and the kinetic equations often refer to diester concentration (c_{DE}), meaning concentration of repeating units.

The concentrations of end groups and low-MW species can be expressed in any of the following units: mol/volume or mol/mass or mol/mol of repeating unit; all three are used in the scientific literature. Mol/mol of repeating unit may be preferable, as it is not affected by volume and mass changes when the removal of volatile products different from glycol or water can be ignored. We use this notation below even though the others could be used without significant errors, as the change in mass and volume due to glycol and water removal is low.

As already stated, reactions and diffusion of low-MW by-products occur in the amorphous domains; as a consequence, the concentration of the terminal reactive groups in the amorphous regions is higher than that of the overall sample, depending on the fraction of amorphous and crystalline domains. If ϕ_w is taken as the weight fraction of crystalline domains, it will be easy to transform the overall bulk concentration (c_{bulk}) into the effective concentration (c_{eff}) in the amorphous domains, by dividing the bulk concentration, c_{bulk}, by the weight fraction of amorphous domains ($1 - \phi_w$):

$$c_{eff} = \frac{c_{bulk}}{1 - \phi_w} \tag{3.8}$$

The degree of crystallinity and the size of crystallites increase significantly only within the preheating time, after which they stabilize, showing only very small increases during SSP [61–63]. Accordingly, ϕ_w can be considered constant over all the SSP process (at least for PET, PBT, and other fast-crystallizing polyesters).

a. Transesterification Reactions   The most important reaction for the molecular-weight increase during SSP is reaction (3.1a), which leads to the formation of glycol (this is often the only reaction considered). It is an equilibrium reaction with K_{eq} typically lower than 1 (frequently assumed to be 0.5). Reaction (3.1a) is usually assumed to be second order with respect to hydroxyl terminal group concentration. The kinetic and equilibrium equations that are commonly used to express the contribution of this reaction to the glycol and hydroxyl terminal group balances are as follows:

$$v_{trans(1)} = \frac{dc_{Gly}}{dt} = -\frac{dc_{OH}}{2dt} \tag{3.9}$$

$$\frac{dc_{Gly}}{dt} = k_1 c_{OH}^2 - k_{-1} c_E \cdot 2c_{Gly} = k_1 c_{OH}^2 - k_{-1} \cdot 2c_{DE} \cdot 2c_{Gly} \tag{3.10}$$

$$\frac{dc_{Gly}}{dt} = k_1 \left(c_{OH}^2 - \frac{4c_{DE}c_{Gly}}{K_1} \right) \tag{3.11}$$

with

$$K_1 = \frac{k_1}{k_{-1}} = \frac{4(c_{DE})_{eq}(c_{Gly})_{eq}}{(c_{OH}^2)_{eq}} \tag{3.12}$$

where c_{Gly}, c_{OH}, c_E, and c_{DE} are the glycol, hydroxyl end group, ester, and diester concentrations, respectively, while k_1 and k_{-1} are the kinetic constants of the forward and backward reactions, respectively.

In the absence of any catalyst, this reaction is very slow, and protic acids or metal-based catalysts are typically used to increase the reaction rate. Of course, the reaction rate depends on the concentration of the catalyst (typically, a first-order dependence is assumed), and therefore the kinetic equation should include the catalyst concentration. Metal-based catalysts are typically used for the industrial polymerizations of polyesters; however, it has been reported that carboxyl groups can also catalyze this reaction [1,20]. By considering the fact that during the SSP of polyesters, carboxyl groups are typically present along with hydroxyl groups, an autocatalytic contribution cannot be ruled out. If more than one metal-based catalyst were used, as often happens in industrial polymerizations, it should be considered that each catalyst could give a specific contribution to the kinetic equation.

As a consequence, assuming that the various contributions are independent (which is not necessarily true), a general equation including all possible contributions should be used for a better description of the glycol formation and the hydroxyl terminal group depletion due to reaction (3.1a). The rate constant k_1 in equations (3.10) and (3.11) should be replaced by

$$k_1 = k_{o,1} + k_{H^+,1}c_{COOH} + \sum_i k_{cat(i),1}c_{cat(i)} \qquad (3.13)$$

where c_{COOH} is the concentration of carboxyl end groups and the subscripts o, H^+, and $cat(i)$ represent uncatalyzed, proton-catalyzed, and i-metal catalyzed, respectively.

However, use of this comprehensive equation would require knowledge of a number of parameters (kinetic constants, activation energies, and concentrations of the different catalytic species) that are usually unknown. In addition, it has been demonstrated that the presence of COOH groups along with titanium-based catalysts [27,28,64] or some specific combination of different metal-based catalysts [65,66] may lead to synergic or antisynergic effects in a way that is not easily predictable. This means that kinetic constants can change during the polymerization in a manner that is not easily predictable.

To overcome these problems a simplified equation such as (3.11) is typically used; of course, the kinetic constant k_1 (and also k_{-1} in the case that the backward reaction is also considered) is an apparent kinetic constant including all possible effects discussed above (the kinetic constants of all the different catalyzed reactions and the concentrations of all the different catalysts). The lack of data regarding these kinetic parameters and the use of commercial prepolymers with an unknown type and amount of catalyst is one of the reasons that a comparison of kinetic data from different authors is difficult.

b. Esterification Reactions   Esterification [reaction (3.2a)] is another important reaction that has often been considered in describing the MW increase during SSP. It leads to the formation of water and is an equilibrium reaction with a low equilibrium constant, typically close to 1. Reaction (3.2a) is generally assumed to be third order: first order with respect to hydroxyl groups and second order with respect to carboxyl groups in the absence of catalyst, or first order with respect to each hydroxyl or carboxyl group, and with respect to catalyst concentration when metal-based catalysts are used [1,20]. Often, the concentration of catalyst is omitted (included in the kinetic constant) and second-order kinetic equations are used [21,58,67]. The corresponding kinetic and equilibrium equations that are commonly used to describe the contribution of this reaction to balance equations of water, hydroxyl, and carboxyl terminal groups, are given below.

Second-order reaction:

$$v_{est(2)} = \frac{dc_W}{dt} = -\frac{dc_{OH}}{dt} = -\frac{dc_{COOH}}{dt} \qquad (3.14)$$

$$v_{trans(1)} = k_1 c_{OH}^2 \qquad (3.15)$$

$$v_{est(2)} = k_2 c_{OH} c_{COOH} \tag{3.16}$$

$$\frac{dc_W}{dt} = k_2 \left(c_{OH} c_{COOH} - \frac{2c_{DE} c_W}{K_2} \right) \tag{3.17}$$

with

$$K_2 = \frac{k_2}{k_{-2}} = \frac{2(c_{DE})_{eq}(c_W)_{eq}}{(c_{OH})_{eq}(c_{COOH})_{eq}} \tag{3.18}$$

where k_2 (and k_{-2}) includes the catalyst concentration (considered constant).

Third-order reaction:

$$v_{est(2)} = \frac{dc_W}{dt} = -\frac{dc_{OH}}{dt} = -\frac{dc_{COOH}}{dt} \tag{3.14'}$$

$$\frac{dc_W}{dt} = k_2 c_{OH} c_{COOH}^2 - k_{-2} c_E c_W c_{COOH} \tag{3.15'}$$

$$\frac{dc_W}{dt} = k_2 c_{OH} c_{COOH}^2 - \frac{k_2}{K_2} 2 c_{DE} c_W c_{COOH} \tag{3.16'}$$

$$\frac{dc_W}{dt} = k_2 c_{COOH} \left(c_{OH} c_{COOH} - \frac{2c_{DE} c_W}{K_2} \right) \tag{3.17'}$$

with

$$K_2 = \frac{k_2}{k_{-2}} = \frac{2(c_{DE})_{eq}(c_W)_{eq}}{(c_{OH})_{eq}(c_{COOH})_{eq}} \tag{3.18'}$$

Actually, both COOH groups and the metal-based catalysts typically used in industrial polymerization of polyesters can contribute to an increased rate of this reaction [1,20], so the kinetic equations should more appropriately include two distinct contributions, as in

$$k_2 = k_{H^+,2} c_{COOH} + \sum_i k_{cat(i),2} c_{cat(i),2} \tag{3.19}$$

When the concentration of COOH groups is low, the contribution from COOH catalysis may become negligible, and equation (3.19) reduces to

$$k_2 = \sum_i k_{cat(i),2} c_{cat(i),2} \tag{3.19'}$$

If more that one metal-based catalyst is used, as sometimes happens in industrial polymerizations, distinct contributions from each catalyst should be included in the general equation (3.19) or (3.19'). As already discussed for reaction (3.1a), the use of this equation may be very difficult, due to the lack of knowledge about the single kinetic constants and the concentrations of the different catalytic species. As a consequence, a simplified equation such as equation (3.17) or

(3.17′) is typically used. Of course, the kinetic constant k_2 (and also k_{-2} when the backward reaction is also considered) is an apparent kinetic constant that includes the concentrations of catalyst as in (3.19) or (3.19′).

c. Other Reactions Leading to the Formation of Low-Molecular-Weight Linear Oligomers

Even though frequently ignored, the reactions that lead to the formation of monomers and/or linear oligomers [reactions (3.1b), (3.1c) and (3.3a)] can give a not-negligible contribution during the SSP of some polyesters and had to be included in some kinetic models in order to describe experimental results [21,22], in particular to describe the last stage of the SSP process. The simplest way to take into account the formation of oligomers containing carboxyl groups is to include reaction (3.3a) and the diffusion of terephthalic acid (TA) in the kinetic model [21,22]. As stated above, reaction (3.3a) has been much less studied than reactions (3.1a) and (3.2a), and almost no kinetic data are available. According to its stoichiometry, the kinetic equation can be written as

$$v_{acid(3)} = \frac{dc_{TA}}{dt} = k_3 c_{COOH}^2 - k_{-3} c_E \cdot 2c_{TA} = k_3 c_{COOH}^2 - \frac{k_3}{K_3} \cdot 2c_{DE} \cdot 2c_{TA}$$

(3.20)

where c_{COOH} is the carboxyl-group concentration, not including the contribution from COOH groups of TA.

Again, in the absence of specific data, the equilibrium constant can be assumed to be 1. For a more comprehensive kinetic model, reactions (3.1b) and (3.1c), leading to the formation of oligomers containing one or two hydroxyl groups, should also be considered:

$$H \left(O-R'-O-\overset{O}{\underset{\|}{C}}-R-\overset{O}{\underset{\|}{C}} \right)_x OH \qquad H \left(O-R'-O-\overset{O}{\underset{\|}{C}}-R-\overset{O}{\underset{\|}{C}} \right)_y O-R'-OH$$

where x and $y = 1, 2, 3, \ldots$. Due to the low volatility of these oligomers, only the contribution of those with $x = y = 1$ are expected to be important. These reactions are alcoholysis reactions, and in principle the kinetic constant of reaction (3.1a) can also be used to describe these reactions. However, kinetic study is not available on this type of reaction, and thus it is not clear whether the same kinetic constant is also applicable for these reactions.

d. Other Reactions

Reactions (3.1d) and (3.3b) can only be important in the last part of the process, contributing to the reactive groups' mobility through chemical exchange reactions, whereas reaction (3.1e) can become important when the formation of cyclic oligomers has to be considered (as, e.g., for PLLA).

e. Side Reactions   As stated above, side reactions can play an important role during the SSP of some polyesters, and their contributions to the overall process should be included in the mass balances either to account for changes in molecular weight or to describe the formation of some specific by-products (such as acetaldehyde for PET and THF for PBT). It is impossible to propose general kinetic equations that are valid for all side reactions and for all polyesters. Typically, a kinetic equation based on the reaction stoichiometry is used when no specific kinetic studies have been reported for a given reaction. The kinetic equations that have been used for side reactions of some specific polyesters are discussed later.

f. Use of Model Molecules   Based on the discussion above, it appears that one of the main problems when using kinetic equations is the lack of reliable kinetic parameters (kinetic constants, activation energies, etc.). There are two main reasons for that: One is because many reactions can occur at the same time during the synthesis of polyesters, and it is not easy to separate the effect of one reaction from that of others; the second reason is that analysis of the reaction products is difficult when dealing with polymers, as reaction products may not be easily separable by chromatographic techniques.

As a consequence, the resulting kinetic data are often dependent on the various assumptions made by different authors, and/or less reliable results are obtained when some analytical techniques are applied to polymers. One possible way to overcome these problems is to perform similar reactions by using low-molecular-weight model molecules of suitable molecular structure and functionality. In many cases, the kinetic parameters derived from model molecules have been shown to be the same as those found for the corresponding polymers and can be used in SSP kinetic models with good results in terms of predictable capability [22,27,28,41,42,68–73].

g. Diffusion of Low-Molecular-Weight By-Products   The reverse reactions (and therefore the equilibrium constants) can be ignored when the particle dimensions are so small that the diffusion of low-MW by-products out of the particle is much faster than their formation from chemical reactions. Of course, when particle dimensions increase or when diffusivity is very low, both forward and backward reactions and diffusion of by-products have to be included in the mass balance equations. Diffusion is generally assumed to be of Fickean type. Typical equations for one-dimensional unsteady-state diffusion process and for cubic particles are reported in equations (3.21) and (3.22), respectively, for a generic low-MW by-product (i_{byp}).

$$\frac{\partial c_{byp}}{\partial t} = D_{byp} \left(\frac{\partial^2 c_{byp}}{\partial x^2} + \frac{\lambda}{x} \frac{\partial c_{byp}}{\partial x} \right) \tag{3.21}$$

where D_{byp} is the diffusion coefficient of a volatile by-product, x the position within the particle ($x = 0$ in the center and $x = x_0$ on the surface), and $\lambda = 0$,

1, and 2 for plane sheets, cylinders, and spheres, respectively [11]. For cubic particles,

$$\frac{\partial c_{byp}}{\partial t} = D_{byp} \left(\frac{\partial^2 c_{byp}}{\partial x^2} + \frac{\partial^2 c_{byp}}{\partial y^2} + \frac{\partial^2 c_{byp}}{\partial z^2} \right) \tag{3.22}$$

where x, y, and z represent the distances in the Cartesian coordinate system [11]. The diffusion coefficient, D, depends on the amount and size of crystalline domains. Usually, for most polyester SSP there are no changes in the degree of crystallinity and D can be taken as a constant over the entire SSP process [11,63,74,75]. When diffusion is the SSP-controlling step, an MW gradient is expected within a particle, with the highest MW close to the surface and the lowest in the center [22,76,77].

h. Gas-Phase Mass Transfer   Under certain operating conditions, removal of low-MW by-products from the surface of the particles to the surrounding medium has to be included in the kinetic model (typically, when the gas flow rate or vacuum is low). Inclusion in the kinetic model of this contribution could also account for the effect of different gases and of stirring of particles. A parameter Φ_i and suitable boundary-condition equations have been proposed to take gas resistance into account [11]:

$$\Phi_{i,byp} = \frac{k_{i,byp} x_0}{D_{i,byp}} \tag{3.23}$$

$$-D_{i,byp} \frac{\partial c_{i,byp}}{\partial x} = k_{i,byp}(c_{i,byp,s} - c_{i,byp,gasphase}) \tag{3.24}$$

where $D_{i,byp}$ is the diffusivity of the ith by-product within a particle, $c_{i,byp,s}$ the concentration of the ith by-product at the gas–solid interface, $c_{i,byp,gasphase}$ the concentration of the ith by-product in the gas phase, and $(k_{i,byp})$ a gas-phase mass-transfer coefficient for the ith by-product. An increase in the flow rate will increase $k_{i,byp}$, which also depends on the gas-phase diffusivity of the low-MW by-products in the gas phase.

i. Segmental Mobility   As discussed above, two reactive groups can meet each other either because of translation of whole molecules or by the movement of terminal segments of fixed molecules. If translational mobility dominates in the molten state, segmental mobility may become important in the SSP when many of the functional groups have reacted and the residual reactive groups are present in the terminal segments of macromolecules having most of the chain included in crystalline domains. Intuitively, the rate of approach of two reactive groups by segmental mobility at a suitable distance to react will depend on the residual amount of reactive groups. A kinetic constant that aims to describe the progress of a reaction in regime II (see Fig. 3.1) should decrease with the progress of the SSP process becoming zero in regime III, when no more reactive groups are able to meet other reactive groups by segmental mobility. In Chapter 2, the effect of segmental mobility is incorporated in the kinetic models, where the chain length constitutes a critical parameter.

A previous attempt to consider the effect of segmental diffusion quantitatively by separating reaction and segmental-diffusion contributions to the kinetic constant of transesterification [reaction (3.1a)] and esterification [reaction (3.2a)] was proposed by Chen and Chen [62]. They observed that the kinetic constants usually employed in the kinetic equations [k_1 and k_2 in equations (3.11) and (3.17)] are unaffected by the reactive group concentrations (c_{OH} and c_{COOH}) only in the first stage of the SSP, when the kinetic parameters are the same as those observed during the molten-state polymerization, and the rate of the process is controlled by the chemical reactions (regime I). As SSP proceeds, the SSP becomes more and more segmental diffusion–controlled, and effective kinetic constants, $k_{1,eff}$ and $k_{2,eff}$, changing with time have to be included in the kinetic equations (regime II) in order to describe experimental results.

The authors separated $k_{1,eff}$ and $k_{2,eff}$ in two different contributions, $k_{1,0}$ and $k_{1,1}$ for the transesterification reaction and $k_{2,0}$ and $k_{2,1}$ for the esterification reaction, where the notation $(.,0)$ is referred to the kinetic constants when the SSP is controlled by the chemical reaction (regime I) and the notation $(.,1)$ to the constants related to segmental diffusion. By measuring both hydroxyl and carboxyl end-group concentration (c_{OH} and c_{COOH}, respectively) at various times, they observed that in the temperature range 169 to 199°C the variation of $(1/\Delta c_{OH})_{transest}$ and $(1/\Delta c_{COOH})$ was linear with $1/t$, and as a consequence, $k_{1,eff}$ and $k_{2,eff}$ decreased with c_{OH} and c_{COOH} according to

$$k_{1,eff} = k_{1,0}\left(1 - \frac{1}{k_{1,1}}\frac{(\Delta c_{OH})_{transest}}{c_{OH}^o - (\Delta c_{OH})_{transest}}\right) \tag{3.25}$$

$$k_{2,eff} = k_{2,0}\left(1 - \frac{1}{k_{2,1}}\frac{\Delta c_{COOH}}{c_{COOH}^o - \Delta c_{COOH}}\right) \tag{3.26}$$

where $\Delta(c_{OH})_{transest}$ represents the changes occurring in hydroxyl concentration due to transesterification reactions, $(\Delta c_{OH})_{transest} = c_{OH}^o - c_{OH} - \Delta c_{COOH}$, and Δc_{COOH} is the difference between the initial concentration of COOH, c_{COOH}^o, and that at time t, c_{COOH}.

The values derived for the kinetic constants $k_{1,0}$, $k_{1,1}$, $k_{2,0}$, and $k_{2,1}$ at various temperatures gave a straight line in an Arrhenius plot, and the activation energies were found to be consistent with values reported previously. According to (3.25) and (3.26), the progressive decrease of c_{OH} and c_{COOH} would lead to negative values of $k_{1,eff}$ and $k_{2,eff}$ when $(\Delta c_{OH})_{transest}$ and Δc_{COOH} approach c_{OH}^o and c_{COOH}^o, respectively. Of course, this has no physical meaning, so these equations have to be considered valid for a limited range of molecular-weight increase, and we can assume that the molecular weight corresponding to the values of $(\Delta c_{OH})_{transest}$ and Δc_{COOH} that make $k_{1,eff}$ and $k_{2,eff}$ become zero is the maximum possible value ($M_{n,lim}$) under the given conditions.

More recently, another simple attempt to explain the upper limit value in the MW increase has been proposed by Duh [57,60,78]. Indeed, Duh did not consider segmental diffusion, but he assumed that a fraction of the overall end groups ("inactive") is not able to react because of either being "dead" (unreactive) or being trapped in crystalline domains. Actually, this approach can also be used to account for segmental diffusion, assuming that a fraction of inactive end groups consists of reactive end groups that are not able to react, as they cannot meet other reactive groups, due to too great a distance (regime III).

When only reaction (3.1a) was considered (only the forward reaction, due to the SSP operating conditions), the kinetic equation proposed to account for the MW increase was similar to equation (3.10), modified by including a parameter $c_{OH,i}$, which accounts for inactive hydroxyl groups:

$$-\frac{d(c_{OH} - c_{OH,i})}{dt} = 2k_1(c_{OH} - c_{OH,i})^2 \tag{3.27}$$

This equation can easily be integrated to give

$$\frac{1}{c_{OH} - c_{OH,i}} - \frac{1}{c_{OH}^o - c_{OH,i}} = 2k_1 t \tag{3.28}$$

Duh's model assumes that the kinetic constant k_1 has a finite constant value that does not change until all the reactive groups have reacted, and then becomes zero for the remaining inactive groups. Indeed, according to the discussion above, a more realistic physical model should include a kinetic constant that changes progressively as SSP proceeds within regime II. Nevertheless, due to the simple mathematics, for practical purposes Duh's model can be considered as a good compromise, and as discussed later, it was also extended, in a general way, to the SSP of polyesters [57,60,78].

j. Chemical Diffusion In principle, interchange reactions (3.1d), (3.3b), and (3.4a) can lead to thorough changes in the topology of chain segments in the intercrystalline regions during SSP; one terminal group can move over a distance much longer than that allowed by segmental mobility after a few steps of these reactions (see Fig. 3.3). Few studies have been reported about the rate of these reactions [53,54]. However, recently, it has been claimed that the chemical diffusion of functional groups that these reactions are able to promote can give a relevant contributions to terminal group mobility for both polyamides and polyesters. It has also been reported that chemical diffusion can be described with exactly the same mathematical form as that of classic diffusion of small molecules [55].

3.3. MODELING SOLID STATE POLYMERIZATION OF POLYESTERS

As discussed above, several reactions and diffusion phenomena can be involved in the SSP of polyesters, and many variables can affect the overall rate of the process and the final quality of the resulting products. As a consequence, the optimization of SSP process by a purely experimental approach would require a big effort in terms of expensive and time-consuming experiments. The modeling approach aims to overcome these problems by developing equations able to predict SSP behavior under different conditions by simulation, therefore saving time and costs. Relevant study of PET SSP follows in Chapter 7.

A comprehensive kinetic model should include contributions from all possible reactions and all possible diffusing products. It should also consider changes in degree of crystallinity and in chain mobility and should be able to describe the effects of all the possible variables, such as temperature, gas type and flow rate, and particle size. As a consequence, it should deal with a large number of differential material-balance equations and would require the knowledge of a huge number of kinetic and diffusion parameters (e.g., kinetic constants, activation energies, equilibrium constants, diffusivities) that are not usually available for all the possible reactions and that it would be difficult, expensive, and time consuming to achieve through an experimental approach.

On the other hand, kinetic models developed on the basis of a too-limited number of reactions and solved by fitting a limited set of experimental data by considering kinetic constants as adjusting parameters would lead to results that are useless in terms of predictive capability were the same kinetic model to be applied to samples and process conditions different from those used to evaluate the adjustable parameters. Kinetic models are therefore usually developed as a compromise between an unmanageable number of balance equations and a too limited number of empirical equations, having in mind what type of data they should be able to predict.

To make mathematics easily tractable and to overcome the problem of the lack of kinetic parameters, simplifying assumptions have to be made. Some of the widely accepted assumptions that are used to simplify the kinetic models are listed below:

1. The equal reactivity principle [11] is generally assumed to be valid when dealing with polymerization kinetics, including SSP. It allows us to use a single kinetic constant for each given reaction at a given temperature, independent of the MWs of the chains bearing the reactive functional groups.
2. All the reactive end groups and catalysts are located in the amorphous regions where the chain mobility is high enough to allow reactions to occur [48,79].
3. The crystallization rate is higher than the reaction and diffusion rates, and changes in the degree of crystallinity during SSP can be neglected [11,74,75].
4. Arrhenius-type equations are usually employed to describe changes in the temperature of both kinetic constants and diffusion coefficients.

Other assumptions are often made in the development of kinetic models such as those reported below:

1. Some possible reactions can be ignored when their role is expected to be negligible with respect to the properties of interest.
2. Diffusion of some low-MW species can be ignored when the particle size is very small (very fast diffusion compared to reaction rate).
3. Mass transfer from the particle surface to the gas phase can be neglected when the gas flow rate is high enough [21,40,60,75,76].

These last assumptions may, however, lead inevitably to limited predictive capability for the model.

Various kinetic models have been reported in the scientific literature; some have been developed by considering just one reaction and no diffusion of by-products, while others are based on a set of equations that consider several reactions and diffusing products. Simple kinetic models allow for analytical solutions resulting in simple equations with a limited number of adjustable parameters. Only numerical solutions are possible for a complex set of differential equations, and several fitting parameters are typically used to describe experimental results.

Some of the kinetic models reported in the literature about the SSP of polyesters are discussed critically in the following. Simple empirical rate equations, able to describe and predict the progress of SSP only in terms of IV (or MW) changes with time, have been proposed and can also be used in the kinetics analysis of polyamides post-SSP (Chapter 4). Of course, their predictive capability is limited; however, they may be more useful in day-to-day work rather than more scientifically correct but more complex kinetic models.

The simplest way to describe the MW (or IV) increase during polyester SSP is an empirical approach that makes use of a power-law rate equation:

$$r = kt^n \tag{3.29}$$

where k and n are adjustable parameters and rate means either dM_n/dt or $d(IV)/dt$. For $n = 0$, equation (3.29) leads to a very simple linear dependence of M_n (or IV) with time. According to Duh, this equation had been found reasonable for fixed-bed or tumble dryer reactors at low temperatures ($\leq 190°C$) and for large particle size [57].

Sometimes, these empirical equations possess some theoretical bases and can be considered as semiempirical equations. For instance, kinetic models in which there is a linear dependence of M_n with time can be derived when the M_n increase is due only to reaction (3.1a) and glycol diffusion is much faster than the reaction rate. In fact, starting from a second-order rate equation, a linear dependence of M_n versus time can easily be obtained and used to fit experimental results [80,81]:

$$-\frac{dc_{OH}}{dt} = k_1 c_{OH}^2 \Rightarrow \frac{1}{c_{OH}} = \frac{1}{c_{OH}^o} + k_1 t \tag{3.30}$$

$$M_n = M_n^o + k_1 t \tag{3.31}$$

An n value of -0.5 in equation (3.29) has been used to describe SSP for PET, PA 66, and PA 610 [43,82–85]. It means that M_n (or IV) increases linearly with the square root of the time:

$$M_n = M_n^o + kt^{0.5} \qquad (3.32)$$

A semiempirical approach like this has been used to describe the SSP of PET samples containing a significant fraction of COOH end groups [85]. The authors explained this behavior, assuming that the main reaction during the studied SSP was the esterification reaction (3.2a) due to the fast water diffusion within thin films. However, a decreasing rate of OH end groups higher than that of COOH end groups is evident from the data reported in that paper, suggesting that a not-negligible contribution from reaction (3.1a) was also present.

All the equations above predict a continuous unlimited increase of MW with time, and it is evident that they are not able to describe the MW increase profile observed over long SSP times when an asymptotic limiting value of MW is typically reached. Nevertheless, they can describe SSP reasonably well in the first part of the process, within a limited range of MW increase.

To explain the limiting value of MW, Bamford and Wayne in 1969 [80] suggested that the rate constant decreases as the SSP proceeds, but they did not propose any equation to account for this. To account for the existing limit for MW, Duh [57] has recently proposed the following semiempirical equation:

$$-\frac{dc}{dt} = 2k_a(c - c_{ai})^2 \Rightarrow \frac{1}{c - c_{ai}} = \frac{1}{c^o - c_{ai}} + 2k_a t \qquad (3.33)$$

where k_a is an apparent kinetic constant accounting for both reactions (3.1a) and (3.2a), c is the total end-group concentration (c_{OH} and c_{COOH}) at a given time t, c^o is the total end-group concentration in the initial prepolymer, and c_{ai} represents the inactive end-group concentration.

As discussed above, these inactive end groups were assumed to include only dead and reactive end groups that are firmly trapped in the crystalline structure. However, this concept can be extended to include the reactive end groups that are not able to react due to their inability to meet other reactive groups by segmental mobility in the last stage of the SSP process. C_{ai} and k_a are actually two adjustable parameters that include the effects of all factors, such as temperature, prepolymer IV, carboxyl concentration, particle sizes, diffusion resistance, morphology, degradation reactions, and so on, and that have to be calculated for any change in such factors.

Equation (3.33) seems able to fit the SSP data very well, even for reaction times as long as 30 hours, and seems to fit experimental SSP data better than any other empirical equation, as discussed in detail in a recent paper [57], where the capability of different empirical and semiempirical rate equations to fit IV versus time data provided for a commercial PET sample were compared. It is well known that the SSP rate decreases when the particle size increases over a given value (dependent on the type of polyester and on the temperature). Obviously, this effect has been attributed to the diffusion resistance of by-products to migration to the surface particles. Accordingly, some semiempirical kinetic models were developed by considering that the SSP process is driven only by diffusion phenomena. For instance, Chen et al. [82] and Chang [86] proposed a pure diffusion-controlled model to describe the SSP of PET. However, the boundary conditions used were chosen such that only the diffusion of the EG initially present was considered, without considering the formation of ethylene glycol (EG) by any type of reaction. This is a quite unrealistic model, as most of the diffusing species present in the initial prepolymer are removed during the drying step typically carried out prior to SSP. In agreement with Duh [57], we believe that those papers actually demonstrate only that the shape of the curves of the MW increase with time during SSP is quite similar to the shape of diffusion curves of by-products out of a particle.

The SSP rate can be controlled either by chemical reactions or by the physical diffusion of low-MW by-products within the solid particles or by both of these phenomena at the same time, depending on reaction conditions. Therefore, more realistic models should include a set of reactions containing both reaction rate and diffusion terms; unfortunately, no analytical solutions become possible as soon as the kinetic models take into account more than a single reaction and the diffusion of one or more species.

Several kinetic models have been developed by considering a set of equations that include both reaction kinetic and diffusion equations; several assumptions have usually been made concerning the types of reactions considered and/or the source of kinetic and diffusion parameters in order to make the mathematics simpler and to reduce the number of adjustable fitting parameters. Some kinetic models developed for the SSP of PET have considered only reaction (3.1a), ignoring the presence and the role of carboxyl groups [11,82,86]. A kinetic model including reaction (3.1a) and the diffusion of glycol has been proposed [11] to describe the SSP occurring in a single particle of arbitrary shape (e.g., sphere, cylinder, plane sheet, cube). The model was developed by assuming that the carboxyl concentration is negligible and therefore only reaction (3.1a) and the diffusion of EG were considered. The mass-transfer process was assumed to be isothermal and the diffusion of Fickean type. Two material balance equations with respect to EG and hydroxyl end groups were written and solved for various boundary conditions. Based on this model, the authors discussed the limiting cases of pure kinetic control, pure diffusion control, and an intermediate case, where both reaction and diffusion determine the evolution of the SSP process. Additionally, they analyzed the case where gas-side resistance assumes importance.

Huang and Walsh [87] studied the SPP of PET for various particle sizes at temperatures of 190 to 220°C and under different gas flow rates. They considered a single reaction [reaction (3.1a)] and concluded that the SPP rate of PET is not always determined by a single control mechanism, and that the controlling mechanism changes under different operating conditions. At a given gas flow rate, the SPP reaction mechanism for a large sample changes from chemical reaction control to interior diffusion control with increasing temperature. At a given reaction temperature, the SPP reaction control mechanism for a small sample changes from surface diffusion control to chemical reaction control with increasing gas flow rate. At a given reaction temperature and gas flow rate, the SPP reaction mechanism changes from interior diffusion control to surface diffusion control with decreasing particle size.

Although in theory it is possible to prepare polyester prepolymers with only hydroxyl end groups (no carboxyl groups), most industrially prepared polyesters contain both hydroxyl and carboxyl end groups (typically, commercial PETs contain 10 to 30% of carboxyl end groups, and this percentage may reach 80% in PBT). As a consequence, to describe the effect of different OH/COOH end-group ratios, reactions (3.1a) and (3.2a) both have to be considered. Therefore, more realistic models should include esterification reaction (3.2a), and accordingly, many authors [37,40,49,58,62,75,76,88,89] have proposed kinetic models, including reaction (3.2a).

Duh [58] extended the concept of inactive groups of its previous kinetic model [60] to another kinetic model that considers two reactions, (3.1a) and (3.2a), and he used this model to investigate the role of several SSP parameters. He concluded that (1) in the presence of sufficient catalyst the esterification kinetic constant (k_2) is lower than the transesterification constant (k_1) and, as a consequence, the SSP rate increases monotonically with decreasing COOH end-group concentration; and (2) k_1 decreases with increasing COOH concentration if the catalyst concentration is low. He suggested also that during the SSP of pelletized PET, both reactions and diffusion are important, and as the diffusion rate of water is higher than that of EG, the contribution of esterification may become more important and an optimal OH/COOH end-group ratio should exist.

A mathematical model including nine reactions and the diffusion of water, EG, and AA has been proposed by Kang [37] to describe the SSP of PET spherical particles. Kinetic and diffusion parameters were taken from the literature for melt polymerization, and some of these data were modified with prefactor adjustable parameters in order to fit the model results with the experimental data (IV and carboxyl group content versus time) reported in a previous paper by other authors [62]. The good fit of the model predictions with previous experimental data [82] seems to validate the model. However, the approach has a high number of fitting parameters, and the role of some reactions included in the model with respect to SSP rate and to the predictive capability of the role of variables is not clear.

A particle model to describe the SSP in a single PET pellet has recently been proposed [76]; the model, similar to that proposed by Kang [37], considers nine reactions, including side reactions but ignoring acidolysis, and 10 components (but only water, EG, and acetaldehyde (AA) were considered able to diffuse out of the pellets). This model has been optimized against changes in MW and COOH end groups by including a fitting parameter that should account for the limited mobility in SSP with respect to melt polymerization. Then the model has been used to simulate the effects on the MW increase of variables such as temperature, time, pellet size, and position within the pellets. The model was also used to describe the generation of AA and to simulate the formation of vinyl ester groups and terephthalic acid [76,90]. Chen and Chen [62] concluded that in the SSP of PET fine particles, a higher hydroxyl concentration leads to faster rates of MW increase, whereas a higher carboxyl concentration is preferable for the SSP of PET pellets. These conclusions were confirmed by Duh [58], who found that the SSP rate decreases monotonically with decreasing carboxyl content for fine PET particles, while an optimal carboxyl/hydroxyl end-group ratio exists for PET pellets.

A limiting value in MW is typically observed for high SSP times; it can be accounted for either by the chain-scission side reaction (when its rate equals that of polycondensation reactions) or by a decrease in the kinetic constant due to a limitation in segmental diffusion, or for both reasons. Chen and Chen [62] used a purely kinetically controlled model to take into account segmental diffusion in the SSP of PET. The model assumes a decrease in the effective kinetic constants when the SSP proceeds, and the concentration of reactive end groups becomes smaller and smaller. The model was based on two reactions, (3.1a) and (3.2a), and was tested with respect to the change in concentrations of hydroxyl and carboxyl end groups. The authors considered two different limit cases: in the first, the conditions were chosen so that diffusion was negligible, and in the second, the SSP rate is controlled by the diffusion of by-products only. At low temperature (170 to 199°C) and for small particle sizes, the diffusion of by-products is much faster than the chemical reactions and can be neglected; accordingly, a kinetic model based on equations (3.11) and (3.16′) (only the forward contributions) was used.

To account for the asymptotic limit observed experimentally for the MW ($M_{n,\text{lim}}$), the authors proposed equations (3.25) and (3.26) to account for the effective kinetic constant decrease. $k_{1,0}$ and $k_{2,0}$ were calculated by extrapolating to lower temperatures the kinetic constants derived from polymerization in the molten state, while $k_{1,1}$ and $k_{2,1}$ were derived by experimental data fitting. It appears that these equations would lead to unrealistic negative values of the effective kinetic constants, as both c_{OH} and c_{COOH} tend to zero; this has no physical meaning, and the model has to be considered valid until the OH and COOH concentrations at which $k_{1,\text{eff}}$ and $k_{2,\text{eff}}$ become zero, and the corresponding MW has to be considered the limit value for the increased MW ($M_{n,\text{lim}}$).

It is interesting to use equations (3.25) and (3.26) to calculate the changes predicted for $M_{n,\text{lim}}$ by changing some variables. According to the data reported in that paper [62], by simple calculations we find that:

1. The actual amount of active groups increases by increasing the temperature, which means that $M_{n,\text{lim}}$ increases with temperature.
2. $M_{n,\text{lim}}$ increases by increasing the initial $c_{\text{COOH}}/c_{\text{OH}}$ ratio between 0 and 0.5.

These results are in qualitative agreement with the generally accepted assumption that the role of segmental diffusion decreases as the temperature increases for both reactions (3.1a) and (3.2a) and different values of these features have actually been found by different authors in various experiments [57,58,60]. The activation energies for reactions (3.1a) and (3.2a) ($E_{a,10}$ and $E_{a,20}$) are 100 and 80 kJ mol^{-1}, respectively, in good agreement with literature data for the same reactions in the molten state. The activation energies $E_{a,11}$ and $E_{a,21}$ for the segmental mobility are 57.3 and 64.0 kJ mol^{-1}, close to that of 63 kJ mol^{-1} reported for segmental diffusion well above the glass-transition temperature, T_g [91]. Apparently, the model also seems to be able to account for the SSP rate increase after remelting.

The same approach [62] was extended qualitatively to particles with higher sizes, for which diffusion of water and/or EG cannot be neglected. By considering the decrease in both hydroxyl and carboxyl terminal group concentrations, they arrived at the conclusion that in the temperature range 210 to 240°C, the esterification rate is not affected by the diffusion of water, whereas the diffusional resistance of EG plays an important role with respect to the contribution of reaction (3.1a) to the overall rate of SSP.

An attempt to account for segmental mobility during SSP of PET was proposed by Wu et al. [38]. Their kinetic model considered the diffusion of both water and EG (with no mass-transfer resistance to the gas phase) and included three side reactions in addition to reactions (3.1a) and (3.2a). The model proposed also tries to take account of the effect of segmental diffusion, in analogy with the approach proposed in a previous paper to explain the reduction in the rate of the termination reaction in free-radical polymerization [92]. An empirical constant fitting parameter Θ (with the dimension of time) was used to take account of segmental diffusion. Using kinetic parameters and diffusion coefficients taken from the literature (but not reported in the paper), they used the model to calculate the effect of some variables; a good fit between simulation and experimental results was obtained for the effect of temperature and particle sizes.

Ma et al. [21] examined the influence of reaction environments on the solid state polymerization (SSP) of thin (180-μm) PET particles at 250°C by following the intrinsic viscosity (IV) increase and end-group depletion. Based on their experimental conditions (gas flow rate and particle thickness), they ignored gas-phase mass transfer and diffusion of both EG and water. Hence, the observed rate of SSP was assumed limited by the chemical reaction kinetics, and the experimental data were described in terms of reactions (3.1a) and (3.2a) by ignoring side reactions. They used the Constant_Relative_Variance model (commercial software gPROMS), where kinetic constants are considered as adjustable parameters, to fit experimental results. They observed that under a stream of nitrogen (SSP-N$_2$), the data can easily be represented up to IV < 1.3 dL g^{-1}, even without including inactive group concentrations in the kinetic equations. However, the fitting deviated from experimental values at reaction time $t > 1$ h (IV > 1.3 dL g^{-1}) and they found that experimental results could be described satisfactory by using the Duh model up to IV = 2.3 dL g^{-1}. A comparison of the results obtained for the inactive end-group concentrations with those of Duh [57] extrapolated to the SSP conditions employed by Ma et al. [21] indicated that the validity of the relation proposed by Duh generally cannot be extended to other SSP conditions, suggesting that the inactive end-group concentration actually includes the effects of some variables.

Ma et al. [21] also found that the sublimate collected during the SSP of PET under vacuum consisted of terephthalic acid (TA), monohydroxyethyl terephthalate (MHET), bishydroxyethyl terephthalate (BHET), and cyclic oligomers. This last evidence led the authors to conclude that other reactions, in addition to reactions (3.1a) and (3.2a); have to be used to describe SSP under vacuum at very high IV values. The removal of these aromatic oligomers from the PET particles under high vacuum can therefore contribute to enhancing the rate of rise of IV during SSP-vacuum as compared to SSP-N_2. By considering that BHET, MHET, and TA can be developed by intermolecular glycolysis, acidolysis, and transesterification reactions, these reactions were included in the kinetic scheme along with reactions (3.1a) and (3.2a) assuming that their kinetic constants have the same value of k_1, the kinetic constant of reaction (3.1a). Using the $k_1, k_2, c_{OH,i}$, and $c_{COOH,i}$ values obtained from SSP-N_2 experiments, the authors found that the IV(t) profile predicted in the short range was much faster than the SSP-vacuum experimental observations. Therefore, they concluded that the net contribution of the reactions leading to the formation of low-MW products containing carboxyl groups to the progress of SSP-vacuum has to be assumed controlled by the rate of mass transfer of the aromatic condensates (both through polymer particles and possibly also from the particle surface to the surrounding medium). Making further somewhat gross approximations, the authors concluded that a satisfactory match with the experimental measurements up to IV $= 2.75$ dL g^{-1} can be obtained, suggesting that the enhancement in overall reaction rate during SSP-vacuum as compared to SSP-N_2 can be attributed to the mass transfer–controlled removal of low-MW products containing aromatic rings during the last stage of SSP-vacuum.

The c_{COOH}/c_{OH} end-group molar ratio is typically higher, and side reactions are faster for PBT with respect to PET. These differences can account for the different MW profiles with time observed for PET and PBT. In fact, although a limiting value in MW was typically found for PET (see the discussion above), the MW profiles observed over long times at each point within a 3-mm-thick slab of PBT submitted to SSP at 214°C showed a maximum before leveling off at a significantly lower MW. Accordingly, acidolysis reaction (3.3a) and two side reactions (THF formation and chain-scission reaction) and the diffusion of TA had to be included in the kinetic model in order to describe the experimental results [22]. As discussed in more detail later, SSP is controlled by different reactions and diffusing species at different times, and the increase in MW is expected to be different in different positions, with higher increasing rates on the surface. MW profiles as a function of time and position within solid samples submitted to SSP have both been measured for PBT (up to 96 hours) [22] and calculated from model simulations for both PBT [22] and PET [76]. For PET, the differences in the SSP rates between the surface and the center increase when the temperature increases.

3.3.1. Effects of Variables and Predictions Based on Kinetic Models

The kinetic models discussed above have usually been tested against their capability to describe experimental results. Sometimes they have been used to predict the effects of variables by simulation. Major factors influencing the SSP rate are catalysts (type and concentration), temperature, particle sizes, type and concentration of reactive end groups, and gas flow rate/vacuum. Next, we summarize the main conclusions reported in the literature about the effects of some variables on the kinetic aspects.

a. Effect of Particle Shape and Dimension   Many authors have considered the effect of particle sizes. As a general conclusion, it has been found that their effect depends on temperature and gas flow rate (vacuum) during SSP. At a given gas flow rate, the SPP reaction mechanism for large particles changes from chemical reaction control to interior diffusion control with increasing temperature. At a given reaction temperature, the SPP reaction control mechanism for small particles changes from surface diffusion control to chemical reaction control with increasing gas flow rate. At a given reaction temperature and gas flow rate, the SPP reaction mechanism changes from interior diffusion control to surface diffusion control with decreasing particle size.

In a recent paper, after a preliminary study for the SSP of PET, Duh [60] concluded that at 230°C under a stream of nitrogen at 2.5 cm s^{-1} velocity, there is no effect of the particle size on the SSP rate when the particle diameter is lower than 180 μm. Fortunato et al. [93] found that the SSP of PBT under vacuum at 200°C is not affected by particle size when the particle diameter is below 0.75 μm. To increase the SSP rate, a possible alternative to the reduction of particle size is to use porous samples. Indeed, increasing the porosity decreases the characteristic length scale for the diffusion of by-products and thus favors faster SSP. Porous particles can be generated either by compacting low-MW prepolymers or by using either an inert gas or a foaming agent dispersed in the melt phase [94–96].

For a diffusion-controlled process, Ravindranath et al. [11] used a kinetic model to simulate the effect of particle-shape geometry on the SSP rate of PET, as shown in Figure 3.5. Under the same operating conditions, the MW increase is maximum for spheres ($\lambda = 2$) and cubes (dashed line) and minimum for flat films ($\lambda = 0$). The influence of particle diameter on the polymerization degree (P_n) was also simulated for spherical particles, and the results are shown in Figure 3.6 [11]. $\Theta^* = k_1 c_{0,\text{DE}} t$ and $\xi_0^* = (k_1 c_{0,\text{DE}}/D)^{0.5} x$ ($c_{0,\text{DE}}$ = concentration of diester groups in the prepolymer and x = distance in the direction of diffusion) are nondimensional parameters that account for reaction time and particle size, respectively. For a given reaction time (Θ^* constant), the MW increase

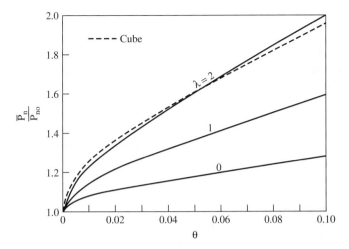

Fig. 3.5. Effect of particle geometry on the degree of polymerization (P_n) (P_{n0} is the initial degree of polymerization). (From Ravidranath and Mashelkar [11] by permission of John Wiley & Sons, Inc.)

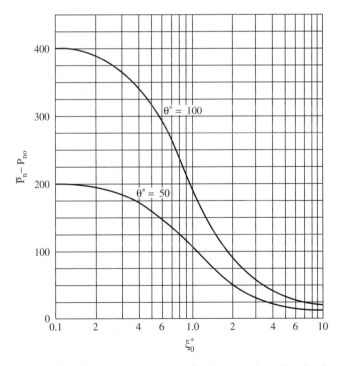

Fig. 3.6. Influence of particle size on degree of polymerization (P_n) for fixed reaction times (P_{n0} is the initial degree of polymerization). (From Ravidranath and Mashelkar [11] by permission of John Wiley & Sons, Inc.)

is higher the smaller ξ_0^* is, or in other words, the smaller the particle size for a given (kinetic constant/diffusivity) ratio, or the higher the diffusion rate with respect to the reaction rate for a fixed particle size. When ξ_0^* is below 0.2, the process is reaction-rate controlled, whereas when $\xi_0^* > 5$, SSP approaches a diffusion-controlled regime. For an intermediate range of ξ_0^* values, the process is controlled by both diffusion and reaction rate.

b. Effect of Temperature   The main effect of temperature is on both kinetic constants and diffusivity, and typically, an Arrhenius equation is used to describe the effect of temperature on both these kinetic parameters. In Tables 3.1 and 3.2, kinetic constants and diffusion coefficients reported in the literature for PET are collected. The range of these data appears to be quite wide. Values reported for the activation energies range from about 63 to 145 kJ mol^{-1} for reactions (3.1a) and (3.2a) and from 17 to 130 kJ mol^{-1} for the diffusion rate of EG. This is not surprising, as many variables are different in different studies, and the assumptions made by different authors can be different. In some studies it was also found that the activation energy is different in different temperature ranges [63,86,87]. This probably reflects changes in the SSP controlling mechanism.

As the activation energies for diffusion of both EG and water are lower than those of reactions (3.1a) and (3.2a), an increase in temperature will increase the reaction rate much more than the diffusion rate, with possible changes from reaction-controlled to diffusion-controlled SSP. As the activation energy of side reactions (3.5) and (3.6) is much higher than that of reactions (3.1a) and (3.2a), the lower the temperature, the less important will be the role of side reactions.

c. Effect of Carboxyl/Hydroxyl End Group Molar Ratio in the Prepolymer   Both —OH and —COOH end groups are involved in reactions that lead to MW increase [mainly reactions (3.1a) and (3.2a)] as well as in side reactions. As their concentration appears in kinetic equations, it is not surprising that the SSP rate depends on the c_{COOH}/c_{OH} ratio. For the SSP of PET, it has been demonstrated that the optimal ratio is in the range 0.3 to 0.8, depending on the SSP temperature and on pellet sizes [38,58,63]. Jabarin [97] and Wu et al. [38] reported that the maximum SSP rate of pelletized PET is achieved when $c_{COOH}/c_{OH} = 0.5$; Schaaf et al. [98,99] found a maximum SSP rate for ratios ranging from 0.3 to 0.6; Chen and Chen [62], based on the different diffusivities of EG and water ($D_W > D_{EG}$), concluded that a high c_{OH} value is better in fine or porous samples, whereas a high c_{COOH} value is preferred in granulated PET.

However, Duh [58] recently found monotonically decreasing SSP rates with increasing acid end groups in starting PET, for fine particles of IV $= 0.25$ dL g^{-1}, regardless of catalyst concentration. Because the diffusion of water is faster than that of EG, a high c_{COOH}/c_{OH} ratio is expected to increase the SSP rate when the process is diffusion controlled. The c_{COOH}/c_{OH} ratio is more critical for PBT than for PET, as fast side reactions give a strong contribution to the change (increase) in the COOH/OH end-group molar ratio during SSP. A low initial c_{COOH}/c_{OH} ratio is generally preferred, as under the same conditions it leads to a faster SSP process and to a higher final MW [22].

TABLE 3.1. Kinetic Constants and Activation Energies Reported for the SSP of PET

Reference	k_1 (kg mmol^{-1} h^{-1})	$E_{a,1}$ (kJ mol^{-1})	k_2 (kg mmol^{-1} h^{-1})	$E_{a,2}$ (kJ mol^{-1})	Notes
[82]	$(*)k = 6.6 \times 10^{17} \exp^{-178/RT}$	178	—	—	$(*)k$ in h^{-1}, not related to a specific reaction
[67]	$9.91 \times 10^4 \exp^{-77.5/RT}$	77.5	—	—	
[84]	$(*)k_{\text{Goodyear}} = 624 \times 10^{10} \exp^{-95.4/RT}$ $(*)k_{\text{FirestoneA}} = 828 \times 10^{10} \exp^{-97.1/RT}$ $(*)k_{\text{Eastman7328}} = 5.70 \times 10^{10} \exp^{-77.0/RT}$	77.0–97.1	—		$(*)k$ = g mol^{-1} min$^{-0.5}$, not related to a specific reaction
[62]	2.034×10^{-3}	100	6.41×10^{-5}	80	At 199°C
[11]	2.3×10^{-5}	—	—	—	At 160°C
[81]	$6.59 \times 10^7 \exp^{-99.6/RT}$	99.6	—	—	Not related to a specific reaction
[88]	2.1×10^{-4}	—	7.9×10^{-4}	—	At 225°C
[37]	1.3×10^{-3}	133.8	4.0×10^{-3}	130.8	At 230°C
[60]	$1.029 \times 10^9 \exp^{-0.0068[Cl]_0 - 98.65/RT}$	98.65	—	—	Where $[Cl]_0$ is the initial concentration of end groups (mmol kg^{-1})
[57]	$653.044 \exp^{-80.9/RT}$	80.9	—	—	Not related to a specific reaction
[58]	1.63×10^{-3} to 1.03×10^{-3}	—	1.38×10^{-3} to 1.07×10^{-3}	—	At 230°C, dependent on COOH prepolymer concentration
[63]	—	63–67 80–145	—	—	At 200–230°C At 160–200°C
[21]	1.15×10^{-2}	—	3.4×10^{-2}	—	At 250°C
[89]	—	135.4	—	125.2	PET modified with isophthalic acid (ter/iso molar ratio 98.8/1.2)

TABLE 3.2. Diffusion Coefficients and Activation Energies for EG and Water Diffusion in the SSP of PET

Reference	D_{EG} (cm² s⁻¹)	$E_{a,EG}$ (kJ mol⁻¹)	D_W (cm² s⁻¹)	$E_{a,W}$ (kJ mol⁻¹)	Notes
[49]	0.005×10^{-6} at 210°C	126			
	0.009×10^{-6} at 220°C				
	0.019×10^{-6} at 230°C				
[62]	1×10^{-6} at 230°C	22.6			
[11]	1.95×10^{-6} at 220°C	130.8			
	3.6×10^{-6} at 230°C				
	6.0×10^{-6} at 240°C				
	9.5×10^{-6} at 250°C				
[88]	2.6×10^{-6} at 225°C	—	5.8×10^{-6} at 225°C		
[37]	3.1×10^{-6} at 230°C	16.7	5.7×10^{-6} at 230°C	10.0	
[50]	2.18×10^{-6} at 230°C	113			
[89]	2.60×10^{-6} at 200°C	17.6	5.80×10^{-6} at 200°C	9.57	PET modified with isophthalic acid
	2.86×10^{-6} at 210°C		6.09×10^{-6} at 210°C		(ter/iso molar ratio 98.8/1.2)
	3.12×10^{-6} at 220°C		6.38×10^{-6} at 220°C		

d. Effect of Catalysts Even though it is well known that reaction (3.1a) occurs at a significant rate only in the presence of suitable catalysts [100] and that reactions (3.1a) and (3.2a) are both affected by type and concentration of catalyst, its role has been ignored in most SSP studies. In a recent paper, Duh [101] made an extensive study of the effect of catalyst combined with carboxyl group content [54]. The reaction time required to achieve a 0.60 dL g^{-1} IV for PET prepolymers containing different amounts of Sb-based catalyst (from 0 to 300 ppm) showed that even the addition of a small amount of catalyst leads to a strong increase in the SSP rate, especially for prepolymers with a low carboxyl group content. When the catalyst concentration reaches a certain limit (about 150 ppm of Sb for PET), a further increase in catalyst concentration does not influence the SSP rate. These results are in contrast with those reported by Kokkalas et al. [102], who found that the SSP rate increases up to 2000 ppm of Sb_2O_3.

Duh has also reported that 2 to 5 ppm of Ti-based catalyst can be as effective as 100 ppm of Sb-based catalyst in catalyzing PET prepolymers with a low carboxylic group content. Also, the transesterification rate increases by increasing Sb concentration (and decreasing COOH concentration), and the effect of the catalyst is more pronounced in prepolymer with a low content of COOH groups. This suggests that COOH groups compete with OH end groups for catalyst and can lead to a reduction in the catalytic activity with respect to reaction (3.1a) if they form stronger interactions with the catalyst. Similar results were found previously for Ti-based catalyst from studies on model molecules [27,64].

In another study, Karayannidis et al. [85] found that the decrease in OH end groups was higher than that of COOH end groups when thin PET films were submitted to SSP at 180 to 230°C. When the same PET film was dissolved in *o*-chlorophenol, reprecipitated in methanol, and the PET powder recovered again submitted to SSP, it was found that contrary to what observed for the parent PET film, the rate of change in OH and COOH end-group concentrations was the same in the temperature range 180 to 220°C, and only at 230°C was the decrease in OH groups slightly higher than that of COOH groups. These results can be ascribed to the removal of the transesterification catalyst during the dissolution–reprecipitation treatment and put in evidence once more the role of the catalyst. Without catalyst, reaction (3.2a) is still active due to COOH catalysis, whereas the rate of reaction (3.1a) is strongly reduced and its contribution becomes significant only at higher temperature.

Based on these results, it can be concluded that the kinetic constants used in SSP kinetic models depend on the type and concentration of catalyst, but also on its interaction with the functional groups present in the reaction medium, and in particular with COOH groups. This last effect can also influence the optimal COOH concentration in the initial prepolymer for the maximum SSP rate. Any comparison between SSP reaction rates is useless in the absence of data concerning type and concentration of catalysts.

e. Effect of Crystallinity Degree Usually, after a relatively short preheating step, the degree of crystallinity can be considered constant. Therefore, its main role during SSP is to define the amorphous mass fraction and therefore the concentration of reactive groups. Other effects of the degree of crystallinity on the PET SSP rate have been reported and explained on the basis of different rate-controlling mechanisms [38,41,103]. When the SSP is by-product-diffusion controlled, high crystallinity reduces the SSP rate by increasing the diffusion resistance. On the other hand, when the SSP is reaction controlled, high crystallinity increases the SSP rate by increasing the terminal group concentration in the amorphous regions [63]. The role of crystallinity seems particularly important in the SSP of PLLA, where control of the degree of crystallinity has been used to obtain the final PLLA free of di-L-lactide monomer [104]. In the slow-crystallizing PEN, the control of crystallization is a critical step in the SSP process [17].

f. Effect of Nature and the Flow Rate of Carrier Gas   Several conflicting results are reported about the effects of the method used to remove by-products from the particle surfaces during SSP (vacuum or gas, nature and flow rate of the inert gas). Typically, an increase in the flow rate of the carrier gas leads to an increase in the SSP rate, before leveling out at high flow rates [87,88,105–107]. For the SSP of PET, a nitrogen velocity of 0.8 cm s^{-1} was sufficient to achieve the maximum SSP rate in 1.1- to 2.7-mm particles at 190°C, while a nitrogen velocity of 2.5 cm s^{-1} was required for the maximum reaction rate at 220°C. Ma et al. [21] noticed that the kinetics of SSP under a stream of nitrogen is not affected by increasing the velocity beyond 5 cm s^{-1} for PET particles of average thickness 180 μm. Similar results were reported by Duh [60], who found that the IV of a PET prepolymer (0.35 dL g^{-1} initial IV, 100- to 150-mesh particle size) reached a maximum value after 2 hours at 230°C when the nitrogen flow rate was 1.5 cm s^{-1}.

Ma et al. [21] found that when nitrogen was used as a carrier gas, the reaction rate and the extent of molecular-weight buildup are somewhat lower than those of SSP under vacuum. As it is known that supersaturation of solutes can enhance their diffusive devolatilization from polymer matrices [108,109], it can be supposed that high vacuum levels lead to supersaturation of low-MW products such as aromatic by-products, also containing carboxyl groups. As a consequence, vacuum can be more efficient than a stream of nitrogen for their removal.

For a diffusion-controlled process, Ravindranath and Mashelkar [11] simulated the effect of the gas-phase resistance of ethylene glycol. The results are reported in Figure 3.7 for various values of the parameter Φ [see equation (3.23)]; the gas-phase resistance increases by decreasing Φ. It appears that the higher Φ is, the lower is the effect of gas resistance on the P_n. This effect becomes very important for Φ values lower than 50 and can be neglected for Φ values above 500. As the value of Φ decreases with decreasing particle size, x_0, the gas-side resistance may become important with respect to the MW increase when very small particles are submitted to SSP. In that case it is important to operate SSP under high vacuum or high gas flow rate in order to increase Φ by increasing k_i, the gas-phase mass-transfer coefficient.

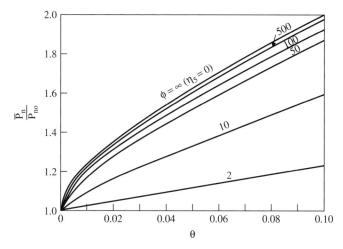

Fig. 3.7. Influence of Φ on degree of polymerization (P_n) (P_{n0} is the initial degree of polymerization). (From Ravidranath and Mashelkar [11] by permission of John Wiley & Sons, Inc.)

The SSP rate can be also affected by the type of gas used to remove volatile by-products. For instance, Hsu [106] found that an SSP value of 0.18- to 0.25-mm PET particles ($M_n = 16{,}500$ g mol^{-1}) at 250°C for 7 h, nitrogen velocity 2 cm s^{-1}, led to an M_n value of 58,000, whereas helium or carbon dioxide under the same conditions led to M_n values of 81,000 and 90,000, respectively. Hsu attributed this to the influence of carrier gas diffusion on the diffusivity of EG in the polymer and to the interaction of the carrier gas with EG. Devotta and Mashelkar [40] found similar results and considered these effects in terms of the influence of the associated free-volume changes on diffusivities. Mallon et al. [110] found no discernible effect among the same carrier gases during the SSP of 0.1-mm PET particles at 226°C.

An interesting discussion on this point has been reported by Ravindranath and Mashelkar [11]. According to these authors, there are two possible ways in which the type of inert gas could influence the SSP rate: the change in the gas-side mass-transfer coefficient (k_i) or plasticization of the amorphous polymer domains by the carrier gas. While an increase in the gas flow rate and by-product gas phase diffusivity can induce increases in k_i, gas absorption can induce plasticization effects in the amorphous domains and can enhance the low-MW by-product diffusivities within the particles, at a constant gas flow rate. This plasticization effect can account for changes due to different carrier gases.

3.4. SOLID STATE POLYMERIZATION OF TYPICAL POLYESTERS

The basic knowledge of the principles of SSP reported above is valid for all types of polyesters, and there are many analogies in terms of molecular structure and reactivity between different polyesters. However, the SSP of different types of polyesters may be characterized by some specific features, and the main features of the SSP of some polyesters—PET, poly(butylene terephthalate) (PBT), poly(ethylene naphthalate) (PEN), poly(trimethylene terephthalate) (PTT), and poly(L-lactic acid) (PLLA)—are discussed below.

3.4.1. Poly(ethylene terephthalate)

Most SSP studies on polyesters have dealt with PET, due to its industrial importance, and most previous discussion on kinetic models was based on the SSP results for PET. So here we summarize only a few concepts and some specific features of the SSP of PET. With the aim of obtaining a high reaction rate while controlling product quality, several studies have been devoted to examining the influence of factors such as temperature [49,62,82,84,86,87,105,106,111], pellet size [38,49,62,82,86,106], crystallinity level [40,49,60,86,103], nature and rate of carrier gas [40,50,87,88,105,106], and catalyst concentration [101,102].

Tables 3.1 and 3.2 present kinetic and diffusion data reported in the literature. The strong differences among results reported by various authors reflect the various assumptions made on developing kinetic models and the different characteristics of the PET prepolymer. As the activation energies of reactions (3.1a) and (3.2a) are higher than those for the diffusivity of EG and water, an increase in temperature will favor the reaction rate more than the diffusion rate, and diffusion will be the controlling step more and more as the temperature increases.

From Table 3.2 it appears that diffusivity of water is higher than that of EG, so that reaction (3.2a) is more important as particle size increases and SSP is diffusion controlled. As discussed above, this is one reason why there is an optimal COOH/OH molar ratio in the starting prepolymer in order to reduce the SSP time required to achieve a high MW value (typically, it ranges from 0.3 to 0.8, depending on SSP conditions) [38,58,63,97–99].

At 230°C, the SSP is controlled in part or completely by diffusion for particles with diameters higher than 180 μm at a nitrogen flow rate of 2.5 cm s^{-1}, and for a flow rate lower than 1.5 cm s^{-1} for particles of 100 to 150 μm [60]. Although the crystallite size does not depend significantly on the initial IV, the level of crystallinity was found to be slightly higher for samples prepared from lower-IV$_0$ prepolymers [63].

Stopping SSP after a given time, remelting the polymer, and then restarting SSP on the particle obtained after solidification of the molten polymer has been proposed as an alternative strategy to reach very high MW. This two- or multistep SSP could help to overcome the problems related to the reduced segmental mobility of end groups as SSP proceeds.

It is well known that several side reactions occur during melt polymerization, leading to chain scission and/or to the formation of acetaldehyde and of diethylene glycol moieties. Chain scission is a well-known reaction for all polyesters with hydrogens in the β position with respect to the ester bond. For PET it leads to the formation of carboxyl and vinyl ester end groups:

$$(3.5_{PET})$$

The kinetic constant at 280°C is 5×10^{-7} s^{-1} and the activation energy is about 180 kJ mol^{-1} [69]. The residual metal catalyst may increase the rate of chain scission [42]. The concentration of vinyl ester group with time has been calculated from kinetic models [38,76,90].

Vinyl end groups can undergo further reactions with the formation of acetalde-hyde (AA) [41,42,72,112,113]:

$$(3.6_{PET})$$

PETs derived from melt polymerization generally contain about 24 to 172 ppm of AA; this amount is reduced to about 1 to 2 ppm after SSP [49,84]. Recently, Kim and Jabarin [90] used a kinetic model, including side reactions, to describe the experimental generation of AA.

Another side reaction that occurs during the PET polymerization leads to the formation of diethylene glycol (DEG) moieties:

$$(3.6b_{PET})$$

The reaction mechanism is not well established [113–116] and no evidence has been reported of the effect of SSP on the final content of DEG moieties.

3.4.2. Poly(butylene terephthalate)

Because of the lower melting point, the SSP of PBT is commonly carried out at a lower temperature (200 to 225°C) than the SSP of PET; nevertheless, its SSP rate is higher. Unlike PET, precrystallization is not required, as PBT crystallizes faster than PET. Relatively few papers have dealt with the SSP of PBT [19,22,77,93,117–119]; some of them disregarded diffusion phenomena and considered only the effect of some variables on the MW increase [77,93,119], while a detailed study including the effects of side reactions and diffusion was reported in papers of Gostoli, Pilati, and others [22,118].

It is well known [19,41,68,69,73,120] that two important side reactions occur during polymerization of PBT: a chain-scission reaction [reaction (3.5$_{PBT}$)],

$$(3.5_{PBT})$$

and the formation of THF from the transformation of a 4-hydroxybutyl ester terminal group into a carboxyl group,

$$(3.6_{PBT})$$

In particular, the chain-scission reaction for PBT is much faster than for PET [69,73], and it has to be included in any kinetic model that aims to be predictive over long SSP times. On the other hand, reaction (3.6$_{PBT}$), which transforms hydroxyl end groups into carboxyl end groups, is also important, due to its effect on the rate of reactions (3.1a) and (3.2a). Other experimental evidence that has to be considered when modeling the SSP of PBT is the presence of oligomers with carboxyl groups, and in particular of terephthalic acid (TA), among the volatile products removed during SSP. To explain the presence of TA, the acidolysis reaction has to be considered.

Accordingly, the SSP of PBT was modeled quantitatively [22,118] by account-ing for five chemical reactions: (3.1a), (3.2), (3.3), (3.5$_{PBT}$), and (3.6$_{PBT}$), and for the diffusion of three volatile species: 1,4-butanediol, water, and TA. The reaction kinetic constants were taken from independent literature data (obtained either from melt polymerization or from model molecules), and the diffusivities were derived from data fitting.

By analyzing the experimental IV profile over long times at 214°C and at different positions within a 3-mm-thick slabs, it was observed that the IV profile showed a maximum (more pronounced for external than for internal layers) and that the time to reach the maximum is nearly the same at each location within the slab. In the initial period, and roughly up to time t_{max}, corresponding to the max-imum in IV, the SSP is diffusion controlled and dominated by reactions (3.1a) and (3.2a) in competition with reaction (3.5$_{PBT}$). The contribution of reaction (3.1a), whose rate depends on the square of c_{OH}, decreases suddenly in the early stages of the process, due to the fast disappearance of OH end groups accord-ing to reactions (3.1a), (3.2a), and (3.6$_{PBT}$). On the contrary, reaction (3.2a) is important up to the maximum, due to the increasing concentration of carboxyl groups; at t_{max}, the OH end groups have disappeared almost completely and the rate of reaction (3.2a) decreases strongly. For times higher than t_{max}, the slope of IV versus time is dominated by reaction (3.5$_{PBT}$) and a decrease of IV is observed. A marked and continuous MW decrease would occur and the experimentally observed asymptotic behavior would not be observed in the absence of reaction (3.3a), which, at very long times, is in competition with the degradation reaction (3.5$_{PBT}$).

The best fit of the experimental data was obtained with the following diffusion coefficients: $D_{Gly} = 10^{-5}$ cm^2 s^{-1}, $D_W = 1.2 \times 10^{-5}$ cm^2 s^{-1}, and $D_{TA} = 4 \times 10^{-6}$ cm^2 s^{-1}, were D_{Gly}, D_W, and D_{TA} are the diffusion coefficients of 1,4-butanediol, water, and terephthalic acid, respectively. As expected, due to faster diffusion from the outer layers, there is a MW gradient within the slab, with higher MWs on the surface. The predictive capability of the model was tested on results reported previously for PBT powders [119], in which diffusion is not a rate-controlling process. With no adjustable parameters the model was able to describe quite satisfactorily the experimental results for different initial MWs and c_{OH}/c_{COOH} ratios, and this good agreement is further support for the validity of the reactions considered and of the numerical values of the kinetic constants. The model was then used to analyze the effect of some parameters and to derive predictive information about the effect of some variables. Under the experimental conditions used (slabs 3 mm thick) reactions (3.1a) and (3.2a) are both diffusion controlled, whereas reaction (3.3a) is controlled by both chemical kinetics and diffusion. The main effect on the MW increase in the first part of the SSP process is related to the values of D_G and D_W, and it has been reported that the influence of D_G is limited, whereas the water diffusion coefficient, D_W, has a primary role. For PBT fine particles (no diffusion considered), the kinetic model predicts that small values of $r = c_{OH}/c_{COOH}$ allow for only small increases in MW; the SSP becomes very attractive for all initial MW when the hydroxyl/carboxyl terminal group ratio r is sufficiently large, typically above 0.5, and for very large values of r, the maximum MW attainable tends to become almost independent on the initial MW value (of course, the time necessary to obtain the maximum MW becomes larger as the initial MW decreases). In the presence of diffusive resistances, both the maximum MW and the time needed to reach it decrease.

3.4.3. Poly(ethylene naphthalate)

Poly(ethylene naphthalate) [PEN; also called poly(ethylene naphthalene dicarboxylate] differs from PET in that the acid and aromatic component of its polymer chain is naphthalene 2,6-dicarboxylic acid instead of terephthalic acid. The repeating unit of PEN is

PEN has a higher glass-transition temperature than that of PET (about 121°C compared to 80°C for PET), a higher melting temperature (melting range 250 to 290°C), and a reduced tendency to crystallize [121]. The intrinsic viscosity (IV) of PEN obtained from the melt polymerization is typically in the range 0.50 to 0.60 dL g^{-1}, and the SSP processes are used primarily to achieve higher molecular weights. Due to the relatively low reactivity [122], SSP temperatures are typically in the range 230 to 240°C. As PEN crystallizes at a slower rate and at higher temperature than those of PET, more complex conditions of annealing are necessary for the crystallization stage to reduce the risk of pellet sintering. In particular, the rate of PEN crystallization is significant only above 180°C (with a maximum at ca. 200°C), but at 195 to 200°C the tendency for particles to stick together is very strong. As consequence, the useful temperature range for this critical stage is very narrow. Furthermore, prior to crystallization, PEN pellets have to be devolatilized at a lower temperature to avoid rapid volume expansions of volatile products (water and EG), which can produce a phenomenon known as "popcorning" [123,124].

Sun and Shieh [125] investigated the kinetics of the SSP of PEN prepolymer using a simple empirical approach. They found that the number-average molecular weight is related linearly to the square root of the reaction time. SSP was carried out at a temperature between 200 and 245°C for prepolymers with initial MWs between 12,000 and 15,000 and with particle diameters between 0.25 and 0.50 mm. For PEN containing Sb_2O_3 as catalyst, the apparent rate constants were calculated and an activation energy of 33.2 kJ mol^{-1} was derived using an Arrhenius-type equation.

3.4.4. Poly(trimethylene terephthalate)

Poly(trimethylene terephthalate) (PTT) is obtained by the melt polycondensation of terephthalic acid or dimethyl terephthalate with 1,3-propanediol. The chemical structure of the repeating unit of PTT is

The polycondensation process is basically the same as that for PET but with some major differences [126]:

- Because of its lower thermal stability, PTT is melt-polymerized at a lower temperature, between 250 and 275°C, in order to prevent degradation.
- The lower reactivity of propanediol needs a larger amount of active catalyst (usually, titanium alkanolates).
- The side reaction of ester-bond scission leads to the formation of allyl-ester end groups, which can react with water or hydroxyl end groups to generate allyl alcohol, which, in turn, in the presence of oxidizers, can lead to the

formation of acrolein [as described schematically in reaction (3.6$_{PTT}$), where X can be either H or R, a polymer chain or oligomer or glycol]:

$$\text{(structure)} \xrightarrow{k_{5(PTT)}} \text{(structure)}$$

$$(3.5_{PTT})$$

$$\text{(vinyl ester)} \xrightarrow[\text{X-OH}]{k_{6(PTT)}} \text{(structure)} + CH_2{=}CH{-}CH_2{-}OH \quad \text{(allyl alcohol)}$$

$$\downarrow \text{[oxid.]}$$

$$CH_2{=}CH{-}\overset{O}{\underset{\|}{C}}{-}H \quad \text{(acrolein)}$$

$$(3.6_{PTT})$$

- For steric reasons, the favored cyclic oligomer is a low-melting cyclic dimer, which has a melting point of 254°C and tends to sublimate as crystals giving rise to deposits in the chemical plant.
- During the polymerization, propanediol can dimerize into dipropylene glycol, which can be incorporated in the polymer chains (like diethylene glycol in the case of PET). Solid state polymerization at 180 to 220°C under nitrogen allows us to obtain PTT with a IV higher than 1.0 dL g^{-1} [127,128]; it also prevents yellowing and reduces the amount of both acrolein and cyclic dimer.

The process has been described in some recent patents [128–130] and articles [78,131]. Taking advantage of the fast crystallization and the low tendency for particles to stick, the PTT can be solid state–polymerized at a temperature as high as 225°C (only 3°C below T_m), while a safe temperature for PET is about 215°C (about 40°C below T_m) [129]. A simplified continuous SSP process has been proposed in which annealing, preheating, and crystallization steps are combined into one single step [130].

By considering that PTT and PET have the same SSP mechanism, Duh has proposed for PTT [78] a kinetic model similar to that already used from the same researcher to describe the SSP behavior of PET [57]. It describes the SSP of PTT with a modified second-order kinetic model that included an apparent inactive end-group concentration, as discussed before [equation (3.33)]. As for PET, the SSP rate of PTT increases with increasing temperature, increasing the prepolymer IV and decreasing the pellet size. The result is that the PTT solid state polymerizes about 2.2 times faster than PET, and that in the range between 200 and 225°C the average rate about doubles with each 10°C (apparent activation energy of about 110 kJ mol^{-1}). The SSP rate increases by about 30% when the PTT particle size is decreased from 2.5 to 1.5 g for each 100 pellets.

3.4.5. Poly(L-lactic acid)

Poly(L-lactic acid) (PLLA) is a semicrystalline biodegradable aliphatic polyester that can be derived from renewable resources. PLLA has a glass-transition temperature between 50 and 80°C and a melting temperature between 170 and 180°C, depending on the amount of residual monomer [25,26,104]. It can be prepared by the ring-opening polymerization of L,L-lactide (also known as L-lactide) [132–135] or by direct polycondensation of L-lactic acid (HO—CHCH$_3$—COOH) [136]. Tin-based catalysts are typically used in both cases, either alone or in combination with p-toluenesulfonic acid (PTSA) [137–139]. The ring-opening polymerization of dilactide has been studied widely [135,140–143]; this polymerization involves a thermodynamic monomer/polymer equilibrium, and the final polymer contains a residual amount of monomer, depending on the equilibrium constant:

$$
\mathrm{H}\!\!\left(\!\!\begin{array}{c} \mathrm{O} \\ \parallel \\ \mathrm{O\text{-}CH\text{-}C} \\ \mid \\ \mathrm{CH_3} \end{array}\!\!\right)_{\!\!n}\!\!\mathrm{OH} \quad + \quad \begin{array}{c} \mathrm{CH_3} \\ \\ \end{array} \quad \rightleftharpoons \quad \mathrm{H}\!\!\left(\!\!\begin{array}{c} \mathrm{O} \\ \parallel \\ \mathrm{O\text{-}CH\text{-}C} \\ \mid \\ \mathrm{CH_3} \end{array}\!\!\right)_{\!\!n+2}\!\!\mathrm{OH}
$$

(3.34)

When the polymerization is carried out in the molten state, the residual amount of monomer is typically a few percent of the final polymerization product and increases with increasing temperature. This remaining monomer is detrimental for mechanical properties, causes corrosion of processing machines, and leads to an unexpected increase in the degradation rate of PLLA. Shinno et al. [104] showed that the concentration of L-lactide monomer can be reduced to zero after SSP carried out in either a one- or a two-step process in closed systems. Segregation and concentration of both monomer and catalyst in amorphous domains, promoted by crystallization, allow the polymer to reach 100% under well-defined conditions. Molecular weights from 40,000 to 110,000 were obtained independent of the amount of residual monomer. A kinetic analysis has been proposed to account for the effect of crystallization on the rate of monomer consumption.

In a different approach, a melt/solid state polycondensation process starting from L-lactic acid was used to prepare a PLLA with high molecular weights [137–139,144,]. In this method the PLLA prepolymer obtained by melt polycondensation of L-lactic acid was subjected to SSP around the T_c of PLLA (about 130 to 150°C).

$$
\mathrm{H}\!\!\left(\!\!\begin{array}{c} \mathrm{O} \\ \parallel \\ \mathrm{O\text{-}CH\text{-}C} \\ \mid \\ \mathrm{CH_3} \end{array}\!\!\right)_{\!\!n}\!\!\mathrm{OH} + \mathrm{H}\!\!\left(\!\!\begin{array}{c} \mathrm{O} \\ \parallel \\ \mathrm{O\text{-}CH\text{-}C} \\ \mid \\ \mathrm{CH_3} \end{array}\!\!\right)_{\!\!m}\!\!\mathrm{OH} \rightleftharpoons \mathrm{H}\!\!\left(\!\!\begin{array}{c} \mathrm{O} \\ \parallel \\ \mathrm{O\text{-}CH\text{-}C} \\ \mid \\ \mathrm{CH_3} \end{array}\!\!\right)_{\!\!n+m}\!\!\mathrm{OH} + \mathrm{H_2O}
$$

(3.2$_{\mathrm{PLLA}}$)

Water has to be removed in order to shift the equilibrium toward high MW [reaction (3.2_{PLLA})] and either high vacuum or a stream of dry N_2 in open systems [137–139], or dehydratants (CaO) in a closed reactor [144] can be used for this purpose. Molecular weights as high as 500,000 [138,139] and 250,000 [137] were obtained.

A ring-chain equilibrium with the formation of cyclic monomer can occur:

$$H\left(O\text{-}CH\text{-}\overset{\overset{O}{\|}}{C}\right)_n OH \rightleftharpoons H\left(O\text{-}CH\text{-}\overset{\overset{O}{\|}}{C}\right)_{n-2} OH \quad + \quad \text{(cyclic lactide)}$$

$$(3.1e_{PLLA})$$

and the residual amount of cyclic monomer can again be controlled by a suitable choice of SSP temperature. The removal of cyclic lactide along with water leads to a reduction in the overall yield (about 90%) when high vacuum or flushing with inert gas is used [138,139].

3.5. CONCLUSIONS

Kinetic models should predict quantitatively the progress of SSP and the final characteristics of the resulting polyesters as well as the effects of various factors. For this purpose, a comprehensive kinetic model is required. It should take into account all possible reactions and the diffusion of all possible diffusing species, and it should include in the kinetic rate equations possible contributions from segmental mobility, chemical diffusion, catalysts, and related synergistic or antisynergistic effects. However, the development of such a model for the SSP of polyesters is at present a formidable task, almost impossible to solve, as it would require the solution of quite a large set of differential equations and the knowledge that a large number of kinetic and diffusion parameters are not available. Therefore, several simplifying assumptions are usually made with regard to reactions and diffusing species to be included in the model.

Several simplified kinetic models have been proposed, and a general conclusion that can be derived from the analysis of literature data is that almost all the kinetic models proposed to describe SSP of polyesters are able to fit the experimental results, despite significant differences in terms of reactions and diffusion equations considered and of kinetic and diffusion parameters used. The main reason is probably because most of the experimental results fitted consist of monotonically increasing or decreasing data (M_n and IV, or —OH and —COOH concentrations, respectively) changing over a limited range of variables. In this case, few adjustable parameters are usually adequate to fit, reasonably well, experimental data for any type of model, independent of the assumptions made for its development.

A kinetic model (and the relative set of kinetic parameters) has a higher value if it is able to describe changes occurring over long times or under different experimental conditions. Unfortunately, this is seldom achievable. Testing a model's predictive capability against a different set of experimental data (e.g., the MW profile resulting within a particle during diffusion-controlled SSP) would be a useful test to validate the model. However, it has rarely been done. Validation against previously published results would also be a good test for a given model, but unfortunately, it is often made impossible, owing to the absence of enough data concerning initial sample characteristics or SSP conditions.

On the other hand, for practical purposes, the best kinetic model depends on the answers expected from the model. So even the simplest empirical model can be useful if the answer requested from a model is just a description of the MW increase in a limited range of time and for a given prepolymer. A simple empirical analytical equation with just one or two fitting parameters can be what researchers need to manage daily an SSP process that is fed with a prepolymer that always has the same initial characteristics.

Of course, if the aim is to describe the SSP over long times, too simple a model can fail. In this case a semiempirical approach such as that proposed by Duh could work better. In this approach, all the effects of the various factors that can influence SSP rate are included in two adjustable parameters that have to be calculated every time a different prepolymer or different SSP conditions are used.

In general, it is not easy to take a kinetic model developed for a given polyester, and to use it and the relative set of kinetic parameters to predict SSP results under different operating conditions, even for the same type of polyester. So, at present, the kinetic models proposed in the literature should be considered as possible tools that can be used to derive a specific set of kinetic parameters for a given sample rather than instruments with general predictive capability.

REFERENCES

1. Pilati F. Polyesters. In: *Comprehensive Polymer Science*, Vol. 5, Eastmond GC, Ledwith A, Russo S, Sigwalt P, eds. Pergamon Press, Oxford, UK, 1988, pp. 275–315.

2. Scheirs J, Long TE. *Modern Polyesters: Chemistry and Technology of Polyesters and Copolyesters*. Wiley, Chichester, UK, 2003.

3. Fakirov S. *Handbook of Thermoplastic Polyesters*. Wiley, Chichester, UK, 2001.

4. Gupta SK, Kumar A. *Reaction Engineering of Step Growth Polymerization*. Plenum Press, New York, 1987.

5. Pilati F, Toselli M, Messori M. Principles of step polymerisation. In: *Waterborne Solvent Based Saturated Polyesters and Their End Applications*, Sanders D, ed. Wiley, New York, 1999, pp. 3–71.

6. Lenz RW. *Organic Chemistry of Synthetic High Polymers*. Interscience, New York, 1967.

7. Goodman I. In: *Developments in Polymerization*, Vol. 2, Haward RN, ed. Applied Science, London, 1984.

8. Korshak VV, Vinogradova SV. In: *Polyesters*. Burdon J, ed. Pergamon Press, Oxford, UK, 1965.

9. Polyesters. In: *Encyclopedia of Polymer Science and Engineering*, Vol. 12, Mark HF, Bikales NM, Overberger CG, Menges G, eds. Wiley-Interscience, New York, 1988.

10. Goodman I. In: *Encyclopedia of Polymer Science and Technology*, Vol. 11, Bikales NM, ed. Interscience, New York, 1969, p. 62.

11. Ravindranath K, Mashelkar RA. Modeling of poly(ethylene terephthalate) reactors. IX. Solid-state polycondensation process. *J. Appl. Polym. Sci.* 1990;39(6):1325–1345.

12. Mallon FK, Ray WH. Modeling of solid-state polycondensation: II. Reactor design issues. *J. Appl. Polym. Sci.* 1998;69(9):1775–1788.

13. Algeri C, Rovaglio M. Dynamic modeling of a poly(ethylene terephthalate) solid-state polymerization reactor: I. Detailed model development. *Ind. Eng. Chem. Res.* 2004;43(15):4253–4266.

14. Rovaglio M, Algeri C, Manca D. Dynamic modeling of a poly(ethylene terephthalate) solid-state polymerization reactor: II. Model predictive control. *Ind. Eng. Chem. Res.* 2004;43(15):4267–4277.

15. Vouyiouka S N, Karakatsani E K, Papaspyrides CD. Solid state polymerization. *Prog. Polym. Sci.* 2005;30(1):10–37.

16. Gantillon B, Spitz R, McKenna TE. The solid state postcondensation of PET: 1. A review of the physical and chemical processes taking place in the solid state. *Macromol. Mater. Eng.* 2004;289(1):88–105.

17. Culbert B, Christel A. Continuous solid-state polycondensation of polyesters. In: *Modern Polyesters, Chemistry and Technology of Polyesters and Copolyesters*, Scheirs J, Long TE, eds. Wiley Chichester, UK, 2003, pp. 143–194.

18. Goltner W. Solid-state polycondensation of polyesters resins. In: *Modern Polyesters, Chemistry and Technology of Polyesters and Copolyesters*, Scheirs J, Long TE, eds. Wiley Chichester, UK, 2003, pp. 195–242.

19. Pilati F. Solid-state polymerization. In: *Comprehensive Polymer Science*, Vol. 5, Eastmond GC, Ledwith A, Russo S, Sigwalt P, eds. Pergamon Press, Oxford, UK, 1988, pp. 201–216.

20. Fradet A, Marechal E. Kinetics and mechanisms of polyesterifications: 1. Diols with diacids. *Adv. Polym. Sci.* 1982;43:51–142.

21. Ma Y, Agarwal US, Sikkema DJ, Lemstra PJ. Solid-state polymerization of PET: influence of nitrogen sweep and high vacuum. *Polymer*. 2003;44(15):4085–4096.

22. Gostoli C, Pilati F, Sarti GC, Di Giacomo B. Chemical kinetics and diffusion in poly(butylene terephthalate) solid-state polycondensation: experiments and theory. *J. Appl. Polym. Sci.* 1984;29(9):2873–2887.

23. Wick G, Zeitler H. Cyclic oligomers in polyesters from diols and aromatic dicarboxyl acids. *Angew. Makromol. Chem.* 1983;112:59–94.

24. Maravigna P, Mondaudo G. Formation of cyclic oligomers. In: *Comprehensive Polymer Science*, Vol. 5, Eastmond GC, Ledwith A, Russo S, Sigwalt P, eds. Pergamon Press, Oxford, UK, 1988, pp. 63–90.

25. Suter UW. Ring-chain equilibria. In: *Comprehensive Polymer Science*, Vol. 5, East-mond GC, Ledwith A, Russo S, Sigwalt P, eds. Pergamon Press, Oxford, UK, 1988, pp. 91–96.

26. Semlyen JA. Ring-chain equilibria and the conformations of polymer chain. *Adv. Polym. Sci.* 1976;21:41–75.

27. Pilati F, Manaresi P, Fortunato B, Munari A, Monari P. Models for the formation of poly(butylene terephthalate): 2. Kinetics of the titanium tetrabutylate–catalysed reactions. *Polymer*. 1983;24:1479–1483.

28. Pilati F, Manaresi P, Fortunato B, Munari A, Passalacqua V. Formation of poly(butylene terephthalate): growing reactions studied by model molecules. *Polymer*. 1981;22:799–803.

29. Pilati F, Munari A, Manaresi P. A reappraisal of catalytic effects of tin(IV) compounds on transesterification and esterification reactions. *Polym. Commun.* 1984;25:187–189.

30. Nondek L, Málek J. Esterification of benzenecarboxyl acids with ethylene glycol: 11. Kinetics of the initial stage of metal-ion catalyzed polyesterification of isophthalic acid with ethylene glycol. *Makromol. Chem.* 1977;178:2211–2221.

31. Habib OMO, Malek J. Esterification of benzenecarboxyl acids with ethylene glycol: VIII. The activity of metal ions in catalytic esterification of aromatic carboxyl acids with aliphatic glycols. *Coll. Czech. Chem. Commun.* 1976;41(9):2724–2736.

32. Otton J, Ratton S. Investigation of the formation of poly(ethylene terephthalate) with model molecules: kinetics and mechanism of the catalytic esterification and alcohol-ysis reactions: I. Carboxyl acid catalysis (monofunctional reactants). *J. Polym. Sci. A.* 1988;26(8):2183–2197.

33. Otton J, Ratton S, Vasnev VA, Markova GD, Nametov KM, Bakhmutov VI, Komarova LI, Vinogradova SV, Korshak VV. Investigation of the formation of poly(ethylene terephthalate) with model molecules—kinetics and mechanisms of the catalytic esterification and alcoholysis reactions: II. Catalysis by metallic derivatives (monofunctional reactants). *J. Polym. Sci. A.* 1988;26(8):2199–2224.

34. Otton J, Ratton S, Markova GD, Nametov KM, Bakhmutov VI, Vinogradova SV, Korshak VV. Investigation of the formation of poly(ethylene terephthalate) with model molecules: III. Metal-catalyzed esterification and alcoholysis reactions: influ-ence of the structure of the reactants and of the nature of the reaction medium. *J. Polym. Sci. A.* 1989;27(11):3535–3550.

35. Otton J, Ratton S. Investigation of the formation of poly(ethylene terephthalate) with model molecules: IV. Catalysis of the esterification of ethylene glycol with benzoic acid and of the condensation of ethylene glycol monobenzoate. *J. Polym. Sci. A.* 1991;29(3):377–391.

36. Hamb FL. Copolyesters of glycols and bisphenols: new preparative process. *J. Polym. Sci. Polym. Chem. Ed.* 1972;10(11):3217–3234.

37. Kang C-K. Modeling of solid-state polymerization of poly(ethylene terephthalate). *J. Appl. Polym. Sci.* 1998;68(5):837–846.

38. Wu D, Chen F, Li R, Shi Y. Reaction kinetics and simulations for solid-state polymerization of poly(ethylene terephthalate). *Macromolecules*. 1997;30(22): 6737–6742.

39. Wang X-Q, Deng D-C. A comprehensive model for solid-state polycondensation of poly(ethylene terephthalate): combining kinetics with crystallization and diffusion of acetaldehyde. *J. Appl. Polym. Sci.* 2002;83(14):3133–3144.

40. Devotta I, Mashelkar RA. Modeling of polyethylene terephthalate reactors: X. A comprehensive model for solid-state polycondensation process. *Chem. Eng. Sci.* 1993;48(10):1859–1867.

41. Buxbaum LH. Degradation of poly(ethylene terephthalate). *Angew. Chem. Int. Ed. Engl.* 1968;7(3):182–190.

42. Zimmerman H. In: *Developments in Polymer Degradation*, Vol. 5, Grassie N, ed. Applied Science, London, 1984, pp. 79–119.

43. Droscher M, Wegner G. Poly(ethylene terephthalate): a solid state condensation process. *Polymer.* 1978;19:43–47.

44. Loontjens T, Pauwels K, Derks F, Neilen M, Sham CK, Seré M. The action of chain extenders in nylon-6, PET, and model compounds. *J. Appl. Polym. Sci.* 1997;65(9):1813–1819.

45. Akkapeddi MK, Gervasi J. Chain extension of PET and nylon in an extruder. *Polym. Prepr.* 1988;29(1):567–570.

46. Karayannidis GP, Psalida EA. Chain extension of recycled poly(ethylene terephthalate) with 2,2′-(1,4-phenylene)bis(2-oxazoline). *J. Appl. Polym. Sci.* 2000;77(10):2206–2211.

47. Cardi N, Po R, Giannotta G, Occhiello E, Garbassi F, Messina G. Chain extension of recycled poly(ethylene terephthalate) with 2,2′-bis(2-oxazoline). *J. Appl. Polym. Sci.* 1993;50(9):1501–1509.

48. Zimmerman J. Equilibria in solid phase polyamidation. *J. Polym. Sci. Polym. Lett.* 1964;2:955–958.

49. Chang S, Sheu MF, Chen SM. Solid-state polymerization of poly(ethylene terephthalate). *J. Appl. Polym. Sci.* 1983;28(10):3289–3300.

50. Mallon F, Ray W. Modeling of solid-state polycondensation. I. Particle models. *J. Appl. Polym. Sci.* 1998;69:1233–1250.

51. Flory PJ. *Principles of Polymer Chemistry*. Cornell University Press, Ithaca, NY, 1953.

52. Pilati F, unpublished results.

53. McAlea KP, Schultz JM, Gardner KH, Wignall GD. Ester interchange reactions in poly(ethylene terephthalate): observation using small-angle neutron scattering. *Polymer.* 1986;27(10):1581–1584.

54. Kugler J, Gilmer JW, Wiswe D, Zachmenn HG, Hahn K, Fisher EW. Study of transesterification in poly(ethylene terephthalate) by small-angle neutron scattering. *Macromolecules.* 1987;20:1116–1119.

55. Srinivasan R, Almonacil C, Narayan S, Desai P, Abhiraman AS. Mechanism, kinetics and potential morphological consequences of solid-state polymerization. *Macromolecules.* 1998;31(20):6813–6821.

56. Gaymans R, Amirtharaj J, Kamp H. Nylon 6 polymerization in the solid state. *J. Appl. Polym. Sci.* 1982;27:2513–2526.

57. Duh B. Semiempirical rate equation for solid state polymerization of poly(ethylene terephthalate). *J. Appl. Polym. Sci.* 2002;84(4):857–870.

58. Duh B. Effects of the carboxyl concentration on the solid state polymerization of poly(ethylene terephthalate). *J. Appl. Polym. Sci.* 2002;83:1288–1304.

59. Li LF, Huang NX, Tang ZL, Hagen R. Reaction kinetics and simulation for the solid-state polycondensation of nylon 6. *Macromol. Theory Simul.* 2001;10(5):507–517.

60. Duh B. Reaction kinetics for solid-state polymerization of poly(ethylene terephthalate). *J. Appl. Polym. Sci.* 2001;81(7):1748–1761.

61. Przygocki W. Nucleation in poly(ethylene terephthalate). *Acta Polym.* 1982;33(12): 729–735.

62. Chen SA, Chen FL. Kinetics of polyesterification. III. Solid-state polymerization of poly(ethylene terephthalate). *J. Polym. Sci. A.* 1987;25(2):533–549.

63. Kim T Y, Lofgren E A, Jabarin S A. Solid-state polymerization of poly(ethylene terephthalate): I. Experimental study of the reaction kinetics and properties. *J. Appl. Polym. Sci.* 2003;89(1):197–212.

64. Fortunato B, Manaresi P, Munari A, Pilati F. Models for the formation of poly(butylene terephthalate): new insights on the catalytic activity of tetrabutyl titanate: 4. *Polym. Commun.* 1986;7:29–31.

65. Colonna M, Banach TE, Berti C, Fiorini M, Marianucci E, Messori M, Pilati F, Toselli M. New catalysts for poly(butylene terephthalate) synthesis: 3. Effect of phosphate co-catalysts. *Polymer.* 2003;44(17):4773–4779.

66. Banach E, Berti C, Colonna M, Fiorini M, Marianucci E, Messori M, Pilati F, Toselli M. New catalysts for poly(butylene terephthalate) synthesis: 1. Titanium–lanthanides and titanium–hafnium systems. *Polymer.* 2001;42(18):7511–7516.

67. Ravindranath K, Mashelkar RA. Finishing stages of PET synthesis: a comprehensive model. *AIChE J.* 1984;30:415–423.

68. Pilati F, Manaresi P, Fortunato B, Munari A, Passalacqua V. Formation of poly(butylene terephthalate): secondary reactions studied by model molecules. *Polymer.* 1981;22:1566–1573.

69. Devaux J, Godard P, Mercier JP. Etude cinétique de la dégradation du poly(oxytetramethyleneoxyterephtaloyle). *Makromol. Chem.* 1978;179:2201–2209.

70. Apicella B, Di Serio M, Fiocca L, Po R, Santacesaria E. Kinetic and catalytic aspects of the formation of poly(ethylene terephthalate) (PET) investigated with model molecules. *J. Appl. Polym. Sci.* 1998;69(12):2423–2433.

71. Pilati F, Marianucci E, Berti C. Study of the reactions occurring during melt mixing of poly(ethylene terephthalate) and polycarbonate. *J. Appl. Polym. Sci.* 1985;30:1267–1275.

72. Goodings EP. *SCI Monogr.* 1961;13:211.

73. Passalacqua V, Pilati F, Zamboni V, Fortunato B, Manaresi P. Thermal degradation of poly(butylene terephthalate). *Polymer.* 1976;17(12):1044–1048.

74. Yoon KH, Kwon MH, Jeon MH, Park OO. Diffusion of ethylene glycol in solid state poly(ethylene terephthalate). *Polym. J. (Tokyo).* 1993;25(3):219–226.

75. Mallon FK, Ray WH. Modeling of solid-state polycondensation. I. Particle models. *J. Appl. Polym. Sci.* 1998;69(6):1233–1250.

76. Kim TY, Jabarin SA. Solid-state polymerization of poly(ethylene terephthalate): II. Modeling study of the reaction kinetics and properties. *J. Appl. Polym. Sci.* 2003;89(1):213–227.

77. Buxbaum LH. Solid-state polycondensation of poly(butylene terephthalate). *J. Appl. Polym. Sci. Appl. Polym. Symp.* 1979;35:59–66.

78. Duh B. Solid-state polymerization of poly(trimethylene terephthalate). *J. Appl. Polym. Sci.* 2003;89(12):3188–3200.

79. Meyer K. Postcondensation of polyamides in partially crystalline condition. *Angew. Makromol. Chem.* 1973;34:165–175.

80. Bamford CH, Wayne RP. Polymerization in the solid phase. *Polymer.* 1969;10:661–681.

81. Karayannidis G, Sideridou I, Zamboulis D, Stalidis G., Bikiaris D, Lazaridis N, Wilmes A. Solid-state polycondensation of poly(ethylene terephthalate) films. *Angew. Makromol. Chem.* 1991;192:155–168.

82. Chen FC, Griskey RG, Beyer GH. Thermally induced solid state polycondensation of nylon 66, nylon 6–10 and polyethylene terephthalate. *AIChE J.* 1969;15:680–685.

83. Griskey RG, Lee BI. Thermally induced solid-state polymerisation in nylon 66. *J. Appl. Polym. Sci.* 1966;10:105–111.

84. Jabarin SA, Lofgren EA. Solid state polymerization of PET: kinetic and property parameters. *J. Appl. Polym. Sci.* 1986;32:5315–5335.

85. Karayannidis GP, Kokkalas DE, Bikiaris DN. Solid-state polycondensation of poly(ethylene terephthalate) recycled from postconsumer soft-drink bottles: I. *J. Appl. Polym. Sci.* 1993;50(12):2135–2142.

86. Chang TM. Kinetics of thermally induced solid state polycondensation of poly(ethylene terephthalate). *Polym. Eng. Sci.* 1970;10:364–368.

87. Huang B, Walsh JJ. Solid-phase polymerization mechanism of poly(ethylene terephthalate) affected by gas flow velocity and particle size. *Polymer.* 1998;39(26):6991–6999.

88. Tang Z, Qiu G, Huang N, Sironi C. Solid-state polycondensation of poly(ethylene terephthalate): kinetics and mechanism. *J. Appl. Polym. Sci.* 1995;57(4):473–485.

89. Zhao J, Xiao H, Qiu G, Zhang Y, Huang N, Tang Z. Solid-state polycondensation of poly(ethylene terephthalate) modified with isophthalic acid: kinetics and simulation. *Polymer.* 2005;46(18):7309–7316.

90. Kim TY, Jabarin SA. Solid-state polymerization of poly(ethylene terephthalate): III. Thermal stabilities in terms of the vinyl ester end group and acetaldehyde. *J. Appl. Polym. Sci.* 2003;89(1):228–237.

91. Bueche F. *Physical Properties of Polymers.* Interscience, New York, 1970, Chap. 4.

92. Chiu WY, Carrat GM, Soong DS. A computer model for the gel effect in free-radical polymerization. *Macromolecules.* 1983;16:348–357.

93. Fortunato B, Pilati F, Manaresi P. Solid-state polycondensation of poly(butylene terephthalate). *Polymer.* 1981;22:655–657.

94. Rinehart VR. Solid state polymerization of polyester prepolymers (Goodyear Tire & Rubber Company). U.S. Patent 4,876,326, 1989.

95. Kremer RA. Solid state polycondensation with porous polyester prepolymer particles (Mobil Oil Corporation). U.S. Patent 3,586,647, 1971.

96. Duh B. Solid state polymerization process for foamed poly(ethylene naphthalate) (Shell Oil Company). U.S. Patent 5,449,701, 1995.

97. Jabarin SA. In: *Polymeric Materials Encyclopedia*, Vol. 8, Salomone JC, ed. CRC Press, Boca Raton, FL, 1996, p. 6091.

98. Schaaf E, Zimmermann H, Dietzel W, Lohmann P. Nachpolykondensation von Polyethylenterephthalat in fester Phase. *Acta Polym.* 1981;32(5):250–256.

99. Schaaf E, Zimmermann H. High-molecular-weight polyester by solid phase repoly-condensation (Akademie der Wissenschaften der DDR). East German Patent 139129, 1979.

100. Schumann HD. Stand der Technologie zur Herstellung von Polyester-Polymer. *Chemiefasern Text. Ind.* 1990;40:1058–1062.

101. Duh B. Effect of antimony catalyst on solid-state polycondensation of poly(ethylene terephthalate). *Polymer.* 2002;43(11):3147–3154.

102. Kokkalas DE, Bikiaris DN, Karayannidis GP. Effect of the Sb_2O_3 catalyst on the solid-state postpolycondensation of poly(ethylene terephthalate). *J. Appl. Polym. Sci.* 1995;55(5):787–791.

103. Medellin-Rodriguez FJ, Lopez-Guillen R, WAAaldo-Mendoza MA. Solid-state poly-merization and bulk crystallization behavior of poly(ethylene terephthalate) (PET). *J. Appl. Polym. Sci.* 2000;75(1):78–86.

104. Shinno K, Miyamoto M, Kimura Y, Hirai Y, Yoshitome H. Solid-state post-polymerization of L-lactide promoted by crystallization of product polymer: an effective method for reduction of remaining monomer. *Macromolecules.* 1997;30(21):6438–6444.

105. Schaaf E, Zimmermann H, Dietzel W, Lohmann P. Nachpolykondensation von Polyethylenterephthalat in fester Phase. *Acta Polym.* 1981;32(5):250–256.

106. Hsu L-C. Synthesis of ultrahigh-molecular-weight poly(ethylene terephthalate). *J. Macromol. Sci. Phys.* 1967;1(4):801–813.

107. Gao Q, Nan-Xun H, Zhi-Lian T, Gerking L. Modeling of solid state polycondensa-tion of poly(ethylene terephthalate). *Chem. Eng. Sci.* 1997;52(3):371–376.

108. Biesenberger JA, Sebastian DH. *Principles of Polymerization Engineering.* Wiley, New York, 1983, p. 573.

109. Ravindranath K, Mashelkar RA. Analysis of role of stripping agents in polymer devolatilisation. *Chem. Eng. Sci.* 1988;43:429–442.

110. Mallon F, Beers K, Ives A, Ray WH. The effect of the type of purge gas on the solid-state polymerization of polyethylene terephthalate. *J. Appl. Polym. Sci.* 1998;69(9):1789–1791.

111. Karayannidis GP, Kokkalas DE, Bikiaris DN. Solid-state polycondensation of poly(ethylene terephthalate) recycled from postconsumer soft-drink bottles: II. *J. Appl. Polym. Sci.* 1995;56(3):405–410.

112. Halek GW. The zero-order kinetics of acetaldehyde thermal generation from polyethylene terephthalate. *J. Polym. Sci. Polym. Symp.* 1986;74:83–92.

113. Reimschuessel HK. Poly(ethylene terephthalate) formation: mechanistic and kinetic aspects of direct esterification process. *Ind. Eng. Chem. Prod. Res. Dev.* 1980;19(1):117–125.

114. Hornof V. Influence of metal catalysts on the formation of ether links in polyethylene terephthalate. *J. Macromol. Sci. A.* 1981;15(3):503–514.

115. Renwen H, Feng Y, Tinzheng H, Shiming G. The kinetics of formation of diethylene glycol in preparation of polyethylene terephthalate and its control in reactor design and operation. *Angew. Makromol. Chem.* 1983;119(1):159–172.

116. Chen L-W, Chen J-W. Kinetics of diethylene glycol formation from bishydroxyethyl terephthalate with zinc catalyst in the preparation of poly(ethylene terephthalate). *J. Appl. Polym. Sci.* 2000;75(10):1229–1234.

117. Borman WFH. Solid state polymerization of poly(1,4-butylene terephthalate) (General Electric Company). U.S. Patent 3,953,404, 1976.

118. Pilati F, Gostoli C, Sarti GC. A model description for poly(butylene terephthalate) solid-state polycondensation. *Polym. Process Eng.* 1986;4:303–319.

119. Dinse HD, Tucek E. Untersuchungen zur Festphasenpolykondensation von Polytetramethylenterephthalat. *Acta Polym.* 1980;31:108–110.

120. Lum RM. Thermal decomposition of poly(butylene terephthalate). *J. Polym. Sci. Polym. Chem. Ed.* 1979;17:203–213.

121. Buchner S, Wiswe D, Zachman H. Kinetics of crystallization and melting behaviour of poly(ethylene naphthalene-2,6-dicarboxylate). *Polymer.* 1989;30:480–488.

122. Po R, Occhiello E, Giannotta G, Pelosi P, Abis L. New polymeric materials for containers manufacture based on PET/PEN copolyesters and blends. *Polym. Adv. Technol.* 1996;7:365–373.

123. Stouffer JM, Blanchard EN, Leffew KW. Production of poly(ethylene 2,6-naphthalate) (E.I. du Pont de Nemours & Company). WIPO Patent WO 97/025364, 1997.

124. Duh B. Process for the crystallization of polyethylene naphthalate (Goodyear Tire & Rubber Company). U.S. Patent 4,963,644, 1990.

125. Sun Y-M, Shieh J-Y. Kinetic and property parameters of poly(ethylene naphthalate) synthesized by solid-state polycondensation. *J. Appl. Polym. Sci.* 2001;81(9):2055–2061.

126. Chuah HH. Synthesis, properties and applications of poly(trimethylene terephthalate). In: *Modern Polyesters: Chemistry and Technology of Polyesters and Copolyesters*, Scheirs J, Long TE, eds. Wiley, Chichester, UK, 2003, pp. 361–397.

127. Schauhoff S, Schmidt DW. New development in the production of polytrimethylene terephthalate (PTT). *Chem. Fibers Int.* 1996;46(4):263–264.

128. Stouffer JM, Blanchard EN, Leffew KW. Production of poly(trimethylene terephthalate) (E.I. du Pont de Nemours & Company). U.S. Patent 5,763,104, 1998.

129. Duh B, Corey AM. High temperature solid state polymerization of poly(trimethylene terephthalate) (Shell Oil Company). U.S. Patent 6,441,129, 2002.

130. Duh B. Solid state polymerization process for poly(trimethylene terephthalate) utilizing a combined crystallization/preheating step (Shell Oil Company). U.S. Patent 6,403,762, 2002.

131. Boehme F, Komber H, Jafari SH. Synthesis and characterization of a novel unsaturated polyester based on poly(trimethylene terephthalate). *Polymer.* 2006;47(6):1892–1898.

132. Jamshidi K, Eberhart RC, Hyon SH, Ikada Y. Characterization of polylactide synthesis. *Polym. Prepr.* 1987;28(1):236–237.

133. Mecerreyes D, Jerome R, Dubois P. Novel macromolecular architectures based on aliphatic polyesters: relevance of the "coordination-insertion" ring-opening polymerization. *Adv. Polym. Sci.* 1999;147:1–59.

134. Duda A, Kowalski A, Libiszowski J, Penczek S. Thermodynamic and kinetic polymerizability of cyclic esters. *Macromol. Symp.* 2005;224:71–83.

135. Duda A, Penczek S. Thermodynamics of L-lactide polymerization. Equilibrium monomer concentration. *Macromolecules*. 1990;23:1636–1639.

136. Ajioka M, Enomoto K, Suzuki K, Yamaguchi A. The basic properties of poly(lactic acid) produced by the direct condensation polymerization of lactic acid. *J. Environ. Polym. Degrad.* 1995;3(4):225–234.

137. Xu H, Luo M, Yu M, Teng C, Xie S. The effect of crystallization on the solid state polycondensation of poly(L-lactic acid). *J. Macromol. Sci. B.* 2006;45(4):681–687.

138. Moon S-I, Lee CW, Taniguchi I, Miyamoto M, Kimura Y. Melt/solid polycondensation of L-lactic acid: an alternative route to poly(L-lactic acid) with high molecular weight. *Polymer.* 2001;42(11):5059–5062.

139. Moon S-I, Taniguchi I, Miyamoto M, Kimura Y, Lee CW. Synthesis and properties of high-molecular-weight poly(L-lactic acid) by melt/solid polycondensation under different reaction conditions. *High Perform. Polym.* 2001;13(2):S189–S196.

140. Kricheldorf HR, Dunsing RK. Polylactones: 8. Mechanism of the cationic polymerization of L,L-dilactide. *Makromol. Chem.* 1986;187:1611–1625.

141. Kricheldorf HR, Saunders IK. Polylactones: 19. Anionic polymerization of L-lactide in solution. *Makromol. Chem.* 1990;191:1057–1066.

142. Kricheldorf HR, Boettcher C, Tonnes KU. Polylactones: 23. Polymerization of racemic and meso-lactide with various organotin catalysts—stereochemical aspect. *Polymer.* 1992;33:2817–2824.

143. Kricheldorf HR, Lee S-R. Polylactones: 32. High-molecular-weight polylactides by ring-opening polymerization with dibutylmagnesium or butylmagnesium chloride. *Polymer.* 1995;36:2995–3003.

144. Qian G, Zhou X-G, Zhu L-B, Yuan W-K. Increasing the molecular weight of poly(L-lactic acid) by solid state polycondensation in a closed system. *J. Polym. Eng.* 2003;23(6):413–422.

4

KINETIC ASPECTS OF POLYAMIDE SOLID STATE POLYMERIZATION

S. N. Vᴏᴜʏɪᴏᴜᴋᴀ ᴀɴᴅ C. D. Pᴀᴘᴀsᴘʏʀɪᴅᴇs

School of Chemical Engineering, National Technical University of Athens, Athens, Greece

4.1. INTRODUCTION

As discussed in Section 1.1.3, solid state polymerization (SSP) can be employed in crystalline monomers as well as in prepolymers. In particular, solid state polyamidation consists of an important finishing technique to prepare high-molecular-weight polyamides (number-average molecular weight,

Solid State Polymerization, Edited by Constantine D. Papaspyrides and Stamatina N. Vouyiouka
Copyright © 2009 John Wiley & Sons, Inc.

$\overline{M_n} > 25,000$ g mol^{-1}), suitable for spinning, extrusion, and injection. It is applied to both polycaproamide (PA 6) and poly(hexamethylene adipamide) (PA 66) monomers and precursors, and its significance is enhanced for specialty polyamides such as poly(tetramethylene oxamide) (PA 42) [1] and poly(tetramethylene adipamide) (PA 46) [2]. Relevant polyamides with a high amide content present high crystallinity, excellent fiber properties, and high melting points ($T_m > 290°C$), but these properties make it difficult to increase their molecular weight sufficiently through the melt technique.

Because SSP involves both chemical and physical attributes, it presents a complex reaction process. Based on the reversible character of the polyamidation reaction and the restrictions set by the nature of SSP, one can identify four process steps and the relevant critical parameters, apart from the starting material composition related to salt or prepolymer, amide group concentration, and prepolymer chemical modification. Each step can be the slowest and thus the rate-determining, without excluding combinations of more than one step [3]:

1. The intrinsic kinetics of the chemical reaction. The key parameters are the reaction temperature and the presence of catalyst.
2. The diffusion of functional end groups correlated to the reactive chain-end mobility. This step is dependent mainly on the reaction temperature, initial prepolymer molecular weight, and crystallinity.
3. The diffusion of the condensate in the solid reacting mass (interior diffusion), which is affected by the reaction temperature, particle size, gas flow rate, and catalyst presence.
4. The diffusion of the condensate from the surface of the reacting mass to the surroundings (surface diffusion). The parameters are similar to those in the preceding step, but gas flow rate dominates as the most important.

The SSP kinetics cover all possible heat and mass transfer phenomena associated with the aforementioned steps. The relevant SSP models are classified into two categories, experimental and analytical, depending on their simplicity and the assumptions used (Table 4.1). Typical examples were discussed in Chapter 3 for polyesters (Section 3.3).

The first category focuses on the treatment of polyamide SSP data: for example, intrinsic viscosity (IV), average molecular weight (MW), and end-group concentrations (C) versus reaction time (t). Usually, main irreversible reactions are considered and simple (empirical) rate equations are tested for their fitting to experimental measurements, revealing the process rate-limiting step. The intrinsic or apparent rate constants (k) are determined based on the most suitable rate expression (represented by r), which may be one of the following:

- A chemical reaction-rate model, according to the mass action law:

$$r = kf(C)$$

TABLE 4.1. Classification of Kinetics Models for Solid State Polyamidation

Simple Rate Expressions

I. Chemical reaction-rate models

$$\text{second order:} \quad -\frac{d[\text{COOH}]}{dt} = k_2[\text{COOH}][\text{NH}_2]$$

Example [4]: third order: $\quad -\dfrac{d[\text{COOH}]}{dt} = k_3[\text{COOH}]^2[\text{NH}_2]$

where [COOH] and [NH$_2$] are the concentrations of carboxyl and amine end groups (mmol kg^{-1}), k_2 is the second-order rate constant (kg mmol^{-1} h^{-1}), and k_3 is the third-order rate constant (kg^2 mmol^{-2} h^{-1})

II. Diffusion models

Example [5]: $\qquad \dfrac{d\overline{M_n}}{dt} = kt^n$

where $\overline{M_n}$ is the number-average molecular weight (kg mol^{-1}) and k is the rate constant (kg mol^{-1} h$^{-(n+1)}$)

Simulation Equations

Water concentration (C_w):

$$\frac{\partial C_w}{\partial t} = D_{wp}\left[\frac{2}{x}\frac{\partial C_w}{\partial x} + \frac{\partial^2 C_w}{\partial x^2}\right] + r_p + \frac{\dot{m}_p}{\rho_p a_c(1-\varepsilon)}\frac{\partial C_w}{\partial z} + D_b\frac{\partial^2 C_w}{\partial z^2}$$

Concentrations of end groups (C_c, C_a) and amide linkages (C_l):
Reactor model (plug flow reactor) main equations [6]:

$$\frac{\partial C_c}{\partial t} = -r_p + \frac{\dot{m}_p}{\rho_p a_c(1-\varepsilon)}\frac{\partial C_c}{\partial z} + D_b\frac{\partial^2 C_c}{\partial z^2}$$

$$\frac{\partial C_a}{\partial t} = -r_p + \frac{\dot{m}_p}{\rho_p a_c(1-\varepsilon)}\frac{\partial C_a}{\partial z} + D_b\frac{\partial^2 C_a}{\partial z^2}$$

$$\frac{\partial C_l}{\partial t} = -r_p + \frac{\dot{m}_p}{\rho_p a_c(1-\varepsilon)}\frac{\partial C_l}{\partial z} + D_b\frac{\partial^2 C_l}{\partial z^2}$$

where C_w is the water concentration in the polymer phase (mol kg^{-1}), D_{wp} the diffusivity of water in the polymer phase (m^2 s^{-1}), x the radial distance from the center of a spherical polymer particle (m), r_p the polymerization rate (mol kg^{-1} s^{-1}), ρ_p the polymer density (kg m^{-3}), a_c the cross-sectional area of the reactor (m^2), ε the voidage, $\dot{m}_p$ the mass flow rate of polymer (kg s^{-1}), z the reactor height from the bottom (m), and D_b the dispersion coefficient (m^2 s^{-1}) (accounting for particles that do not flow at the average velocity).

- A diffusion model, usually a power-law rate expression:

$$r = kf(t)$$

- A combined model:

$$r = kf(C, t)$$

The rate expressions of the experimental approach serve as a tool to predict achievable ranges of molecular weight and to compare different prepolymer reactivities during SSP processes, keeping the same critical experimental conditions, however, so as to be in the same kinetic regime (Section 3.2.2). The kinetic analysis based on simple models can also follow a reverse procedure. The critical experimental conditions can first be determined so as to perform the SSP reaction within a specific kinetic regime. The experimental plan shown in Figure 4.1 is suggested to define critical SSP parameter values for which the process is reaction controlled, including end-group diffusion. It can be seen that the first step is to define the critical gas flow rate (v_{N_2}) and/or particle size (mean diameter, $\bar{d}$) value, above or below which, respectively, surface and interior condensate diffusion is not the controlling step (i.e., the change of v_{N_2} and/or $\bar{d}$ has no effect on the final molecular weight). In particular, these scouting runs should be carried out at a high reaction temperature (T_{max}), so that chemical reaction and end-group diffusion are favored and to exclude the possibility of their control of the process. The second runs set should be carried out at the critical v_{N_2} value or, better, above this value, at at least three reaction temperatures lower than T_{max}. A selected reaction rate model can be used so as to define intrinsic SSP rate constants (k), process activation energy (E_a), and equilibrium constant values (K_{eq}).

In an intrinsic chemical reaction–controlled regime, a further critical experimental condition is the reaction time. In particular, it is suggested that the process be divided into two or even three stages in terms of kinetics (Section 3.2.2). In the first stage the distribution of end groups in the solid polymer is considered homogeneous, as in the melt process, and consequently, the reaction kinetics and mechanisms will be similar. The end groups with the smallest end-to-end distances do not need to diffuse to react, and the reaction rate constant is the intrinsic constant. In the second stage, the diffusion of polymer chain ends begins to be the limiting step and the reaction rate is strongly affected by end-group diffusion limitations. As a result, due to the solid state character of the process and to the restrictions set by the segmental or translational mobility, the rate constant may change versus reaction time at a stable reaction temperature, deviating from Arrhenius theory [7–11]. For the same reason, an asymptotic molecular-weight value is reached at long SSP reaction times. Therefore, short reaction times are usually preferred in intrinsic SSP kinetic analysis, so as to exclude limitations regarding end-group diffusion [7–9].

This SSP analysis, based on experimental data, is of great importance and depicts the system's apparent reactivity, which deviates from the melt or solution, where in general mass-transfer restrictions can be dissipated. The kinetic SSP

I. Eliminate interior and surface condensate diffusion limitations

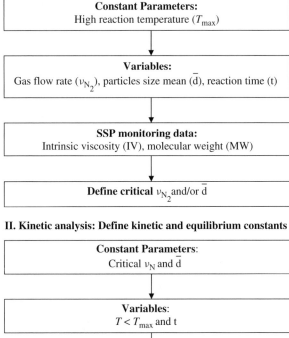

Constant Parameters:
High reaction temperature (T_{max})

↓

Variables:
Gas flow rate (v_{N_2}), particles size mean ($\bar{d}$), reaction time (t)

↓

SSP monitoring data:
Intrinsic viscosity (IV), molecular weight (MW)

↓

Define critical v_{N_2} **and/or** $\bar{d}$

II. Kinetic analysis: Define kinetic and equilibrium constants

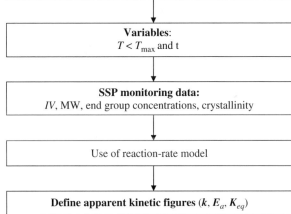

Constant Parameters:
Critical v_N and $\bar{d}$

↓

Variables:
$T < T_{max}$ and t

↓

SSP monitoring data:
IV, MW, end group concentrations, crystallinity

↓

Use of reaction-rate model

↓

Define apparent kinetic figures (k, E_a, K_{eq})

Fig. 4.1. Preliminary experimental plan to study reaction-controlled SSP process, including end-group diffusion effect.

figures calculated can also be used in more analytical (comprehensive) models. Accordingly, the second, analytical approach applies simulation, and the relevant equations involve both physical and chemical steps, based on assumptions regarding one or more controlling mechanisms. These models differ in the number of main and side chemical reactions considered, the number of controlling steps assumed, and the mathematical technique used. They involve solution of the full system of partial differential equations, which describe the change with time and position of all chemical species within the particle (particle model) and/or in a reactor (reactor model) (Table 4.1).

TABLE 4.2. Indicative Physicochemical Parameters Used in PA 66 SSP Modeling

Parameter	Value or Correlation
Heat capacity of polymer	
C_p (J kg^{-1}°C^{-1}) = 7.437T + 1928	25°C < T ≤ 111°C
= 0.344T + 2717	111°C < T ≤ 200°C
Diffusivity of water in polymer phase	
D_{wp} (m^{2}s^{-1}) = 1.2 × 10^{-10} at 202°C	
E_a (kJ mol^{-1}) = 76.5 kJ mol^{-1}	
Reaction activation energy	
$-E_a$ (kJ mol^{-1}) = 48.07	
Polymer density	
ρ_p (kg m^{-3}) = 1129 − 0.52T	

Source: Yao et al. [6].

The analytical models contain a significant number of physicochemical parameters (Table 4.2), such as rate constants and diffusion coefficients, one or more of which must be adjusted to fit the experimental data to the model. In Chapter 7, fundamentals of SSP modeling and product design are discussed especially for PET and PA 6 SSP, aiming to predict the dependence of intrinsic viscosity on gas flow rate, pellet radius, and polymer feed temperature.

4.2. SIMPLE KINETIC MODELS OF SOLID STATE POLYAMIDATION

4.2.1. Fundamental Chemistry in Solid State Polyamidation

The main reaction considered in simple polyamide SSP models is the polyamidation: coupling the free amine ends with carboxyl groups accompanied by the formation of water as condensate:

$$\underset{\overset{\|}{C}}{\overset{O}{\|}}-OH + H_2N \underset{k_r}{\overset{k_f}{\rightleftharpoons}} \overset{O}{\underset{\underset{H}{|}}{\overset{\|}{C}}}-N- + H_2O \tag{4.1}$$

Such reactions consist of equilibrium systems, and in case of continuous by-product (water) removal, the equilibrium is shifted toward polymer formation. The main difference between the two important condensation polymers, polyamides and polyesters, is the higher equilibrium constant of the former ($K_{eq} = k_f/k_r$, where k_f and k_r, the rate constants of the forward and reverse reaction, respectively), indicating that the equilibrium for the amidation reaction is very favorable, much more favorable than that for the esterification and ester interchange polymerization. For this reason, the amidation can be performed in the beginning in aqueous environment, which is not the case, however, for

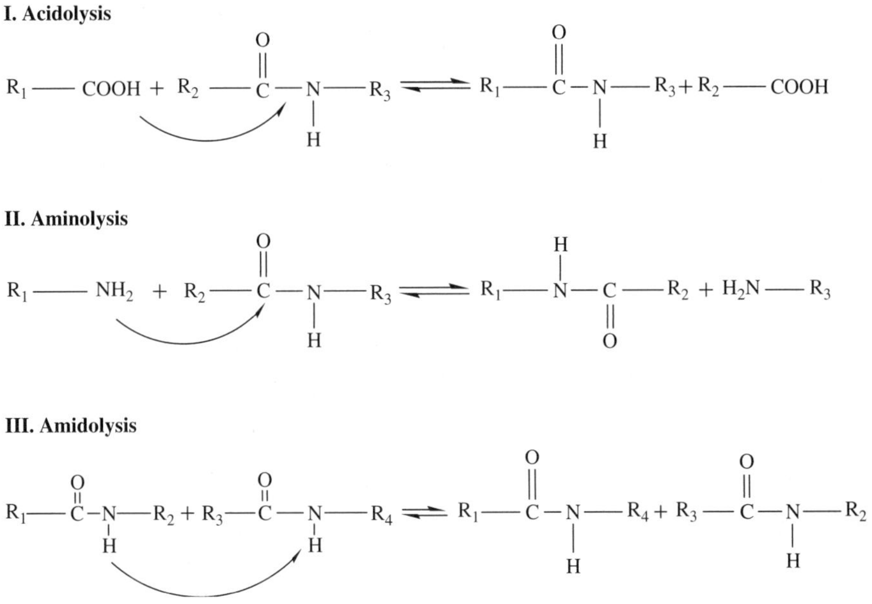

I. Acidolysis

$$R_1 — COOH + R_2 — \overset{\overset{O}{\|}}{C} — \underset{\underset{H}{|}}{N} — R_3 \;\rightleftharpoons\; R_1 — \overset{\overset{O}{\|}}{C} — \underset{\underset{H}{|}}{N} — R_3 + R_2 — COOH$$

II. Aminolysis

$$R_1 — NH_2 + R_2 — \overset{\overset{O}{\|}}{C} — \underset{\underset{H}{|}}{N} — R_3 \;\rightleftharpoons\; R_1 — \underset{\underset{\underset{O}{\|}}{C}}{\overset{\overset{H}{|}}{N}} — R_2 + H_2N — R_3$$

III. Amidolysis

$$R_1 — \underset{\underset{H}{|}}{\overset{\overset{O}{\|}}{C} — N} — R_2 + R_3 — \underset{\underset{H}{|}}{\overset{\overset{O}{\|}}{C} — N} — R_4 \;\rightleftharpoons\; R_1 — \overset{\overset{O}{\|}}{C} — \underset{\underset{H}{|}}{N} — R_4 + R_3 — \overset{\overset{O}{\|}}{C} — \underset{\underset{H}{|}}{N} — R_2$$

Fig. 4.2. Exchange reactions of polyamides.

polyester manufacture held directly in the melt state. In other words, amidation is carried out without concern for shifting the equilibrium until the last stages of reactions.

Apart from amidation reaction, intramolecular and/or intermolecular exchange reactions (Fig. 4.2) may occur during SSP. Intramolecular reactions involve the formation of cyclic macromolecules. On the other hand, intermolecular reactions result in linear products, which can be polyamides, oligomers, or by-products. More specifically, acidolysis is the reaction between an alkyl carboxyl group and an amide linkage, aminolysis is the similar exchange between an alkyl amine and an amide group; meanwhile, amidolysis between two amide groups is also possible [12].

When the terminal group attack the amide linkages of the polymer backbone, there is no significant effect on the final average molecular weight, but the fractional composition is changed, as discussed in Section 3.2.1 for polyesters. A rearrangement of the macromolecule length occurs, the segmental mobility is chemically favored in the amorphous regions, and especially for the SSP, the latter accelerates the end-group diffusion by offering a migration mechanism [13]. Further, the longest chains are the most susceptible to exchange reactions, resulting in narrowing the molecular-weight distribution in the forming polymer [12].

Side reactions can also occur during SSP, especially after prolonged heating. In the case of PA 66, the main cause of polyamide cross-linking is the formation of a secondary amine group, which reacts further with carboxyl end groups and creates branched structures [12]. Both by-products yield upon hydrolysis bis(hexamethylene triamine):

$$(4.2)$$

Finally, in the case of PA 46, the extent and rate of molecular-weight buildup are both inhibited by the irreversible pyrrolidine formation reaction. Pyrrolidine end groups can subsequently react with water, resulting in carboxyl-terminated chains which act as terminating agents [1,2]:

$$(4.3)$$

4.2.2. Direct Solid State Polyamidation

Direct solid state polyamidation starting from dry polyamide monomer (salt or amino acid) has been studied using different techniques [14–19], including heating in sealed vessels under an inert atmosphere, in open vessels while inert gas passes, and in inert liquid media. It presents considerable practical interest, since all the problems associated with the high temperature of melt technology are avoided [14]. However, it is applied mainly on the laboratory scale; therefore, its kinetics studies count a much lower number than post-SSP.

The characteristic of direct SSP is the topotacticity of the process, as discussed in Chapter 6. In most cases, the reaction occurs across the crystallographic axis of the monomer (e.g., of the polyamide salt [Fig. 4.3(a)] [20]), permitting the formation of single crystals with high melting points and perfect hydrogen bonding [Fig. 4.3(b)].

Regarding the mechanisms prevailing in direct SSP, a number of studies [21] report that solid state polyamidation proceeds through nucleation and growth along the crystallographic directions of the monomer. The kinetic data fit well with typical nucleation and growth models (i.e., the process comprises two steps and the kinetic curves are S-shaped, respectively). More specifically:

1. The nucleation stage occurs on the surface of the crystallites and/or at internal surfaces (defects) within the crystallites. Especially in the case of a volatile diamine such as hexamethylenediamine (HMD), the escape of the salt component is found to occur early and precedes water formation,

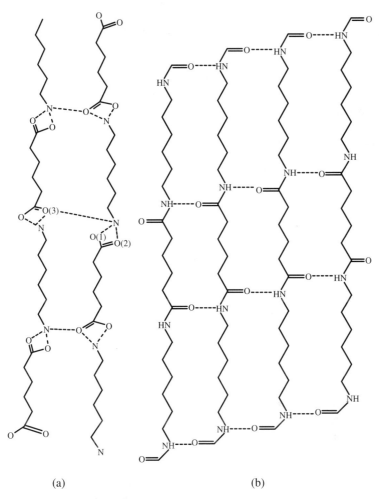

(a)   (b)

Fig. 4.3. PA 66 salt and polymer unit cell. (a) PA 66 salt: the hydrogen bonds N— H— O (1) and N— H— O (2) are formed intramolecularly and N— H— O (3) intermolecularly. (b) PA 66: intersheet hydrogen bonds between parallel chains in the α-form.

creating defects in crystallites which act as nucleation sites. However, this volatilization reduces the reaction rate, due to the disturbance caused in the stoichiometric equivalence of the functional end groups and impedes proper kinetic study of the process.

2. Regarding the growth stage, sintering occurs due to hydration of the reacting mass. As a result, the growth stage may be also performed in a quasi-melt state, depending on the reaction conditions and the salt polar sites, thus increasing the mobility of the reacting species. The relevant solid–melt transition has been described in Section 1.3.1a.

In particular, Volokhina et al. [22] studied the SSP kinetics of three aliphatic ω-amino acids—aminoenanthoic, aminopelargonic, and aminoundecanoic—and three hexamethylenediamine salts—adipic, thiodivaleric, and terephthalic acids—with respect to temperature. Investigation of the amino acid polyamidation process led to the conclusion that the reaction is distinctly autocatalytic. Another Russian group arrived at the same conclusion by studying the SSP of ω-aminoenanthic acid [23]. They suggested that "the autocatalytic acceleration of this reaction is due to increased reaction of the surface of the monomer–polymer boundary." This autocatalysis feature is also supported by Khripkov et al. [24] for hexamethylenediammonium adipate polymerized in the solid state in the presence of boric acid as catalyst.

Finally, the activation energy of polycondensation in solid phase was found to be 251 kJ mol^{-1} for ω-aminoundecanoic acid, 385 kJ mol^{-1} for ω-aminoenanthoic acid, and 752 kJ mol^{-1} for ω-aminopelargonic acid. In the liquid phase, the activation energy is lower: 180 kJ mol^{-1} for ω-aminoenanthoic acid and 159 kJ mol^{-1} for aminopelargonic acid [12].

4.2.3. Prepolymer Solid State Polyamidation

The post-SSP simple kinetics models can be classified into two categories: the first is based on the Flory equations (1.2) and (1.3) and the second on power-law models, without excluding combinations of these. In these models, only the polyamidation reaction is considered, first due to the low temperatures, which impede side reactions, and second, due to the continuous condensate and/or oligomer removal through convection caused by passing inert gas.

Regarding the Flory-based models, it should be mentioned that condensation reactions would be extremely difficult to analyze kinetically if the rate constant for the coupling depended on the degree of polymerization of the reacting species. Fortunately, kinetic studies of simple esterifications involving acids of increasing chain lengths have shown that the rate constant is effectively independent of chain length. These studies led to the Flory equal reactivity principle for step-growth polymerization, which states that the intrinsic reactivity of all functional groups is constant and independent of molecular size. The theoretical justification for this assumption is based on the fact that the collision frequency (which determines the reactivity of a functional group) is independent of molecular mobility (i.e., although the rate of diffusion of the larger molecules depends on their size, the collision frequency of a functional group attached to the end of those chains does not) [4].

Many different but almost equivalent sets of integrated equations of Flory rate expressions have been developed by several researchers focusing on irreversible polyamidation. More specifically, Vouyiouka et al. [25] used the Flory equations and integrated them based on the concentration of the reacted end groups (x) and the polymerization conversion (p_t) of PA 66 SSP. Depending on the prepolymer end group concentrations (amine or carboxyl moieties excess), the expressions for second- and third-order kinetics vary:

Carboxyl group excess:

$$-\frac{d[COOH]}{dt} = k_2([COOH]_0 - x)([NH_2]_0 - x) \xrightarrow{x = [NH_2]_0 p_t}$$

$$A_2 = \frac{1}{D_0}\left(\ln\frac{1}{[COOH]_0} + \ln\frac{[COOH]_0 - [NH_2]_0 p_t}{1 - p_t}\right) = k_2 t \quad (4.4)$$

$$-\frac{d[COOH]}{dt} = k_3([COOH]_0 - x)^2([NH_2]_0 - x) \xrightarrow{x = [NH_2]_0 p_t}$$

$$A_3 = \frac{1}{D_0^2}\ln\frac{[COOH]_0 - [NH_2]_0 p_t}{[COOH]_0(1 - p_t)}$$

$$-\frac{1}{D_0}\left(\frac{1}{[COOH]_0 - [NH_2]_0 p_t} - \frac{1}{[COOH]_0}\right) = k_3 t \quad (4.5)$$

Amine group excess:

$$-\frac{d[COOH]}{dt} = k_2([COOH]_0 - x)([NH_2]_0 - x) \xrightarrow{x = [COOH]_0 p_t}$$

$$A_2 = \frac{1}{D_0}\left(\ln\frac{1}{[NH_2]_0} + \ln\frac{[NH_2]_0 - [COOH]_0 p_t}{1 - p_t}\right) = k_2 t \quad (4.6)$$

$$-\frac{d[COOH]}{dt} = k_3([COOH]_0 - x)^2([NH_2]_0 - x) \xrightarrow{x = [COOH]_0 p_t}$$

$$A_3 = \frac{1}{D_0^2}\ln\frac{[NH_2]_0 - [COOH]_0 p_t}{[NH_2]_0(1 - p_t)}$$

$$-\frac{1}{D_0}\left(\frac{1}{[NH_2]_0 - [COOH]_0 p_t} - \frac{1}{[NH_2]_0}\right) = k_3 t \quad (4.7)$$

where $[COOH]_0$ and $[NH_2]_0$ are the initial concentrations of carboxyl and amine end groups (mmol kg^{-1}), k_2 (kg mmol^{-1} h^{-1}) and k_3 (kg^2 mmol^{-2} h^{-1}) are the rate constants for second- and third-order kinetics, and D_0 is the initial carboxyl or amine end-group excess (mmol kg^{-1}).

Further, Gaymans [26] studied the SSP of unbalanced PA 46 and proposed a kinetic expression in terms of the product of the concentrations of the end groups (P) and of the reaction order (n):

$$-\frac{d[COOH]}{dt} = k(\sqrt{P})^n \Rightarrow \left(\frac{1}{\sqrt{P_t}}\right)^{n-1} - \left(\frac{1}{\sqrt{P_0}}\right)^{n-1} = (n-1)kt \quad (4.8)$$

where k is the rate constant (kg^{n-1} mmol$^{-(n-1)}$ h^{-1}), n the reaction order, and $P = [-COOH][-NH_2]$ (mmol^2 kg^{-2}). Under the specific experimental limits, the SSP reaction did not follow third-order kinetics, but the apparent order varied between 3.5 and 5.2, revealing that with a decreasing reaction temperature, the apparent order increases.

According to Srinivasan et al. [27], who assumed stoichiometric equivalence in PA 66 prepolymer, the reaction rate is expressed through the number-average molecular weight:

$$-\frac{d[COOH]}{dt} = k[COOH]^n \xrightarrow{\overline{M_n} = 1/[COOH]} (\overline{M_n})^{n-1} - (\overline{M_{n_0}})^{n-1} = (n-1)kt$$

$$(4.9)$$

In addition, Jabarin and Lofgren [28] used the following equation and related the number-average molecular weight during SSP to the square root of time:

$$\overline{M_n} = \overline{M_{n_0}} + k\sqrt{t} \qquad (4.10)$$

where k is the rate constant for (4.9) (g^{n-1} mmol$^{-(n-1)}$ h^{-1}) and n is the reaction order for (4.10) (g mmol^{-1} h$^{-0.5}$).

Alternatively, the kinetics model of Duh [29,30] is based on the assumption that two categories of end groups exist during SSP: active and inactive ones (Section 3.2.3i). It was suggested that the overall SSP follows second-order kinetics and the proposed rate equation involves the apparent reaction rate constant and the apparent inactive end-group concentration, which was found to decrease linearly with the SSP temperature. Such a rate expression was proposed to be adequate in cases where SSP is controlled jointly by reaction and end-group diffusion.

$$-\frac{d[C]}{dt} = 2k_2([C] - [C_{ai}])^2 \Rightarrow \frac{[C]_0 - [C]_t}{t}$$

$$= 2k_2([C]_0 - [C_{ai}])[C]_t - 2k_2([C]_0 - [C_{ai}])[C_{ai}] \qquad (4.11)$$

where $[C]_0$ is the initial total end-group concentration (mmol kg^{-1}), $[C]_t$ the total end group concentration at any given time (mmol kg^{-1}), $[C_{ai}]$ the constant apparent inactive end-group concentration (mmol kg^{-1}), and k_2 is the apparent second-order rate constant (kg mmol^{-1} h^{-1}).

The Flory-based models may be transformed so as to take into consideration the two-phase model suggested by Zimmerman et al. (Section 1.3.2a) [31,32]. The end groups and low-molecular-weight substances (condensate, oligomers) are exclusively in the amorphous regions, where the equilibrium is the same as for a completely amorphous or molten polymer at the same temperature. Accordingly, the concentrations of the end groups involved in the SSP reaction should be corrected properly as follows:

$$[C]_a = \frac{[C]}{1 - \varphi_w} \qquad (4.12)$$

where $[C]_a$ is the end-group concentration in amorphous regions (mmol kg^{-1}), $[C]$ the end-group concentration in the polymer mass (mmol kg^{-1}), and φ_w the weight fraction crystallinity.

The two-phase model was used in the study of the SSP of PA 66 fibers [33]. Assuming a balanced prepolymer and a polydispersity index of 2 ($\overline{M_v} \sim \overline{M_w} = 2\overline{M_n}$) during SSP, a rate expression in terms of the reaction order n of the viscosity-average molecular weight and of the degree of crystallinity φ_w has been developed:

$$-\frac{d[C]_a}{dt} - k[C]_a^n \Rightarrow (\overline{M_n})^{n-1} - (\overline{M_{n_0}})^{n-1} = \frac{n-1}{(1-\varphi_w)^{n-1}}kt_{\Rightarrow} \quad (4.13)$$

$$(\overline{M_v})^{n-1} - (\overline{M_{v_0}})^{n-1} = \frac{(n-1)2^{n-1}}{(1-\varphi_w)^{n-1}}kt$$

where k is the rate constant (g^{n-1} mmol$^{-(n-1)}$ h^{-1}), n the reaction order, and φ_w the weight fraction crystallinity. The relevant data gave equally good degrees of fitting with second- and third-order reaction kinetics.

Turning now to the power-law models, a widely used equation is that of Walas [34], who pointed out that the rate (r) of a solid state process, which is controlled by the chemical reaction and diffusion, usually varies as some power of the time (n):

$$r = kt^n \quad (4.14)$$

Griskey and Lee [5] used a modified form of Walas' equation, which assumes that the number-average molecular weight in solid state polymerization varies as a power of the time equal to -0.49 for PA 66 SSP:

$$\frac{d\overline{M_n}}{dt} = kt^n \Rightarrow \ln(\overline{M_n} - \overline{M_{n_0}}) = \ln\frac{k}{n+1} + (n+1)\ln t \quad (4.15)$$

where k is the rate constant (g mmol^{-1} h$^{-(n+1)}$) and n is the power of the time.

During their study of PA 66 SSP, Fujimoto et al. [35] formulated a rate equation according to which the relative viscosity (RV) increases as some power of the time, and it was found that the RV of their samples increased linearly with heating time during SSP:

$$\frac{d(RV)}{dt} = kt^n \Rightarrow (RV)_t = (RV)_0 + \frac{k}{n+1}t^{n+1} \quad (4.16)$$

where k is the rate constant $h^{-(n+1)}$. However, it should be emphasized that such an expression serves only as a tool to get some idea of the rheological behavior of the polymer during SSP. According to Zimmerman and Kohan [32], if the course of the reaction was followed by RV measurements rather than by end-group concentrations, the results could become confusing because of the possibility of branching or cross-linking reactions, which would decrease or increase, respectively, the RV over that expected for a given end-group concentration.

Finally, for the analysis of their kinetic data of PA 6 SSP, Gaymans et al. [36] suggested that the process is limited by the diffusion rate of the autocatalyzing acid end and that this is not only dependent on the concentration and the temperature but also on the changing end group-to-end group distance distribution. Therefore, they developed a kinetic expression that associates the rate of reaction to the concentration of the catalyzing end groups [COOH] and to a power of time (n), thus combining the Flory theory with the Walas equation:

$$-\frac{d[COOH]}{dt} = k[COOH]t^n \Rightarrow \ln\left(-\frac{d[COOH]}{dt}/[COOH]\right) = \ln k + n\ln t$$

$$(4.17)$$

where k is the rate constant $h^{-(n+1)}$. It was found that the SSP kinetics had more than one region and for conversions higher than 30% the reaction rate was first order regarding carboxyl end-group concentration and reciprocal to the reaction time ($n = -1$).

4.3. SIMULATION OF SOLID STATE POLYAMIDATION

Because of the industrial importance of post-SSP, mathematical modeling and process simulation have been employed to gain a better understanding of the relevant mechanism and to predict the influence of different parameters on SSP rate. The majority of these modeling studies involve solution of the full system of partial differential equations (PDEs), which describe the changes with time and position of all chemical species within the particle (particle model) and/or in a reactor (reactor model). Such models have been developed for the SSP of a variety of polymers, belonging to the families of both polyesters and polyamides, including PET, PBT, BPA-PC, PA 6, and PA 66. The principal differential reactor model equations for PA 66 SSP in a continuous moving-bed reactor are given in Table 4.1 [6].

Focusing on particle models [37], the variables can be divided into two types: those that are subject to diffusion and mass transfer to the gas phase (e.g., condensate) and those that are not (e.g., amide linkages). The set of variables subject to diffusion is composed of the concentrations of volatile compounds:

$$\frac{\partial C_v}{\partial t} = R_v + D_v\nabla^2 C_v - \frac{C_v}{M}\frac{dM}{dt} \qquad (4.18)$$

where D_v is the diffusivity of the volatile compound in the polymer mass, C_v the concentration of a volatile, R_v the rate of generation of C_v by reaction, and M the system mass. The diffusion term differs according to the geometry of the reacting particle (x distance from the center of a polymer particle):

$$\text{For flakes (plane sheets)} : D_v\nabla^2 C_v = D_v\frac{\partial^2 C_v}{\partial x^2}$$

$$\text{For pellets (cylinder)} : D_v \nabla^2 C_v = D_v \left(\frac{\partial^2 C_v}{\partial x^2} + \frac{1}{x} \frac{\partial C_v}{\partial x} \right)$$

$$\text{For powder (sphere)} : D_v \nabla^2 C_v = D_v \left(\frac{\partial^2 C_v}{\partial x^2} + \frac{2}{x} \frac{\partial C_v}{\partial x} \right)$$

The second set of variables for nonvolatile compounds is described by a similar equation without the diffusion term:

$$\frac{\partial C_n}{\partial t} = R_n - \frac{C_n}{M} \frac{dM}{dt} \tag{4.19}$$

where C_n is the concentration of a nonvolatile, R_n the rate of generation of C_n by reaction, and M the system mass. In many cases the concentrations are adjusted to account for the end groups and condensate in the amorphous regions through (4.12). An example of PA 66 particle model equations [37] is presented below, and for reactor SSP models, the relevant expressions also include terms for mass and energy transport across the SSP reactor:

$$\frac{\partial \lambda_0}{\partial t} = \frac{\partial \lambda_{1,1}}{\partial t} = \frac{\partial \lambda_{1,2}}{\partial t}$$

$$= (1 - \varphi_w) \left[-k_1 \frac{C_A C_B^2}{(1 - \varphi_w)^3} + \frac{k_1}{K_{eq}} L W_f \frac{C_B}{1 - \varphi_w} \right] - \frac{(\lambda_0, \lambda_{1,1} \text{ or} \lambda_{1,2})}{M} \frac{\partial M}{\partial t}$$

$$\frac{\partial W}{\partial t} = -(1 - \varphi_w) \left[-k_1 \frac{C_A C_B^2}{(1 - \varphi_w)^3} + \frac{k_1}{K_{eq}} L W_f \frac{C_B}{1 - \varphi_w} \right] + D_w \nabla^2 W_f - \frac{W}{M} \frac{\partial M}{\partial t}$$

$$K_{eq} = \frac{[W/(1 - \varphi_w)] - W_f}{(A_{tot} - 2[W/(1 - \varphi_w)] + 2W_f)^2 W_f} \qquad A_{tot} = L + \frac{C_B}{1 - \varphi_w}$$

$$\times \frac{\partial M}{\partial t} = 0.018 D_w \nabla^2 W_f$$

where λ_i is the ith moment of all polymer molecules, $\lambda_{1,1}$ the concentration of total hexamethylenediamine (free and polymerized), $\lambda_{1,2}$ the concentration of total adipic acid (free and polymerized), C_A the amine end concentration, C_B the carboxyl end concentration, L the amide linkage concentration in the amorphous phase, A_{tot} the total carbonyl groups (carboxyls and amides), W the water concentration, W_f the free water concentration, D_w the diffusivity of water, k_1 a rate constant, K_{eq} an equilibrium constant, φ_w the weight fraction crystallinity, and M the system mass.

Regarding PA 6 SSP particle models, Kaushik and Gupta [38] suggested an equation between the overall rate constant of the reaction k_p in the presence of diffusional limitations, and the intrinsic rate constant of reaction k_{p0}:

$$\frac{k_p}{k_{p0}} = \frac{1 - (r_m^2/3D)k_p'W[1 - (\lambda_1/\lambda_0)]}{1 + (r_m^2/3D)k_{p0}\lambda_0}$$

$$\frac{r_m^2}{3D} \equiv \frac{\theta_t}{D/D_0} \tag{4.20}$$

$$\frac{D}{D_0} = \exp\left[\frac{2.303\upsilon_f}{A(T) + B(T)\upsilon_f}\right]$$

where D_0 is the diffusivity at some reference state, υ_f the free volume fraction, and $A(T)$ and $B(T)$ are empirically determined functions of temperature used as parameters in the SSP model. The relevant ratio is a function of the reverse rate constant k_p' of the diffusivity of the polymer molecule D of the the distance of diffusion r_m of the the concentration of water [W] and the ith moment of all polymer molecules, λ_i. Through the term $r_m^2/3D$, the characteristic migration time (θ_t) is introduced as a parameter accounting for segmental mobility and diffusion; meanwhile, the dependence of D/D_0 on the reaction temperature and polymer concentration is assumed to be given by Fujita–Doolitle theory.

The simulation above explained PA 6 SSP experimental data quite well and predicted qualitative trends observed experimentally in the SSP of PA 6 chips with intermediate remelting. More specifically, the authors set critical values for the water diffusivity inside the reacting particles, for the particles' surface water concentration, and for the particles' radius, for which no considerable increase of $\overline{M}_n$ occurs.

Kulkarni and Gupta [39] later developed an improved model that uses the Vrentas–Duda theory for diffusion coefficients. The effects of changing the important operating conditions on SSP (e.g., intermediate remelting of PA 6 powder, water concentration in the vapor phase, minimizing the monomer and water contents before SSP, size and degree of crystallinity of polymer particles) have been studied. The same effects are also considered in the comprehensive particle model for the SSP of PA 6 developed by Li et al. [40]. Xie [41] also developed a particle model for the SSP of PA 6, which explores the effect of different parameters on number-average chain length and polydispersity index.

Yao et al. [6,42] developed a reactor model for PA 66 to describe its SSP in a moving bed and to simulate process startup, shutdown, and different disturbances during operation. A model of PA 66 SSP has been developed by Li et al. [43] in which the variations of molecular weight and water content at different positions inside the chip indicate that the major polymerization occurs in a thin shell near the periphery of the chip and the molecular weight at the core increases more slowly, because of the limited diffusion of water. Furthermore, in agreement with experimental data, in their simulation Yao and Ray [44] found that the residence time in the SSP reactor increases rapidly with a decrease in the DP in the feed line, and the size of PA particles does not have a considerable effect on the SSP rate. In addition, it was shown that the voidage of the inlet polymer flow has little effect on the properties of the final product insofar as the ratio between gas flow and polymer flow is high. In general, with an increase in voidage, the $\overline{M_n}$ drops.

4.4. SIMPLE SSP KINETICS: THE CASE OF POLY(HEXAMETHYLENE ADIPAMIDE)

In the following sections, emphasis is placed on PA 66 SSP kinetics, due to its commercial significance (Section 1.5.2), since it consists one of the most popular polyamide resins, accounting with PA 6 for more than 90% of nylon uses. In particular, empirical SSP models found in the literature are tested based on experimental data. Special interest presents the impact of starting material composition (salt, prepolymer, and comonomer presence) on the process rate.

4.4.1. Solid State Polymerization of Hexamethylenediammonium Adipate

Diamine–diacid salts are important intermediates in the polyamide industry, ensuring stoichiometric portions of polymerization monomers. They are formed through the creation of ionic bonds between the end groups of the reactants:

$$HOOC{-}R_1{-}COOH + H_2N{-}R_2{-}NH_2 \rightarrow {}^-OOC{-}R_1{-}COO^- \cdot {}^+H_3N{-}R_2{-}NH_3{}^+$$

$$(4.21)$$

The volatility of the diamine in combination with the exothermic character of the salt formation demands a reduction in the reacting system temperature to reach high mass yields.

In particular, PA 66 salt is prepared from hexamethylenediamine (HMD) and adipic acid. It is a distinct chemical compound, and it forms white diamond-shaped monoclinic crystals. It is hygroscopic, readily soluble in water (47% w/v at 18°C), and when heated, it melts in the range 190 to 191°C with partial polymer formation. The salt is stable dry or in solution at room temperature, but above 140°C it polymerizes [21,45]. PA 66 salt can be isolated as a solid directly from an alcoholic (i.e., ethanolic or methanolic) solution of the reactants or through precipitating it from their aqueous solution by adding a nonsolvent [e.g., isopropanol or methanol (solubility 0.4% w/v at 25°C)]. Alternatively, Papaspyrides and Bletsos [46,47] suggested a simple technique for reacting adipic acid and hexamethylenediamine in the presence of a cryogenic medium in conventional dry blending equipment.

A thermogravimetric analysis (TGA) chamber can be used as a small-scale direct SSP reactor. The SSP progress can be evaluated by continuous monitoring of the weight of the reacting mass, which decreases due to the loss of hexamethylenediamine and polycondensation water. Even better, the TGA can be coupled to a titrator so as to monitor the effluent gas composition regarding the two escaping components [48]. Use of this technique permits correct calculation of critical species concentrations and kinetics parameters, such as water formed and polymerization conversion, taking into account the volatilization of HMD (Table 4.3).

More specifically, direct PA 66 SSP was analyzed kinetically in a pertinent TGA assembly in the temperature range 160 to 190°C under both dry static and flowing (50 mL min^{-1}) nitrogen. The reduced weight loss [$(\Delta W)_r = \Delta W / m_0$ in %] due to HMD and H_2O (Fig. 4.4) was plotted against reaction time and temperature. The relevant values for water formation remained below the theoretical value [for $p_t = 1$, $(\Delta W)_r = 13.72\%$]; the diamine loss varied between 2.8 and 3.7 wt%.

Based on the water weight loss, the polymerization conversion was calculated through the relevant expression in Table 4.3. The plots developed of p_t against time were S-shaped, showing that the kinetics of the process were characterized by two stages: induction and propagation (i.e., nucleation and growth). The process kinetics were investigated further based on the HMD loss. First, the Flory-based integrated expressions (4.4) and (4.5) were tested by fitting the end-group concentration data, but they failed to describe the process for the pure PA salt SSP under flowing nitrogen. The latter indicates the role of condensate diffusion as a rate-limiting step; therefore, a typical power-law model was

TABLE 4.3. Critical PA 66 Monomer SSP Kinetic Parameters

Parameter	Expression
Polymerization conversion (p_t)	$$p_t = \frac{[H_2O]_t}{[NH_2]_0} = \frac{m_{H2O,t}}{18[NH_2]_0 m_t}$$
	where $[H_2O]_t$ is the concentration of water formed (mol kg^{-1}), $m_{H2O,t}$ the amount of the water that escaped at any given time (g), $[NH_2]_0$ the initial concentration of amine ends (mol kg^{-1}), and m_t the reacting mass (kg).
Reacting mass (m_t)	$$m_t = m_0 - m_{H2O,t} - m_{HMD,t}$$
	where $m_{HMD,t}$ is the amount of the diamine that escaped at any given time (kg), m_0 the initial weight of the salt (kg), and $m_{H2O,t}$ the amount of the water that escaped at any given time (kg).
End-group concentrations	$$[NH_2]_t = [NH_2]_0 - [NH_2]_{lost} - [H_2O]_t$$ $$[COOH]_t = [COOH]_0 - [H_2O]_t$$
	where $[COOH]_t$ and $[NH_2]_t$ are the concentrations of the carboxyl and amine end groups (mol kg^{-1}), $[COOH]_0$ and $[NH_2]_0$ the initial concentrations (mol kg^{-1}), and $[NH_2]_{lost}$ the concentration of amine end groups (mol kg^{-1}) lost due to the volatilization of HMD.

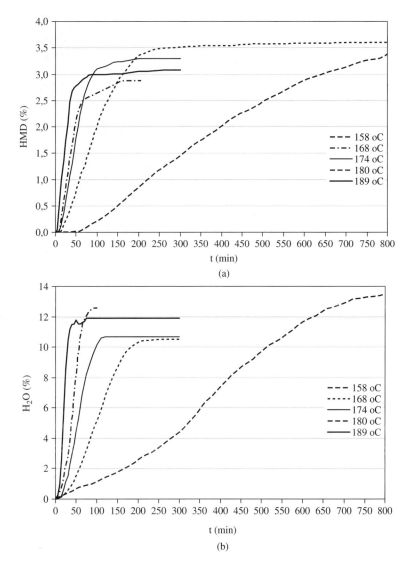

Fig. 4.4. Reduced weight loss due to (a) HMD (%) and (b) H_2O (%) during SSP of PA 66 salt under flowing nitrogen. (From Papaspyrides et al. [48] by permission of Elsevier.)

examined. Based on Walas' rate expression (4.14), equation (4.22) was deduced, including initial amine end concentration and polymerization conversion as calculated through the proper equations of Table 4.3:

$$r = \frac{d[H_2O]}{dt} = kt^n \Rightarrow \ln[H_2O] = \ln([NH_2]_0 p_t) = \ln \frac{k}{n+1} + (n+1)\ln t$$

$$(4.22)$$

where n is the power of the time and k is the rate constant (mol kg^{-1}) min$^{-(n+1)}$. Equation (4.20) was tested by fitting the experimental data depicted in Figure 4.4 at each reaction temperature. Accordingly, $\ln([NH_2]_0 p_t)$ versus $\ln t$ was plotted so as to determine the power of the time from the line slope $(n + 1)$ and the rate constant from the intercept $\ln[k/(n + 1)]$ at $\ln t = 0$. It was found that this simple model describes the overall process satisfactorily and fits all data adequately, based on high correlation coefficients ($r^2 = 0.9810$ to 0.9955) (Fig. 4.5). The power of the time and the SSP rate constants are shown in Table 4.4.

The n values were found to increase linearly ($r^2 = 0.9177$) with the reaction temperature T:

$$n = 0.0475T - 20.215 \qquad (4.23)$$

where T is the absolute temperature in K. This probably shows the effect of T on the diamine diffusion step: As temperature increases, the HMD volatilization is favored, the nucleation stage is shortened, and the SSP is accelerated.

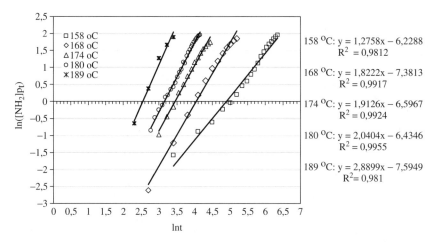

Fig. 4.5. Fitting of the power-law model [equation (4.22)] to the experimental data of pure PA 66 salt SSP under flowing nitrogen. (From Papaspyrides et al. [48] by permission of Elsevier.)

TABLE 4.4. Reaction Order (n) and Rate Constants (k) for the SSP of Pure PA 66 Salt Under Flowing Nitrogen (158–189°C)

$T(°C)$	n	k
158	0.2758	0.0025 mol kg^{-1} min$^{-1.2758}$
168	0.8222	0.0011 mol kg^{-1} min$^{-1.8222}$
174	0.9126	0.0026 mol kg^{-1} min$^{-1.9126}$
180	1.0404	0.0033 mol kg^{-1} min$^{-2.0404}$
189	1.8899	0.0014 mol kg^{-1} min$^{-2.8899}$

Finally, combining equations (4.22) and (4.23), (4.24) was deduced, where the polymerization conversion can be calculated through a power-law model and the effect of diamine loss is also included:

$$\ln p_t = (0.045T - 19.215) \ln t + \ln \frac{k}{0.045T - 19.215} - 2.03 \qquad (4.24)$$

4.4.2. Solid State Polymerization of Poly(hexamethylene adipamide)

The majority of published SSP rate expressions as stated in Section 4.2.3 were assessed for their validity based on PA 66 prepolymer SSP data at 160°C ($\overline{M_{n0}} = 17,000$ g mol^{-1}, pellets size 1.4 to 1.7 mm, $D_0 = 20$ mmol kg^{-1} with COOH excess, 260 mL min^{-1} N$_2$, fixed-bed reactor, short reaction time), to result in the most adequate rate expression for kinetics (regime I, Section 3.2.2) [25]. Use of each kinetic model led to values of the rate constants at any given reaction time, based on which a mean value (k) was determined. The standard deviation of the mean (SDM) was also calculated and indicated the model fitting. In addition, the rate constants were derived by the slope of the lines (k_{plot}) of the rate expressions versus reaction time, and the correlation coefficient (r^2) was deduced. The deviation between k and k_{plot} was also calculated [$\Delta S^2 = (k - k_{plot})^2$] and showed how well the model fits.

The results of the kinetics analysis are presented in Table 4.5. In general, it can be said that the Flory rate expressions fit better than the power-law models, revealing that the SSP rate depends primarily on the concentrations of the end groups under the specific experimental conditions (i.e., the process is reaction controlled). The poor fit of the power-law models indicates that by-product diffusion limitations may be disregarded. This observation may be attributed to the low reaction temperatures studied, since by-product diffusion limitations are generally predominant at much higher operating temperatures ($T > 210°$C), where chemical reaction is no longer the controlling step [10,49,50].

More specifically, the Flory-based expression introduced by Gaymans [26], equation (4.6), does not describe the SSP process successfully. It leads to high SDM values (17 to 18%) and, more important, to significantly high deviations ΔS^2. According to Gaymans, the SSP reaction does not follow third-order kinetics, but the apparent order varies between 3.5 and 5.2, as the SSP temperature decreases from 280°C to 190°C, revealing that the apparent order increases with lower reaction temperature. These SSP data were also tested for $n = 5$ and 6 in (4.8), considering the low operating temperature (160°C), but the fit was still poor (r^2: 0.7461 and 0.7450, respectively).

A similar result was observed in the Srinivasan et al. [27] approach [equation (4.7)]. The high deviation resulted obviously from the nonstoichiometric equivalence of the end groups in the prepolymer sample, which is, however, an assumption for the relevant kinetic expression. For the same reasons, equation (4.8), introduced by Jabarin and Lofgren [28], revealed a poor fit (r^2: 0.5311). The Duh [29,30] model [equation (4.9)] was found to be completely inadequate.

TABLE 4.5. Fitting PA 66 Post-SSP Data at 160°C to Various Kinetic Models

Flory-Based Simple Models

Equation	$100k_2$ (kg mmol^{-1} h^{-1})	$1000k_3$ (kg^2 mmol^{-2} h^{-1})	$100k_{2\text{plot}}$ (kg mmol^{-1} h^{-1})	$1000k_{3\text{plot}}$ (kg^2 mmol^{-2} h^{-1})	$\Delta S^2(k_2)$ (kg^2 mmol^{-2} h^{-2})	$\Delta S^2(k_3)$ (kg^4 mmol^{-4} h^{-2})
(4.4), (4.5) ($t = 1, 2, 3,$ 4 h)	$0.046175 \pm 8.38\%$	$0.006929 \pm 8.88\%$	$0.05,$ $r^2 = 0.9883$	$0.008,$ $r^2 = 0.9894$	1×10^{-9}	1×10^{-12}
(4.8), ($t = 2, 3,$ 4 h)	$0.039791 \pm 16.57\%$	$0.0073003 \pm 17.67\%$	$1.1108001,$ $r^2 = 0.7496$	$0.20783,$ $r^2 = 0.7484$	1×10^{-4}	$1 \times 4 \times 10^{-8}$
(4.9), ($t = 2, 3,$ 4 h)	$0.03475 \pm 21.94\%$	$0.0062747 \pm 23.00\%$	$0.0427,$ $r^2 = 0.8085$	$0.01,$ $r^2 = 0.7978$	6×10^{-9}	1×10^{-11}
(4.10)	$\overline{M_n} = 16{,}884 + 833.15\sqrt{t},\ r^2 = 0.5311,\ k = 833.15$ g mol^{-1} h$^{-0.5}$					
(4.11)	$\dfrac{C_0 - C}{t} = -0.3316C + 37.866$; as derived, different from the model proposed					

Power-Law Models

(4.15) ($t = 2,$ 3, 4 h)	$\dfrac{d\overline{M_n}}{dt} = 303t^{0.3722},\ r^2 = 0.6513,\ k = 303$ g mol^{-1} h$^{-1.3722}$
(4.16)	$\dfrac{dRV}{dt} = 6.3699t$
(4.17) ($t = 2,$ 4 h)	$-\dfrac{d[COOH]}{dt} = 0.0039[COOH]t^{1.2358},\ k = 0.0039$ h$^{-2.2358}$; poor fitting, as negative rate constants for 1 and 3 h were derived

Equation (4.4) and (4.5) presented the lowest deviation (SDM: 8 to 9%) among the aforementioned kinetic models. In the pertinent rate expressions, only the initial carboxyl end-group concentrations $[COOH]_0$ were involved. The latter is really important considering that the —COOH analysis has lower reproducibility and is less accurate than —NH_2 determination.

Regarding the power-law models, the Fujimoto approach [equation (4.14)] had a better fit to the experimental data and showed the linearity of relative viscosity values (RV) versus reaction time, as was found in the Fujimoto et al. study [35]. On the other hand, (4.15) resulted in a low correlation coefficient (r^2: 0.6513), and the reaction order was found equal to 0.3722, deviating significantly from the findings of the previous study ($n = -0.49$) [5]. Regarding (4.17), the fit was very poor, since some data led to negative rate constant values, and only values for selected reaction times (2 and 4 h) could be deduced.

Overall, SSP was described successfully through (4.4) and (4.5), which were examined further for other polyamide prepolymers. Accordingly, the SSP data for PA 6 chips at 200 and 220°C (N_2: 60 mL min^{-1}) provided by Xie [51] were first used. The rate expressions for second- and third-order kinetics (Figs. 4.6 and 4.7) were plotted against the reaction time and showed that in all cases the equations proposed describe SSP behavior appropriately (r^2: 0.9614 to 0.9859). Similarly, Gaymans' data [26], regarding PA 46 SSP at 220°C were used and showed an excellent fit to the model proposed (r^2: 0.9953 to 0.9989).

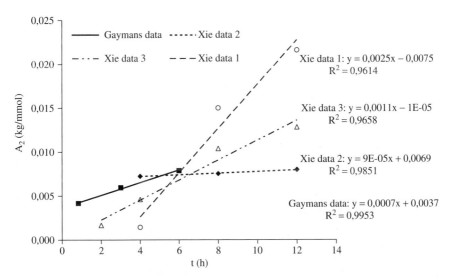

Fig. 4.6. Fitting of the second-order rate expression (4.4) to published post-SSP data. ■, Gaymans data: PA 46 at $T_{SSP} = 220°C$; ○, Xie data 1: PA 6 chips with 0.03% regulator at $T_{SSP} = 220°C$; ◆, Xie data 2: PA 6 chips (cylinder, $\bar{d} = 1.2$ mm); △, Xie data 3: PA 6 chips (cylinder, $\bar{d} = 1.4$ mm). (From Vouyiouka et al. [25] by permission of John Wiley & Sons, Inc.)

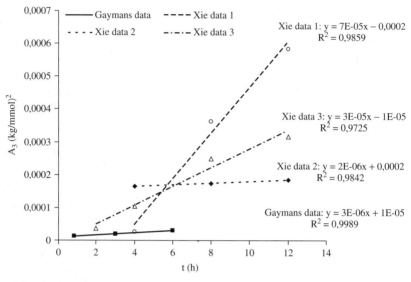

Fig. 4.7. Fitting of the third-order rate expression [equation (4.5)] to published post-SSP data. ■, Gaymans data: PA 46 at $T_{SSP} = 220°C$; ○, Xie data 1: PA 6 chips with 0.03% regulator at $T_{SSP} = 200°C$; ◆, Xie data 2: PA 6 chips (cylinder, $\overline{d} = 1.2$ mm); △, Xie data 3: PA 6 chips (cylinder, $\overline{d} = 1.4$ mm). (From Vouyiouka et al. [25] by permission of John Wiley & Sons, Inc.)

Based on (4.4) and (4.5) and for the temperature range 160 to 200°C, the relevant mean values of the rate constants are presented in Table 4.6 for second- and third-order kinetics. Third-order kinetics presented a slightly better fit than second-order kinetics, which has also been observed for the SSP of PA 66 fibers in the temperature range 220 to 250°C [33]. As expected, the rate constants increased with increasing temperature. The average SSP rate constant for third-order kinetics increased by 63% with each 10°C increase in temperature.

The activation energy (E_a) for PA 66 post-SSP was found to be 70.41 kJ mol^{-1}, which lies within findings reported in the literature (42 to 318 kJ mol^{-1}) (Table 4.7). The Arrhenius equation deduced can be incorporated in

TABLE 4.6. Reaction Rate Constants for PA 66 Post-SSP

	Eq. (4.4)		Eq. (4.5)	
T (°C)	$100k_2$ kg mmol^{-1} h^{-1}	%SDM (k_2)	$1000k_3$ kg^2 mmol^{-2} h^{-1}	%SDM (k_3)
160	0.051	1	0.008	0
180	0.113	11	0.018	9
200	0.221	13	0.040	11

TABLE 4.7. Reaction Kinetic Data Related to the Irreversible Post-SSP of Polyamides

Reference	Starting Material	Operating Conditions	k	E_a (kJ mol^{-1})
[5]	PA 66 ($\bar{d} = 0.18$ cm)	90–135°C, 0–10 h, N$_2$	$k = 1.53 \times 10^{10} \exp(-54.34/RT)$, k: h$^{-0.51}$	54.3
[52]	PA 66 ($\bar{d} = 0.35$–0.20 cm)	120–180°C, 5–20 h, N$_2$	$k = 1.39 \times 10^4 \exp(-43.89/RT)$, k: h$^{-0.5}$	43.9
	PA 610 ($\bar{d} = 0.22$–0.34 cm)	120–180°C, 5–20 h, N$_2$	$k = 1.68 \times 10^4 \exp(-55.18/RT)$, k: h^{-1}	55.2
[35]	PA 66 ($\bar{d} = 0.3$ cm)	160–210°C, 0–80 h, N$_2$	$\log k = 13.8 - \dfrac{5.90 \times 10^3}{T}$, k: h^{-1}	108.7
[27]	PA 66 fibers	220–250°C, 0–4 h, N$_2$	$k = 6.29 \times 10^{40} \exp(-317.68/RT)$, k: g^2 mol^{-2} s^{-1}	317.7
[33]	PA 66 fibers	220–250°C, 0–4 h, N$_2$	$k = 3.06 \times 10^{18} \exp(-175.56/RT)$, k: g mol^{-1} s^{-1}; second order	175.6
			$k = 1.18 \times 10^{31} \exp(-254.98/RT)$, k: g^2 mol^{-2} s^{-1}; third order	255.0
[36]	PA 6 ($\bar{d} = 0.02$–0.05 cm)	110–205°C, 1–24 h, N$_2$	For conversions $\rangle$ 30%: $k = 0.28$	

comprehensive models and used to determine SSP apparent rate constants:

$$k_{hom} = 4.8 \times 10^{-6} \exp\left[\frac{70.41}{R}\left(\frac{1}{423} - \frac{1}{T}\right)\right] \qquad (4.25)$$

where k_{hom} is the rate constant for third-order kinetics for homopolymer PA 66 post-SSP ($kg^2 \, mmol^{-2} \, h^{-1}$), T the absolute reaction temperature (K), and R the universal gas constant ($kJ \, mol^{-1} \, K^{-1}$).

4.4.3. Compositional Effects in Solid State Polymerization of Poly(hexamethylene adipamide)

The effect of sulfur-containing comonomer, sodium 5-sulfoisophthalic acid (NaSIPA) is examined with respect to the kinetics of PA 66 post-SSP [53]. As an aromatic dicarboxylic acid [Fig. 4.8(a)], NaSIPA reacts with the free amine groups of polyamide structure, and an anionically modified copolymer (ionomer) is formed [Fig. 4.8(b)]. The comonomer is usually added during the initial stages of the conventional solution-melt polyamidation or through melt polycondensation of PA 66 homopolymer with a master batch containing NaSIPA. However, the sulfonated PA 66 copolyamides generally have high melt viscosities, which limit the extent of melt polymerization and hinder the effective discharge of the polymerized melt from the reactor. As a result, the conventional solution-melt polycondensation is interrupted at a low- or medium-molecular-weight product. Where higher molecular weights are required, the sulfonated PA 66 prepolymers are further polymerized in the solid phase.

NaSIPA-containing polyamides consist of important commercial grades since in many cases they present improved properties in comparison to homopolymer grades. Their main attribute is the improvement of the polymer dyeability to cationic dyes, resulting in fibers or films with deep and brilliant colors and resistance to stains, fading, and yellowing throughout their life cycle. It has also been found that NaSIPA incorporation contributes in minimizing operational problems related to the use of pigments and stabilizers as well as in avoiding problems during spinning. For this reason, the preferred range of NaSIPA to be used is 1

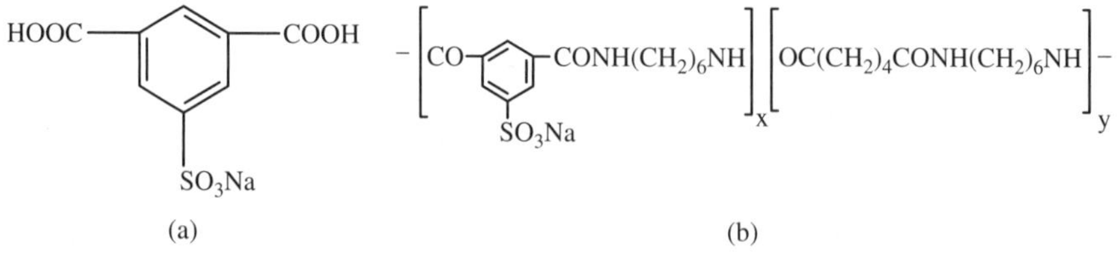

Fig. 4.8. (a) Sodium 5-sulfoisophthalic acid (NaSIPA); (b) PA 66 copolyamide with NaSIPA (x, 0.01–0.05; y, 0.95–0.99).

to 2 wt% (added at the salt stage, i.e., prior to polymerization) for most combinations of pigments and copper; above 4 wt%, the additive itself begins to lower the relative viscosity of the polymer and gives poorer operability [53–58]

The effect of NaSIPA on PA 66 SSP has been examined thoroughly, considering the importance of the additive on an industrial scale and that its presence affects the majority of nylon production lines. NaSIPA-containing copolyamides (0 to 3 wt% NaSIPA) were submitted to SSP runs under flowing and static nitrogen in the temperature range 160 to 200°C. At each reaction temperature, the copolyamides exhibited reduced SSP rates, as indicated by the lower rate in each case, the final RV, $\overline{M_n}$, and p_t values.

The kinetics of the process were studied using the rate expression (4.5), due to the carboxyl end-group excess in the prepolymers. The aforementioned retarding behavior of the copolyamides became even clearer when comparing the apparent rate constants (k_3) for third-order kinetics: PA 66 homopolymer presented the higher rate constant in the temperature range 160 to 200°C (Fig. 4.9), and more specifically, k_3 decreased by 11 to 57% as the amount of NaSIPA increases.

SSP runs were also carried out under static nitrogen using PA 66 homopolymer (PA) and copolymer containing 3% w/w NaSIPA (PA 3) at 160 and 200°C for 4 hours. Regarding NaSIPA's negative effect on the SSP process, the picture remained the same (Table 4.8). In addition, the importance of the by-product removal was revealed, since rate constants under static nitrogen (k_3^s) decreased by 12 to 19% in comparison to the flowing nitrogen processes. However, this decrease under static nitrogen did not vary significantly when comparing PA and

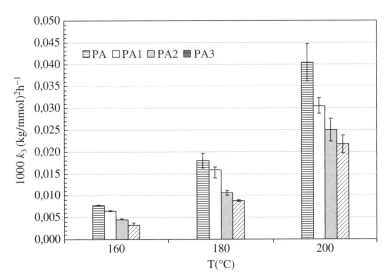

Fig. 4.9. Rate constants (k_3) during PA 66 post-SSP under flowing nitrogen. PA, PA 66 homopolymer, PA1, PA 66 containing 1 wt)% NaSIPA; PA2, PA 66 containing 2 wt % NaSIPA; PA3, PA 66 containing 3 wt % NaSIPA. (From Vouyiouka et al. [53] by permission of Elsevier.)

TABLE 4.8. Rate Constants (k_3^s) During PA 66 Post-SSP Under Static Nitrogen and Percent Rate Decrease Comparing Static and Flowing Nitrogen Processes

Sample	$1000k_3^s$ (kg^2 $mmol^{-2}$ h^{-1})	$\dfrac{k_3^s - k_3}{k_3}$ (%)
At 160°C		
PA	0.006577	−15
PA3	0.003059	−19
At 200°C		
PA	0.031866	−12
PA3	0.016762	−15

PA 3, showing that both grades "experienced" similar difficulty in removing the polycondensation water. This observation seems quite useful, as discussed below.

The SSP behavior of the ionomers was studied further so as to explore the relevant reaction mechanism. The retardation reasons were first sought in the various properties of the copolyamides against the homopolymer. Thus, the retarding behavior may be attributed to the anticipated higher by-product removal resistance in the NaSIPA-modified grades, due to the increased hygroscopic sites of the additive structure. Based on literature findings, the presence of the sodium sulfonate groups hinders drying, but it may initiate and/or catalyze the hydrolysis reaction. This was suggested in the case of sulfonated PET copolymers found significantly susceptible to acidic hydrolysis and exhibiting much higher hydrodegrability than that of pure PET [59]. However, the SSP results under static nitrogen indicated that the hygroscopicity of the additive structure cannot fully explain the dramatic rate decrease (almost 57%) observed in the copolyamide samples, since PA and PA 3 showed similar difficulty in the removal of the polycondensation water during SSP.

On the other hand, if considering the effect of the aromatic comonomer (isophthalic acid) on the SSP slowdown, it may be assumed that this does not apply here, since in the case of PET copolyesters with isophthalic acid, the SSP performance was exactly the opposite: The rate constants for the copolymers increased as a function of the aromatic diacid content, probably due to the less compact structures and thus to the increased diffusion of by-products [60].

Further, the SSP process might have been influenced by the morphology of the sulfonated ionomers. According to the model proposed by Eisenberg et al. [61] for random ionomers, the NaSIPA ionic moieties aggregate into *multiplets*, which, in turn, aggregate themselves into *clusters*, finally creating a contiguous phase of restricted mobility in the polymer mass. Respectively, during the SSP of NaSIPA-containing copolyamides, the ionic groups are localized in the amorphous regions and create a low-mobility area, which obviously impedes diffusion of the functional end groups and/or the water escape, slowing the SSP reaction. However, the rate decrease observed experimentally increased proportionally to the amount of NaSIPA, and this cannot be explained solely by the aforementioned aspects.

Based on the considerations above, another mechanism is proposed regarding the ion-modified PA 66 SSP. The retardation is correlated with the partial deactivation of the protonated amine ends [62,63] by the sulfonate (SO_3^-) units:

$$NH_3^+ + SO_3^- \rightarrow NH_3^+ \cdot SO_3^- \tag{4.26}$$

which exist in the amorphous regions of the copolyamides. Through this reaction, NaSIPA improves the polymer resistance to acid stains, since the amine groups are no longer available to absorb them (e.g., wine, soft drinks), which is really important for applications such as flooring covers [56–58]. As a result, during SSP there are two types of end groups: active ($[NH_3^+]_{act}$) and inactive ($[NH_3^+]_{inact}$) protonated amine ends. The former are able to participate in the polymerization reaction, and their concentration in the copolyamides is lower against PA 66 homopolymer, resulting in reducing the reactant concentration and thus the reaction rate. The inactive amine groups are attached to SO_3^- in the restricted mobility domains (clusters), and their constant concentration during SSP is assumed to be a fraction of the total amine ends, depending on the SO_3^- content, the reaction temperature, and the morphology of the ionomer.

According to this mechanism, the reactive species concentrations are changed and the relevant correlated kinetic figures are formed as in Table 4.9. In particular, it is proposed that the active amine ends are less than the experimentally measured value and more specifically, decrease proportional to a function of the sulfonate groups (C_s), using also a deactivation factor ($1/J$). The latter reflects the influence

TABLE 4.9. Modified Kinetics for the Post-SSP of NaSIPA-Containing PA 66

Kinetics	Expression
Active protonated amine-end concentrations	$[NH_3^+]_{act} = [NH_3^+]_t - [NH_3^+]_{inact} = [NH_3^+]_t - [NH_3']_t \dfrac{C_s}{J}$ where $[NH_3^+]_t$ is the amine-ends concentration measured experimentally (mmol kg^{-1}), C_s the sulfonate group content (mmol kg^{-1}), and $1/J$ the deactivation factor (J in mmol kg^{-1}).
Rate simple model [based on (4.5)]	$\dfrac{-d[COO^- + COOH]}{dt} = k_{cop}[COO^- + COOH]^2[NH_3^+]_t$ $= k_{cop}[COO^- + COOH]^2 \dfrac{[NH_3^+]_{act}}{1 - C_s/J}$ where k_{cop} is the apparent SSP rate constant of the copolyamides for third-order kinetics (kg^2 mmol^{-2} h^{-1}).
Correlation of apparent rate constants	$k_{cop} = k_{hom}\left(1 - \dfrac{C_s}{J}\right) \hfill (4.27)$

of the different ionomers morphology as discussed by Eisenberg et al. [61] and of the reaction temperature on the concentration of the inactive amine groups.

The assumption of the inactive amine-end mechanism can be tested through examining the fitting of SSP experimental data to (4.27) (Table 4.8), which shows the correlation between the SSP rate constants for homopolymer and copolymer, respectively. For all reaction temperatures, the ratio of k_{cop}/k_{hom}, based on the data of Figure 4.9, was plotted versus C_s and the linearity ($r^2 = 0.9538$) verified the validity of the mechanism suggested. The slope of the line represents the deactivation factor and was found equal to—0.0053 kg mmol^{-1} (i.e., J is 189 mmol kg^{-1}). This value of C_S/J may be considered as the fractional decrease of the total amine ends in the copolyamide grades (i.e., for PA 1, the active amine ends are 82% of the experimentally measured value; for PA 2, 65%; and for PA 3, 47%).

Also considering the Arrhenius equation for PA 66 homopolymer SSP [equation (4.25)], the compositional effect of NaSIPA on the SSP apparent rate constant can be assessed when knowing C_s and T:

$$k_{cop} = 4.8 \times 10^{-6} \exp\left[\frac{70.41}{R}\left(\frac{1}{423} - \frac{1}{T}\right)\right]\left(1 - \frac{C_s}{189}\right) \qquad (4.28)$$

where k_{cop} is the SSP rate constant of the copolyamides for third-order kinetics (kg^2 mmol^{-2} h^{-1}), C_s the sulfonate group content (mmol kg^{-1}), T the reaction temperature (K), and R the universal gas constant (kJ mol^{-1} K^{-1}).

A further modification of this rate expression can be made:

$$k_{cop} = 4.8 \times 10^{-6} \exp\left[\frac{70.41}{R}\left(\frac{1}{423} - \frac{1}{T}\right)\right]$$
$$\times \left\{1 - 6 \times 10^{-3} \exp\left[\frac{-4.587}{R}\left(\frac{1}{423} - \frac{1}{T}\right)\right]C_s\right\} \qquad (4.29)$$

where R the universal gas constant (kJ mol^{-1} K^{-1}), in order to include the slight effect of reaction temperature on the deactivation factor. $1/J$ values were calculated for each reaction temperature and were expressed successfully as an exponential function of $1/T$ (r^2: 0.9773). In particular, $1/J$ decreased slightly with temperature, apparently due to the induced increase of reacting species mobility in ionomers, resulting in inactive amine-end partial rejection from clusters and, finally, activation with rise in T. Equation (4.29) may be used successfully in simulation models, especially since the relevant reaction rate constants are usually extrapolated from melt polymerization data, despite the very different morphology of the solid polymer.

A similar kinetics analysis has also been done in case of catalyzed post-SSP of PA 66 [64,65]. The effect of a phosphonate [calcium-bis((3,5-bis (1,1-dimethylethyl)-4-hydroxyphenyl)methyl)ethylphosphonate, Ciba Irgamod 195] (Fig. 4.10) on the post-SSP rate was evaluated and the rate expression was modified so as to involve the phosphorus content.

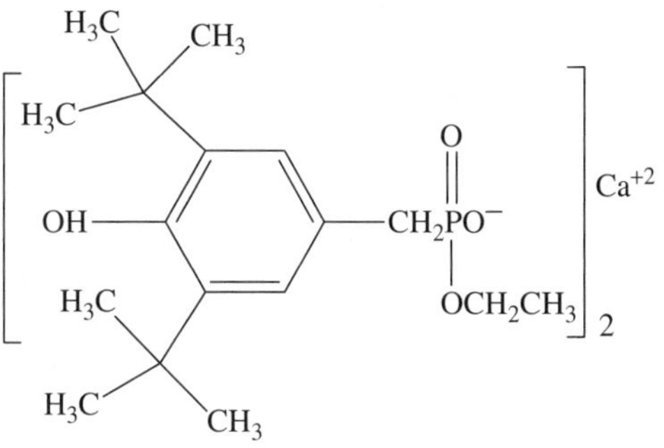

Fig. 4.10. Molecular structure of phosphonate used as catalyst in PA 66 post-SSP.

More specifically, it was assumed that the rate constant of the catalyzed process (k_{cat}) as derived from the power-law model (4.16) ($n = 0$) increases proportionally to a function of the phosphorus concentration ([P]), using a constant b (ppm^{-1}):

$$k_{cat} = k(1 + b[P])$$ (4.30)

where k is the rate constant for the uncatalyzed process at the same temperature (h^{-1}) and [P] is the experimentally measured phosphorus concentration (ppm).

For the reaction temperatures studied, the k_{cat}/k ratio was plotted versus [P] and the linearity ($r^2 = 0.9349$) verified the validity of (4.30). The slope of the line represented the factor b and was found to be 0.0028 ppm^{-1}. Finally, also considering the Arrhenius equation for PA 66 post-SSP [equation (4.31)], the compositional effect of the phosphorus-containing catalyst on the SSP apparent rate constant can, when knowing [P] and T, be assessed through (4.32):

$$k = 3.675 \exp\left[\frac{61.03}{R}\left(\frac{1}{423} - \frac{1}{T}\right)\right]$$ (4.31)

$$k_{cat} = 3.675 \exp\left[\frac{61.03}{R}\left(\frac{1}{423} - \frac{1}{T}\right)\right](1 + 0.0028[P])$$ (4.32)

where T is the absolute reaction temperature (K), [P] the phosphorus content (ppm), and R the universal gas constant (kJ mol^{-1} K^{-1}).

4.5. CONCLUSIONS

A significant amount of literature on SSP kinetics has been accumulated during the last 50 years and may be classified to experimental and theoretical data, including simulation aspects. In both cases the scope is to assess the importance of each possible process factor, to suggest a relative mechanism, to predict SSP behavior, and finally, to optimize the overall procedure. In the chapter, emphasis is placed on progress in determining the intrinsic rate constants of associated main chemical reactions. The corresponding kinetic models are either based on Flory theory, where the rate expressions are in terms of end-group concentrations, or on a power-law description of the rate with respect to reaction time. Recent advances in modeling and simulation on a larger scale of the various physical and chemical processes occurring within a SSP reactor are reported. The goal is to describe the dynamic evolution of all chemical species within the particle and its surroundings, and to assess its dependence on basic process variables.

REFERENCES

1. Gaymans R, Venkatraman V, Schuijer J. Preparation and some properties of nylon 42. *J. Polym. Sci. Polym. Chem. Ed.* 1984;22:1373–1382.

2. Roerdink E, Warnier J. Preparation and properties of high molar mass nylon-4,6: a new development in nylon polymers. *Polymer.* 1985;26:1582–1588.

3. Vouyiouka S, Karakatsani E, Papaspyrides C. Solid state polymerization. *Prog. Polym. Sci.* 2005;30(1):10–37.

4. Flory P. *Principles of Polymer Chemistry*. Cornell University Press, Ithaca, NY, 1975, pp. 75–83, 317–325.

5. Griskey R, Lee B. Thermally induced solid-state polymerisation in nylon 66. *J. Appl. Polym. Sci.* 1966;10:105–111.

6. Yao K, McAuley K, Berg D, Marchildon E. A dynamic mathematical model for continuous solid-phase polymerisation of nylon 6,6. *Chem. Eng. Sci.* 2001;56: 4801–4814.

7. Shi C, DeSimone J, Kiserow D, Roberts G. Reaction kinetics of the solid-state polymerization of poly(bisphenol A carbonate) facilitated by supercritical carbon dioxide. *Macromolecules.* 2001;34:7744–7750.

8. Ma Y, Agarwal U, Sikkema P, Lemstra P. Solid-state polymerisation of PET: influence of nitrogen sweep and high vacuum. *Polymer.* 2003;44:4085–4096.

9. Shi C, Gross S, DeSimone J, Kiserow D, Roberts G. Reaction kinetics of the solid state polymerisation of poly(bisphenol A carbonate). *Macromolecules.* 2001;34: 2062–2064.

10. Kim T, Lofgren E, Jabarin S. Experimental study of the reaction kinetics and properties. *J. Appl. Polym. Sci.* 2003;89:197–212.

11. Chen S, Chen F. Kinetics of polyesterification: III. Solid-state polymerization of polyethylene terephthalate. *J. Polym. Sci. A.* 1987;25:533–549.

12. Korshak V, Frunze T. *Synthetic Hetero-Chain Polyamides*. IPST, Jerusalem, Israel, 1964, pp. 90–95, 119, 128.

13. Almonacil C, Desai P, Abhiraman A. Morphological consequences of interchange reactions during solid state polymerization in oriented polymers. *Macromolecules.* 2001;34:4186–4199.

14. Papaspyrides C. Solid-state polyamidation of dodecamethylenediammonium adipate. *Polymer.* 1984;25:791–796.

15. Papaspyrides C. Solid-state polyamidation of nylon salts. *Polymer.* 1988;29.114–117.

16. Papaspyrides C. Solid state polyamidation processes. *Polym. Int.* 1992;29:293–298.

17. Oya S, Tomioka M, Araki T. Studies on polyamides prepared in the solid state: I. Polymerisation mechanism. *Kobunshi Kagaku.* 1966;23(254):415–421.

18. Katsikopoulos P, Papaspyrides C. Solid-state polyamidation of hexamethylenediammonium adipate: II. The influence of acid catalysts. *J. Polym. Sci. A.* 1994;32: 451–456.

19. Khripkov E, Kharitonov V, Kudryavtsev G. Some features of the polycondensation of hexamethylene diammonum adipinate. *Khim. Volokna.* 1970;6:63–65.

20. Brown C. Further refinement of the crystal structure of hexamethylenediammonium adipate. *Acta Crystallogr.* 1966;21:185–189.

21. Papaspyrides C. Solid state polyamidation. In: *The Polymeric Materials Encyclopedia*, Salamone JC, ed. CRC Press, Boca Raton, FL, 1996, pp. 7819–7831.

22. Volokhina A, Kudryavtsev G, Skuratov S, Bonetskaya A. The polyamidation process in the solid state. *J. Polym. Sci.* 1961;53:289–294.

23. Bagramyants B, Bonetskaya A, Yenikolopyan N, Skuratov. The reason for the high temperature coefficient in solid state polycondensation of ω-aminoacids. *Vysokomol. Soyed.* 1966;8(9):1594–1598.

24. Khripkov E, Kharitonov V, Kudryavtsev G. Solid-phase catalytic polycondensation: I. Polycondensation of hexamethylenediammonium adipate. *Khim. Volokna.* 1970;6:63–65.

25. Vouyiouka S, Papaspyrides C, Weber J, Marks D. Polyamide solid state polymerization: evaluation of pertinent kinetic models. *J. Appl. Polym. Sci.* 2005;97(2): 671–681.

26. Gaymans R. Polyamidation in the solid phase. In: *Polymerization Reactors and Processes*, Henderson JN, Bouton CT, eds. ACS Symp. Ser. American Chemical Society, Washington, DC, 1979.

27. Srinivasan R, Desai P, Abhiraman A, Knorr R, Solid-state polymerisation vis-à-vis fiber formation of step-growth polymers: I. Results from a study of nylon 66. *J. Appl. Polym. Sci.* 1994;53:1731–1743.

28. Jabarin S, Lofgren E, Solid state polymerization of poly(ethylene terephthalate): kinetic and property parameters. *J. Appl. Polym. Sci.* 1986;32:5315–5335.

29. Duh B. Reaction kinetics for solid-state polymerisation of poly(ethylene terephthalate). *J. Appl. Polym. Sci.* 2001;81:1748–1761.

30. Duh B. Semiempirical rate equation for solid state polymerization of poly(ethylene terephthalate). *J. Appl. Polym. Sci.* 2002;84:857–870.

31. Zimmerman J. Equilibria in solid phase polyamidation. *J. Polym. Lett.* 1964;2: 955–958.

32. Zimmerman J, Kohan M. Nylon: selected topics. *J. Polym. Sci. A.* 2001;39: 2565–2570.

33. Srinivasan R, Almonacil C, Narayan S, Desai P, Abhiraman A. Mechanism, kinetics and potential morphological consequences of solid-state polymerization. *Macromolecules.* 1998;31:6813–6821.

34. Walas S. Uncatalysed heterogenous reactions. In: *Reaction Kinetics for Chemical Engineers*. McGraw-Hill, New York, 1959, pp. 126–130.

35. Fujimoto A, Mori T, Hiruta S. Polymerization of nylon-6,6 in solid state. *Nippon Kagaku Kaishi.* 1988;3:337–342.

36. Gaymans R, Amirtharaj J, Kamp H. Nylon 6 polymerization in the solid state. *J. Appl. Polym. Sci.* 1982;27:2513–2526.

37. Mallon F, Ray H. Modeling of solid state polycondensation: I. Particle models. *J. Appl. Polym. Sci.* 1998;69:1233–1250.

38. Kaushik A, Gupta S. A molecular model for solid-state polymerisation of nylon 6. *J. Appl. Polym. Sci.* 1992;45:507–520.

39. Kulkarni M, Gupta S. Molecular model for solid-state polymerization of nylon 6: II. An improved model. *J. Appl. Polym. Sci.* 1994;53:85–103.

40. Li L, Huang N, Tang Z, Hagen R. Reaction kinetics and simulation for the solid-state polycondensation of nylon 6. *Macromol. Theory Simul.* 2001;10:507–517.

41. Xie J. Kinetics and simulation of solid state polymerisation for nylon 6. *Ind. Eng. Chem. Res.* 2001;40:3152–3157.

42. Yao K, McAuley K. Simulation of continuous solid-phase polymerisation of nylon 6,6: II. Processes with moving bed level and changing particle properties. *Chem. Eng. Sci.* 2001;56:5327–5342.

43. Li L, Huang N, Liu Z, Tang Z, Yung W. Simulation of solid-state polycondensation of nylon-66. *Polym. Adv. Technol.* 2000;11:242–249.

44. Yao K, Ray W. Modeling and analysis of new processes for polyester and nylon production. *AIChE J.* 2001;47(2):401–412.

45. Nelson W. *Nylon Plastics Technology.* Published for the Plastics and Rubber Institute, Newnes-Butterworths, Sevenoaks, UK, 1976, pp. 214–215.

46. Papaspyrides C, Vouyiouka S, Bletsos I. New process for the production of dry hexamethylenediammonium adipate. *J. Appl. Polym. Sci.* 2003;88(5):1252–1257.

47. Bletsos I, Papaspyrides C. Preparation of low-water-content, diamine–dicarboxylic acid monomer salts (E.I. du Pont de Nemours & Company). U.S. Patent 5,801,278, 1998.

48. Papaspyrides C, Vouyiouka S, Bletsos I. New aspects on the mechanism of the solid state polyamidation of PA 6,6 salt. *Polymer.* 2006;47(4):1020–1027.

49. Chang T. Kinetics of thermally induced solid state polycondensation of poly(ethylene terephthalate). *Polym. Eng. Sci.* 1970;10(6):364–368.

50. Chang S, Sheu M, Chen S. Solid-state polymerization of poly(ethylene terephthalate). *J. Appl. Polym. Sci.* 1983;28:3289–3300.

51. Xie J. Kinetics of the solid-state polymerization of nylon-6. *J. Appl. Polym. Sci.* 2002;84:616–621.

52. Chen F, Griskey R, Beyer G. Thermally induced solid state polycondensation of nylon 66, nylon 6–10 and polyethylene terephthalate. *AIChE J.* 1969;15(5):680–685.

53. Vouyiouka S, Papaspyrides C, Weber J, Marks D. Solid state post-polymerization of PA 6,6: the effect of sodium 5-sulfoisophthalic acid. *Polymer* 2007;48(17): 4982–4989.

54. Po R. In: *The Polymeric Materials Encyclopedia*, Salamone JC, ed. CRC Press, Boca Raton, FL, 1996, pp. 6100–6106.

55. Muller H, Rossbach V. Dyeing of anionically-modified polyamide fibers with cationic dyes: I. Uptake of the dye and its binding. *Text. Res. J.* 1977;47:44–51.

56. Flamand C, Pensacolz F. Basic dyeable acid dye resistive polyamides containing terminal aryl disulfonated groups (Monsanto Company). U.S. Patent 3,542,743, 1970.

57. Studholme M. Acid dye stain-resistant fiber-forming polyamide composition containing masterbatch concentrate, containing reagent and carrier (Prisma Fibers). U.S. Patent 6,117,550, 2000.

58. Caldwell J. Sulfonated polyamides (Eastman Kodak Company). U.S. Patent 3,296,204, 1967.

59. Gaona O, Kint D, Ilarduya A, Alla A, Bou J, Guerra S. Preparation and hydrolytic degradation of sulfonated poly(ethylene terephthalate) copolymers. *Polymer.* 2003;44:7281–7289.

60. Schiavone R. Solid state polymerization (SSP) of low molecular weight poly(ethylene terephthalate) (PET) copolyesters compared to conventional SSP of PET. *J. Appl. Polym. Sci.* 2002;86:230–238.

61. Eisenberg A, Hird B, Moore R. A new multiplet-cluster model for the morphology of random ionomers. *Macromolecules.* 1990;23:4098–4107.

62. Mizerovskii L, Kuznetsov A, Bazarov Yu, Bykov A. Kinetics of reversible repolyamidation of polycaprolactam below the melting point of the polymer. *Vysokomol. Soyed.* 1983;A25(5):1056–1062.

63. Mallon F, Ray W. A comprehensive model for nylon melt equilibria and kinetics. *J. Appl. Polym. Sci.* 1998;69:1213–1231.

64. Vouyiouka S, Papaspyrides C, Pfaendner R. Catalyzed solid state polyamidation. *Macromol. Mater. Eng.* 2006;291(12):1503–1512.

65. Pfaendner R, Fink J, Simon D, Papaspyrides C, Vougiouka S. Process for the preparation of polyamides in the presence of a phosphonate (Ciba Specialty Chemicals Lampertheim GmbH). WIPO Patent WO 2007/006647, 2007, p. 36.

5

CATALYSIS IN SOLID STATE POLYMERIZATION PROCESSES

RUDOLF PFAENDNER

Ciba Lampertheim GmbH, Lampertheim, Germany

5.1. INTRODUCTION

A number of factors affect the solid state polycondensation (SSP) of polymers (e.g., polyamides or polyesters), such as the reaction temperature and time, particle size, crystallinity and porosity, polymer composition, initial molecular weight, polymer end groups, potential side reactions, and reaction medium (e.g., reduced pressure, inert gas). In addition, as SSP usually involves condensation of the polymer end groups as well as transesterification–transamidation reactions, it is sensitive to catalysts or additives in general. In a positive way, these additives

Solid State Polymerization, Edited by Constantine D. Papaspyrides and Stamatina N. Vouyiouka
Copyright © 2009 John Wiley & Sons, Inc.

can accelerate the SSP process and even improve some of the end-product properties. A number of chemical compounds, which are discussed below in detail, have been described in the literature to act as catalysts and/or to improve the resulting product quality.

Although some SSP review articles refer to selected catalyst categories [1–3], in this chapter we provide a comprehensive overview of the topic, including open as well as patent literature. Only publications where SSP is mentioned explicitly are covered, whereas any use of catalysts in polycondensation melt reactions is generally excluded despite potential overlapping. Moreover, solid state processes not related to polycondensation reactions (e.g., modification, branching, grafting of polyolefins in the solid state) are not considered in this review.

Regarding polymer type, polyesters such as poly(ethylene terephthalate) (PET), poly(butylene terephthalate) (PBT), and polyamides [e.g., polyamide 6 (PA 6) and polyamide 66 (PA 66)] are the main focus of SSP publications; however, related polymer structures and copolymers are mentioned in some cases. The diversity of experimental parameters and materials type with regard to origin, synthetic process, molecular weight, end-group structures, and concentrations derived from many years of publications renders a comparison of results from different authors rather difficult. In particular, industrial SSP processes use mainly polymers or lower-molecular-weight prepolymers as starting materials. SSP processes from monomer salts are described in the area of polyamides; however, these are even more complex, as intermediate melt states could be involved. In addition, catalyst structures might have been modified through the incorporation process or through previous reaction steps (e.g., reacted in the polymer as an end group). Furthermore, it is often not possible to analyze the SSP effect separately, as residual catalysts from previous reaction steps are often not known or analyzed and might result in a mixed effect. Therefore, general statements about the relative efficacy of single additives are necessarily weak.

Potential additives to improve the SSP process can be split into three classes:

1. Catalysts (metals, acids, bases)
2. Reactive additives (chain extenders, branching agents)
3. Inert additives (these influence the process but do not participate in the chemical reaction)

We should mention, however, that some additives might have an inhibition effect as well [e.g., titanium dioxide (TiO_2)] [1], or might reduce the catalytic activity [e.g., phosphites in antimony oxide (Sb_2O_3)-catalyzed SSP]. Accelerating the SSP process presents, first, cost-effective benefits (e.g., energy savings, higher throughput), and second, a shorter reaction time or lower reaction temperature, which improves the polymer quality through reducing side reactions (coloration, branching, gel formation, diethylene glycol structures in PET), since milder reaction conditions can be used.

Dispersion of the additive or catalyst is an essential parameter of SSP. An additive can be introduced as a concentrate; as a solution before polymerization; by improved diffusion via water vapor, slurry, solvent, and evaporation; in the recirculating gas flow [4]; and via reactive carrier [5]. The additive or catalyst can be added to monomers, prepolymers, or polymers at different stages within the manufacturing process. It might be mixed with monomers or prepolymers at an early stage before the reactor is filled, or added to the reactor directly. Alternatively, the catalyst or additive can be added in the course of the melt polycondensation process or before the particle forming–crystallization step, or finally, dry mixed with the solid polymer particle before the mixture is submitted to SSP.

5.2. CATALYSTS IN POLYESTER SOLID STATE POLYMERIZATION PROCESSES

5.2.1. Metal-Type Catalysts

Poly(ethylene terephthalate) (PET) molecules in either the melt state or the solid state are capable of performing intra- or intermolecular interchange reactions between the hydroxyl groups and ester linkages or between two ester linkages. An increase in the molecular weight is possible only if an intermolecular reaction takes place between the hydroxyl end group and a terminal ester group under elimination of ethylene glycol. As stated in the early days of SSP research, any catalyst active in the polycondensation of PET in the melt state should also be efficient in the solid state as long as the reaction temperature is high enough to activate the terminal hydroxyl (and/or carboxyl) groups or those ester linkages close to the chain ends [6].

Antimony oxide (Sb_2O_3) is still the most widely used catalyst in the polycondensation of PET and therefore is naturally present for the SSP processes when PET is, for example, used in bottle or tire cord applications. Alternatively, antimony glycolate or antimony acetate may be used. The simplified mechanism proposed for antimony oxide catalysis is, first, the formation of an antimony glycolate and its reaction with a free ethylene glycol end group of the polymer under release of ethylene glycol [8]. The polymeric bound antimony reacts with an OH end group and forms an ester linkage while the antimony glycolate is regenerated (Fig. 5.1).

Sb_2O_3 offers a high catalytic activity and chemical stability, does not form color, and does not catalyze side reactions. A specific feature of antimony, which is contrary to that of other transesterification catalysts, is that the presence of carboxyl groups is not detrimental for its activity [7]. In the industrial PET process, Sb_2O_3 is dissolved in ethylene glycol to form antimony glycolate before it is added to the polycondensation reactor. The optimized concentration for the melt process (about 150 to 300 ppm) is not changed when going into the SSP (i.e., no additional catalyst is added). However, if stabilizers such as phosphorous acids or phosphites are used to avoid color formation and to improve the thermal

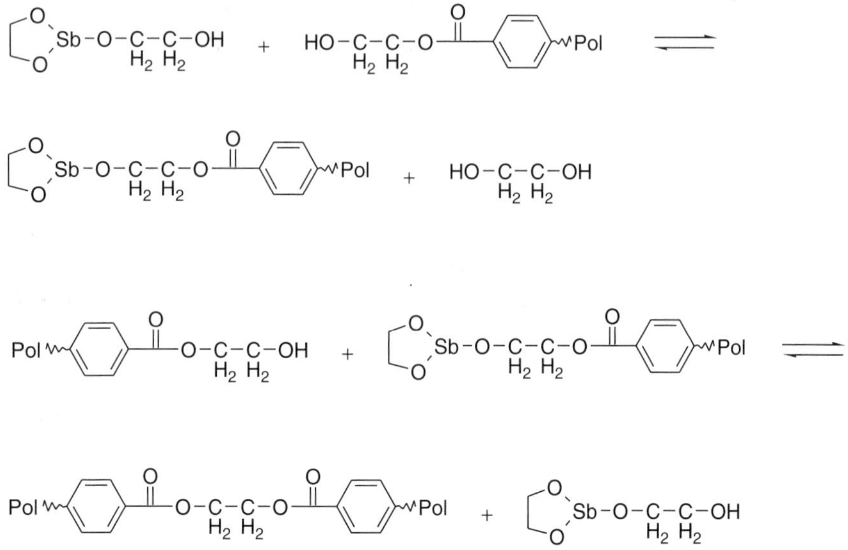

Fig. 5.1. Mechanism of PET polycondensation catalyzed by antimony oxide.

stability, the active concentration of Sb is reduced through reduction of Sb(III) to metallic Sb(0), which is seen as an inert molecule.

The SSP rate and rate constant increase with increasing Sb concentration; however, a critical content is found for the rate constant. It is believed that a macromolecule with an antimony glycolate end group shows higher reactivity but a lower mobility or diffusion rate than those of a normal OH or COOH end group. With increasing Sb concentration, the activation energy is decreasing while the frequency factor of the Arrhenius equation is reduced. A maximum for the SSP rate constant at 150 ppm is published [7], being in the range of contents used on an industrial scale. On the other hand, the maximum achievable molecular weight of PET catalyzed by Sb_2O_3 seems to be limited and is in the range of 40,000 g mol^{-1} ($\overline{M_n}$) at a concentration of 2000 ppm Sb_2O_3 when the starting prepolymer is synthesized by manganese diacetate [9].

Apart from Sb_2O_3 and other soluble antimony compounds or modified antimony derivatives [10], a number of other metals have been evaluated in the melt or in model reactions based on Ti, Sn, Mn, Zn, and Pb [8]; however, there are no data on relevant SSP processes. Germanium oxide (GeO_2) has a certain industrial importance to produce PET resins with minimized coloration, and it can be expected that its influence on the SSP is analogous to that of Sb_2O_3.

For very high-molecular-weight polyesters ("superpolyesters"), titanium catalysts such as sodium titanium butylate are proposed [10]; however, other catalysts (zinc acetate/Sb_2O_3 or cobalt acetate) allow the synthesis of high-molecular-weight polyesters as well [6]: A molecular weight of PET as

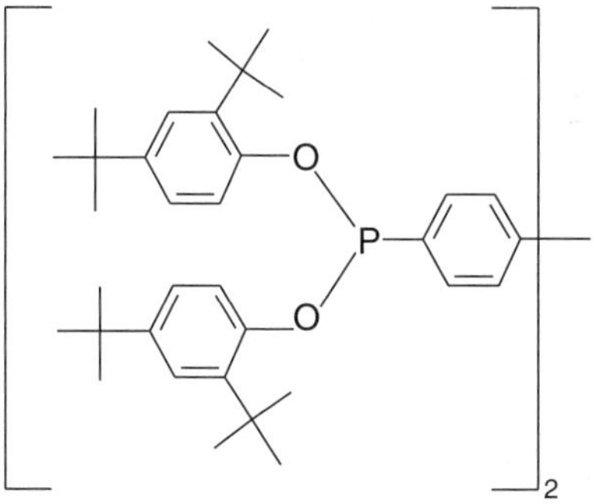

Fig. 5.2. Chemical structure of Ciba Irgafos P-EPQ: Tetrakis(2,4-di-*tert*-butylphenyl) [1,1-biphenyl]-4,4′-diylbisphosphonite.

high as 65,000 g mol^{-1} (zinc acetate/Sb$_2$O$_3$) or 120,000 g mol^{-1} (cobalt acetate) is analyzed after SSP.

More recent publications claim the use of mixed metal catalysts or combinations of traditional catalysts with co-additives. For example, combinations of antimony oxide (or germanium oxide, tetrabutyl titanate) with heterogeneous cocatalysts such as hydrotalcites (5 to 50 ppm) and stabilizers (e.g., P-EPQ, Fig. 5.2) [11] have been investigated. These combinations result in a twofold increase in the intrinsic viscosity (IV, Table 5.1) after SSP versus control without increasing the metal content in the polymer or negatively influencing the processing properties.

Antimony–tin combinations, such as 130 ppm Sb (e.g., as antimony oxide) and 80 ppm Sn (e.g., as monobutyltin tris (2-ethylhexoate)), result in rapid melt polymerization because of the tin catalyst and a small increase in the SSP rate, but in a reduced acetaldehyde formation in subsequent processing steps [12].

TABLE 5.1. Delta IV [Δ(IV) = (IV)$_t$ − (IV)$_0$] Values During Polyester SSP at 210°C[a]

		Δ (IV) (dL g^{-1})		
t_{SSP}(h)	Control (260 ppm Sb)	200 ppm Sb + 50 ppm HT + 100 ppm PEPQ	260 ppm Sb + 25 ppm HT + 150 ppm PEPQ	260 ppm Sb + 50 ppm HT + 100 ppm PEPQ
2	0.058			
3	—	0.112	0.125	0.115
4	0.116			
6	—	0.202	0.204	0.215
7	0.173			

[a]The effect of Sb$_2$O$_3$ as catalyst in combination with hydrotalcite (HT) and stabilizer (Ciba Irgafos PEPQ).

TABLE 5.2. Effect of Zinc-p-Toluenesulfonate as Catalyst During PET SSP^a

Catalyst	Catalyst Metal (ppm)	IV_0 (dL g^{-1})	IV_{final} (dL g^{-1})
Antimony glycolate	280	0.206	0.690
Zinc-p-toluenesulfonate	100	0.216	1.440
Germanium oxide	100	0.189	0.639

^{a}At 225°C for 24 h.

In the SSP of thermoplastic elastomer copolyether esters, an optimized ratio of titanate catalyst and alkaline earth metal cocatalyst results in process improvement: 25% of time saving is achieved at 190°C by a selected ratio of Ti/Mg (1 : 1) accomplished with titanium tetrabutoxide (1000 ppm) and magnesium acetate (630 ppm) [13].

Polyaryl esters have been synthesized by the use of dibutyltin dilaurate and titanium tetraisopropoxide and dimethylaminopyridine. The titanium catalyst (0.5%) results in the best performance, where an increase in the inherent viscosity of 0.47 dL g^{-1} from 0.21 dL g^{-1} versus 0.26 dL g^{-1} (uncatalyzed SSP at 230°C, 6 h, vacuum) is claimed [14].

To reduce the metal content and to use a more environmentally friendly metal than antimony, zinc-p-toluenesulfonate has been identified as an efficient catalyst. In the range 75 to 100 ppm of zinc, the IV increase of a polyester prepolymer at 225°C is decisively higher than by using Sb_2O_3 or GeO_2 as catalyst (Table 5.2) [15]. Furthermore, it is claimed that side reactions producing diethylene glycol units and color formation are minimized.

The synthesis of liquid crystalline polyesters or polyesteramides from 4,4′-biphenol and terephthalic acid/2,6-naphthalene dicarboxylic acid via SSP at temperatures of 320 and 340°C showed the beneficial activity of an alkali metal cation in the range 5 to 1000 ppm (e.g., potassium 4-hydroxybenzoate). At 15 ppm potassium, the melt viscosity after SSP is doubled and tripled at a concentration of 25 ppm [16]. In the presence of an alkali metal cation, the color of the liquid crystalline polyester is lighter and the melting point is higher.

5.2.2. Phosphorus- and Sulfur-Based Catalysts

It has been claimed that similar to the effect of zinc-p-toluenesulfonate, benzenesulfonic acid as a completely metal-free system can act as an SSP catalyst [4]. By adding benzenesulfonic acid and, for comparison, titanium and antimony to catalyst-free PET prepolymer via a solvent process, it could be shown that benzenesulfonic acid results in the highest IV increase within the combinations tested (Table 5.3). The catalysts are introduced via solution or preferably via an inert gas stream and must therefore be volatile enough for this process.

An additional attractive class of SSP accelerators are the hydroxyphenyl-alkylphosphonic esters [17], which were originally developed for the SSP of polyester recyclates but are also found to be active in virgin polyesters and

TABLE 5.3. Benzenesulfonic Acid as Catalyst Compared to Ti and Sb During PET SSP at 240°C

	IV (dL g^{-1})			
t_{SSP}(h)	Control	Ti(O-i-Pr)$_4$[a]	C$_6$H$_5$SO$_3$H[b]	Sb(OC$_4$H$_9$)$_3$[c]
0	0.165	0.162	0.159	0.166
1	—	—	0.263	0.181
2	—	0.285	0.511	0.313
4	0.171			
6	—	0.419		
24	0.215	0.625	0.988	0.457

[a]Titanium tetraisoprionate.
[b]Benzenesulfonic acid.
[c]Antimony tributoxide.

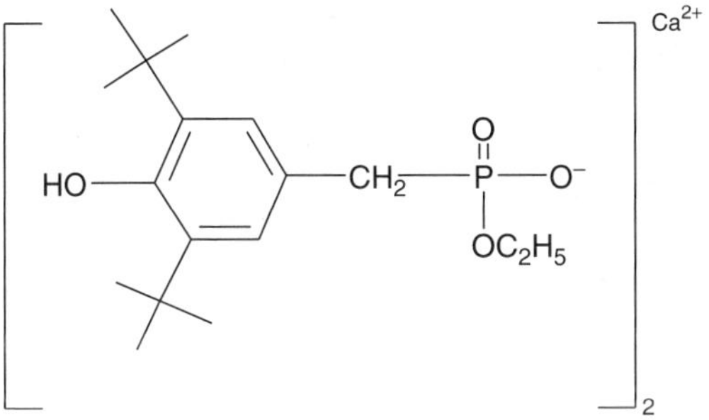

(a)

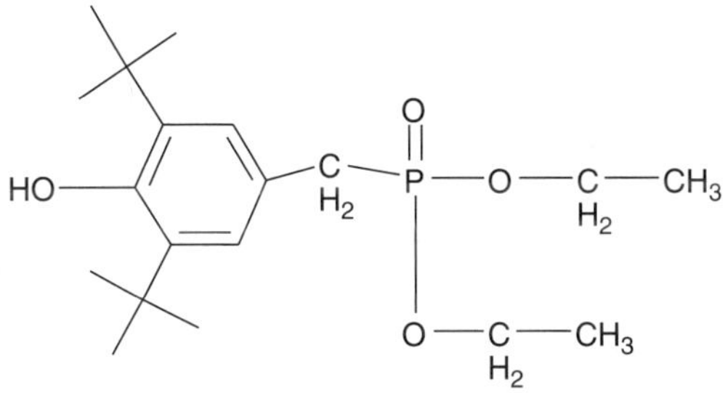

(b)

Fig. 5.3. Chemical structures of hydroxyphenylalkylphosphonic esters: (a) phosphonate 1 (Ciba Irgamod 195), Calciumdiethylbis[[[3,5-bis(1,1-dimethylethyl)-4-hydroxyphenyl]methyl]phosphonate]; (b) phosphonate 2 (Ciba Irgamod 295) diethyl [[3,5-bis(1,1-dimethylethyl)-4-hydroxyphenyl] methyl]phosphonate.

TABLE 5.4. Hydroxyphenylalkylphosphonic Esters as Cocatalyst During PET Recyclate SSP[a]

Cocatalyst	IV_0 (dL g^{-1})	IV_{final} (dL g^{-1})
Control	0.44	0.89
0.25% phosphonate 1	0.47	1.08
0.25% phosphonate 2	0.51	1.18

[a] At 230°C for 23 h.

polyamides, as discussed below. Their combinations with stabilizers (e.g., phenolic antioxidants) and various reactive polyfunctional co-additives are also of interest for adjusting polymer properties. The hydroxyphenylalkylphosphonic esters (Fig. 5.3) act as cocatalysts in accelerating the SSP process of recycled PET (Table 5.4). In these experiments, the cocatalyst is incorporated via extrusion, and 0.25% of the phosphonic esters stabilizes the recyclate against thermal degradation at the extrusion step and accelerates the SSP process. Moreover, at extended SSP times, a much higher final molecular weight can be achieved in the presence of the co-catalyst [e.g., 82,000 g mol^{-1} (IV = 2.25 dL g^{-1}) versus 69,000 g mol^{-1} (IV = 1.95 dL g^{-1}, uncatalyzed SSP)] [18]. The mechanical properties of the PET recyclate (e.g., elastic modulus, tensile impact strength, notched impact strength) are influenced and enhanced as expected at the higher molecular weight. Properties measured on monofilaments derived from catalyzed SSP upgraded recyclate have shown a performance identical to that of the virgin material.

Addition of hydroxyphenylalkylphosphonic esters as cocatalyst to virgin PET results in productivity improvement and better color, and can therefore replace other phosphorus-based stabilizers used for color improvement. At the same time, the concentration of cobalt, which is often used to mask yellowing of PET, can be reduced [19]. The effect of hydroxyphenylphosphonic esters on the polymerization rate of PET was confirmed independently. Furthermore, it was found that the content of cyclic oligomer and cyclic trimer is somewhat reduced when phophonate 1 is used at the polycondensation stage together with phosphoric acid, antimony acetate, and cobalt acetate [20].

A related chemical approach to the hydroxyphenylalkylphosphonic esters described is the incorporation of carboxyphosphonic acids in the PET chain, resulting in a long-chain branching and a permanent thermal and thermooxidative stabilization of a highly brilliant PET [21] with some minor increase in IV when the phosphorus compound (120 to 300 ppm) is present. The preferred carboxyphosphonic acid is carboxyethylenephosphonic acid (Fig. 5.4), which is added at the early stage of synthesis to guarantee incorporation in the polyester chain.

The SSP of PET, and especially of PBT, can be accelerated by the addition of phosphites such as triphenylphosphite or tri-o-cresylphosphite. Addition of 2% of phosphite results in a viscosity number of 145 versus 107 without phosphite at 180°C after 10 hours of SSP [22]. As expected, higher reaction temperatures and increased SSP time result in higher viscosity numbers, contrary to the increase in triphenylphosphite concentration, where no difference is seen within the tested loading band of 1 to 3%. Finally, phosphorous acid is useful to catalyze the synthesis of PET from terephthalic acid and ethylene oxide in the solid phase [23]. This type of process using the SSP of monomers is better known from polyamide, as described below.

5.3. CATALYSTS IN POLYAMIDE SOLID STATE POLYMERIZATION PROCESSES

Compared to polyesters, the literature of polyamide SSP shows numerous references dealing with the synthesis of polymers from the monomer salt in a solid state process. The autocatalytic process takes place at a temperature range close to the melting point of the ω-amino acids (aminoenanthoic, aminopelargonic, aminoundecanoic acid), or starts at about 235°C in the case of the hexamethylene-diamine salts of adipic acid. The need of additional catalysts for this process was recognized early, since the reaction rate of the solid process is not high enough and is significantly low compared to the melt or solvent processes. The addition of catalysts revealed clearly that a number of basic, acidic, or neutral molecules can speed up the SSP process [24]. For example, the addition of 2% MgO completes the polycondensation of aminoenanthoic acid at 184°C within less than 120 minutes, whereas without catalyst after 360 minutes a conversion of only 20% is achieved. According to the effectiveness of catalysts, the following list has been compiled: 1% boric acid, 0.2% magnesium oxide > 0.5% ammonium oxalate > 0.5% zinc acetate > 0.2% sodium carbonate > 1% camphorsulfonic acid > 0.6% acetic acid > 0.5% ammonium sulfate > 1% stanneous chloride.

In the synthesis of polyoxamides in either the melt or the solid phase, the following suitable catalysts have been identified: $SbF_3/As_2O_3 \gg GeO_2 > Sb_2O_3 > Bi_2O_3 \sim PbO$ [25]. The performance differences of the catalysts are explained by steric and/or solubility effects and depend on their ability to form coordination complexes with the oxalic unit. The synthesis of poly(2-methylhexamethyleneoxamide) is carried out at 245°C in the solid phase and high molecular weights, and good heat stability of the polymer is achieved. Nevertheless, the molecular weight is reduced again at extended heating times.

In a profound study of the SSP of dodecamethylenediammonium adipate in the range of 130°C or hexamethylenediammonium adipate at about 140°C, the performance of boric acid versus sulfuric acid and phosphoric acid on one side and a

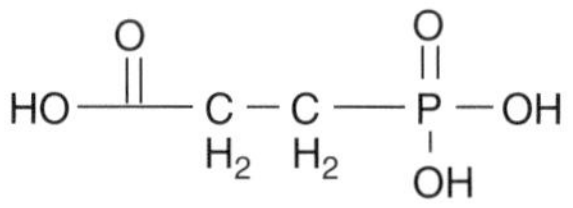

Fig. 5.4. Chemical structure of carboxyethylenephosphonic acid.

large series of group IIA [$Ca(OH)_2$, $Ba(OH)_2$, $Sr(OH)_2$], group VA (Na_2HAsO_4, Na_3AsO_3, As_2O_3, Sb_2O_3, Bi_2O_3), and group VIII ($NiSO_4$, Co_2O_3, Fe_2O_3, Ni_2O_3) metal compounds on the other side was proven [26–30]. The starting materials were synthesized by a coprecipitation technique or by addition of a dispersion agent to guarantee homogeneous distribution of the catalyst. It was found that metal hydroxides are more efficient than metal oxides. Na_2CO_3, $NaHSO_4$, and SiO_2 seem to be inactive, and sodium-5-sulfoisophalic acid slowed down the reaction [31]. Boric acid proved to be the best catalyst for the polyamidation process of hexamethylenediammonium adipate, where a more than tenfold increase in the conversion rate is achieved at 138.5°C compared to the uncatalyzed process [29]. Furthermore, the polycondensation process, which is difficult to control due to the melting–solidification–agglomeration behavior, is easier to run. There is a critical catalyst concentration above which no further increase in the reaction rate is observed, and especially for boric acid, this content is found between 0.03 and 0.07 wt%. The catalyst favors easier removal of the water formed in the reaction (instead of an hydrated salt structure), avoiding water accumulation and inhibiting a transition of solid to the melt state, which are phenomena to be excluded in the process [30]. Finally, it can be stated that from today's point of view, boric acid is still the best solution for a solid state polyamidation process starting from the monomers.

Catalysts used for the polycondensation reaction of polyamides are reported to be beneficial as well for the SSP process of polyamide prepolymers. Therefore, mineral acids, such as sulfuric acid and phosphoric acid, and their salts, such as potassium dihydrogenphosphate and potassium hydrogen sulfate, are described as early examples in the patent literature [32,33]. The final molecular weight achieved in the process depends on the catalyst amount (e.g., the maximum viscosity after 24 h at 190°C of PA 6 is achieved when 1×10^{-4} mol phosphoric acid per mole of monomer is used). At higher and lower concentrations of phosphoric acid, a much lower molecular weight is reported.

The polycondensation of carboxylic acids and amines is autocatalyzed through the acid group. Therefore, the limiting step of the reaction is the acid diffusion. Whereas the autocatalyzed reaction is slowed down with reaction time, the H_3PO_4-catalyzed reaction continues undisturbed. This fact is explained through easier diffusion of H_3PO_4 compared to the polymer chain end at higher molecular weights [34]. For acceleration of the SSP, only a minor fraction of the polymer end-group concentration is needed in the form of the catalyst. Although phosphoric acid is an efficient polyamide catalyst, there is a high risk of reactor fouling due to salt deposits, which will influence the product quality. Therefore, it was proposed to introduce H_3PO_4 via a water vapor–phosphoric acid mixture, which results in improvement in the technical process [35], the preferred concentration of H_3PO_4 in the polyamide being in the range 50 to 100 ppm. The concentration of phosphoric acid can be optimized by adding a chain stabilizer (e.g., acetic acid [33] or aminocaproic acid [36]) to limit the viscosity and to obtain a constant polyamide quality. It might be an advantage to introduce the catalyst via a concentrate [37]. There, the preferred phosphorus concentration is between 20 and

100 ppm, and the SSP temperature of the polyamide-6 examples is performed at 170 or 180°C. By using the catalyst, the final relative viscosity (RV) under comparable conditions can be more than doubled (e.g., 9.9 versus 4.5 at 1 h /170°C, catalyst: H_3PO_4, [P] = 41 ppm). In addition, the process allows the synthesis of very high-molecular-weight polyamides (PA 6, RV = 14.2, [P] = 42 ppm, 30 h at 180°C).

Phosphoric acid (0.1 wt%) has also been used for catalyzing the SSP process of polyamide 46 [38]. Alternatively to H_3PO_4, H_3BO_3 [39] or *p*-aminobenzoic acid [40] are suggested for PA 6 SSP. SSP polyamide process improvements with other phosphorus-based chemical alternatives have been described by using phenylphosphinic acid [41], H_3PO_3, triphenylphosphite, tris(2,4-di-*tert*-butylphenyl)phosphite, ammonium phosphate, or sodium dihydrogen phosphate [37].

Some more complex phosphorus compounds, such as 2-(2'-pyridyl)ethylphosphonic acid have been selected, especially if an additional base is added to reduce thermal degradation, branching, and gel formation [42], which is a risk with phosphoric acid or other multifunctional phosphorus compounds. By choosing 2-(2'-pyridyl)ethylphosphonic acid (other alternative heterocyclic compounds are 4-morpholino, 1-pyrrolidino, 1-piperidino), no loss of catalytic activity with base addition is found, while thermal degradation is decreased and gelation avoided. The relative viscosity of PA 66 is more than doubled after 1 h at 160°C when the catalyst is used at 1×10^{-5} molar concentration in the presence of 2.2×10^{-5} $KHCO_3$. 2-(2'-Pyridyl)ethylphosphonic acid (Fig. 5.5) and, alternatively, potassium tolylphosphinate and sodium and manganese hypophosphite are preferably used by combination with an oxygen-free ultradry gas of low dew point [43].

The catalytic activity of the hydroxyphenylalkylphosphonic esters mentioned above (Fig. 5.3) have been tested more recently in polyamide SSP [44]. The catalysts were incorporated via extrusion in commercial PA 66, and the SSP was performed in a fixed-bed reactor under flowing nitrogen at 160 and 200°C with a reaction time of 4 h. The efficiency of the catalyst was clearly demonstrated, as the RV is tripled after 4 h at 200°C. Compared to an uncatalyzed SSP process, the RV is increased up to 57%. Increase in the SSP temperature speeds up the reaction. The catalytic activity was correlated to the structure of the phosphonate 1 or 2 and the additive mobility, whereas a partial incorporation in the polymer chain is assumed. The incorporation may take place in the compounding step and prevents losing the phosphonate from volatility reasons in the process (the volatility of phosphonate 1 is not an issue even if it is not

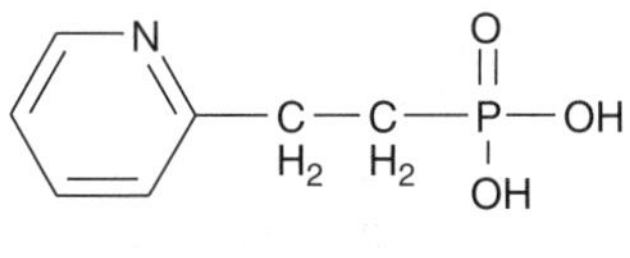

Fig. 5.5. Structure of 2-(2'-pyridyl)ethylphosphonic acid.

TABLE 5.5. Hydroxyphenylalkylphosphonic Esters as SSP Catalysts in PA 66 SSP

		RV				
Catalyst	Concentration (ppm)	Extruded Compound	160°C, 0 h	160°C, 4 h	200°C, 0 h	200°C, 4 h
Control	0	54.8	57.7	66.9	69.4	137.9
Melt compounding						
Phosphonate 1	1000	50.8	54.5	67.7	58.5	151.0
Phosphonate 1	5000	60.3	53.8	72.1	62.8	216.6
Phosphonate 2	1000	52.3	55.2	71.7	65.3	151.4
Phosphonate 2	5000	63.7	70.9	81.9	72.5	161.3
Addition at polycondensation stage						
Phosphonate 1	1000	45			56	185
Phosphonate 2	1000	54			68	262

incorporated in the polymer chain due to the salt structure). At the lower reaction temperature of 160°C, the phosphonate 2 performs better than the phosphonate 1, probably because of its higher mobility and easier attachment to the polymer chain.

The best performance has been achieved with phosphonate 1 at temperatures of 200°C (Table 5.5), where the slow mobility of the salt seems to be compensated through higher reaction temperatures. In addition to the catalytic activity, the multifunctional hydroxyphenylalkylphosphonic esters will serve as antioxidants in the process, with subsequent application preventing the polyamide from oxidation.

The catalytic effect of phosphonates 1 and 2 is even more pronounced when the phosphonate is added during synthesis. A hexamethylenediammonium salt, polycondensated in the presence of the phosphonates (1000 ppm), shows an increase in the RV at 200°C of 230% (phosphonate 1) and 288% (phosphonate 2) in comparison to the uncatalyzed process ([45]; Table 5.5, "addition at polycondensation stage"). Moreover, the color of the resulting polyamide is lighter, which is again an effect of the antioxidant group as part of the molecule [45].

5.4. REACTIVE ADDITIVES IN SOLID STATE POLYMERIZATION PROCESSES

A different possibility for increasing the molecular weight during SSP apart from catalysts is the use of low-molecular-weight reactive additives. These molecules, often referred to as *chain extenders*, react with the end groups of the polymer chain and build up the molecular weight. As the reactive additives are not necessarily bisfunctional, there is a possibility to introduce via multifunctional molecules branching points, which result in a very fast increase not only in the molecular weight but also in modified polymer structures and properties. Furthermore, a combination of reactive additives with suitable catalysts is feasible. Although this process can be applied to all SSP reactions, examples are shown primarily in the patent literature of manufacturing polyesters [17,46–49].

The use of tetracarboxylic acid dianhydride (cycloaliphatic, aliphatic, or aromatic) as such or in combination with a specified crystallization process allows polyester SSP to be carried out efficiently in a reasonable time frame, even at lower temperatures and also for the use of copolyesters of lower melting points. The reduction in SSP time achieved is claimed to be from 15 to 38 h to reach an intrinsic viscosity of 0.8 to 1.1 dL g^{-1} in a continuous plant down to 2 to 5 h [46]. Preferred anhydride is pyromellitic dianhydride (PMDA) or 3,3′,4,4′-benzophenonetetracarboxylic acid dianhydride (Fig. 5.6). The final molecular weight and the molecular-weight increase shown by the IV depend on the concentration of the dianhydride (e.g., PMDA) (Table 5.6). To guarantee homogeneous distribution of the PMDA and to avoid gel formation, it is incorporated in the form of a 20% master batch in crystallized PET powder via twin-screw extrusion at about 300°C before SSP. In addition to pure poly(ethylene terephthalate), copolyesters [e.g., copoly(ethylene terephthalate-isophthalate)] can be used [46–48], whereas the targeted application is the manufacture of mineral water bottles. Unfortunately, the catalyst with which the PET is produced is not mentioned in the patents cited; however, catalyst residues from the polycondensation should have an influence on the reaction as well.

Apart from dianhydrides, aromatic diisocyanates have been used for polyester SSP, where best results are achieved with diphenylmethane 4,4′-diisocyanate

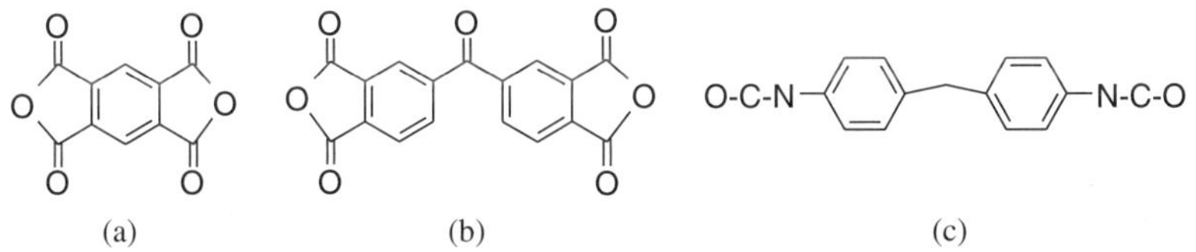

(a)   (b)   (c)

Fig. 5.6. Chemical structures of reactive additives in SSP processes: (a) pyromellitic dianhydride; (b) benzophenone-3,3′,4,4′-tetracarboxylic acid dianhydride; (c) 4,4′-diphenylmethane diisocyanate.

TABLE 5.6. SSP of PET in the Presence of Pyromellitic Dianhydride

PMDA (wt%)	T_{SSP} (°C)	t_{SSP} (h)	IV_0 (dL g^{-1})	IV_{final} (dL g^{-1})	Rate Constant, k (dL g^{-1} h^{-1})
0	202	5	0.57	0.635	0.013
0	216	5	0.57	0.685	0.023
0.3	202	5	0.62	0.794	0.0348
0.45	202	5	0.62	0.885	0.053
0.6	202	5	0.62	1.16	0.108
0.1	216	5	0.62	0.78	0.032

(Fig. 5.6, 0.3 to 0.5%) [50]. An intrinsic viscosity of 2.06 dL g^{-1} after 105 min at 245°C in a fluid-bed reactor is found (control IV = 1.20 dL g^{-1}) from a PET prepared in the presence of antimony oxide as catalyst. As expected, the concentration of free carboxylic end groups of the polyester is very low after reaction with the diisocyanate compared to the control sample without diisocyanate.

To avoid the use of diisocyanate itself, addition of polyurethanes to PET has been proposed, whereas under SSP conditions the polyurethane depolymerizes and produces isocyanate intermediates in situ [51], which react with carboxylic or alcoholic groups of the polyester. The solid state time is reduced and high-molecular-weight material is produced. Examples show the SSP of PBT in the presence of 3% polyurethane and 0.25% antioxidant at 190°C. After 6 h, the mean-number average molecular weight achieved is 20,389 g mol^{-1} in the presence of polyurethane in contrast to only 16,564 g mol^{-1} in its absence.

Another approach to adding reactive molecules before SSP has been published for branched polyesters [52]. These PBT polyesters are synthesized with small amounts of polyols (e.g., trimethylolpropane, pentaerythritol) or tricarboxylic acids (e.g., trimesic acid). By adding bisphenol A–polycarbonate (1%) to this branched polyester, the solid state polymerization time at 207°C to achieve material suitable for blow molding is reduced considerably.

Combinations of hydroxyphenylalkylphosphonic esters and several chemical classes may be used for an accelerated molecular-weight increase and/or branching or cross-linking in the solid state process. As reactive additives, epoxides, oxazolines, oxazolones, oxazines, isocyanates, anhydrides, acyllactams, maleimides, alcohols, carbodiimides, and esters have been suggested [17].

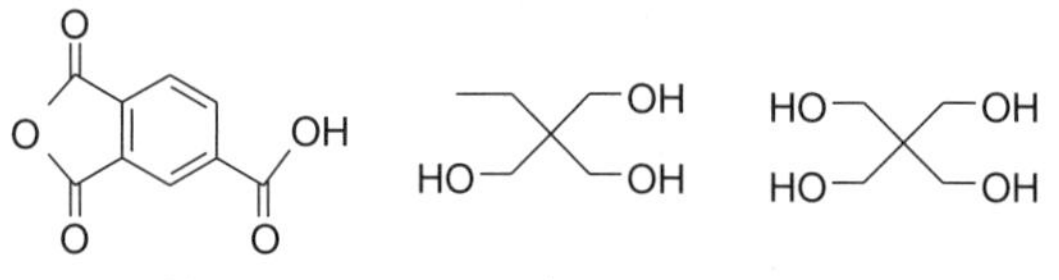

(a)   (b)   (c)

Fig. 5.7. Chemical structures for manufacturing branched polyesters: (a) trimellitic anhydride; (b) trimethylolpropane; (c) pentaerythrithol.

TABLE 5.7. SSP of Reactive Additives in Combination with Selected Catalysts

| | | | IV$_{final}$ (dL g^{-1}) | |
| | Reactive | | $T_{SSP} = 220°C$, | $T_{SSP} = 220°C$, |
Catalyst	Additive	IV$_0$ (dL g^{-1})	$t_{SSP} = 8$ h	$t_{SSP} = 16$ h
Sb$_2$O$_3$	Control	0.60	0.70	0.78
Sb$_2$O$_3$ + phosphonate 1	Trimellitic anhydride	0.60	0.83	0.97
	Trimethylolpropane	0.60	0.82	0.95

In a typical example, a combination of trimethylolpropane, pentaerythritol, or trimellitic anhydride (Fig. 5.7, about 200 ppm) in the presence of Sb_2O_3 (200 ppm) and hydroxyphenylalkylphosphonic ester salt (phosphonate 1, Fig. 5.3, 200 ppm) results in faster SSP rate at 220°C, as shown by the IV values (Table 5.7) [53]. These data confirm what has been found analogously for the melt reaction process of PET with tetracarboxylic acid dianhydrides [54] and the combinations of tetracarboxylic acid dianhydrides with multifunctional alcohols [55] in the presence of phosphonates.

5.5. INERT ADDITIVES IN SOLID STATE POLYMERIZATION PROCESSES

Additives that are inert in the sense that they do not react with the polymer can influence the SSP process. As the SSP rate increases with increasing crystallinity due to enhanced end-group concentration in the amorphous phase [56], additives that influence the crystallinity will modify the SSP process. Also, additives that influence the surface of the granules that are to be solid state–polymerized (e.g., by preventing the beads from sticking together or by modifying the porosity of the granules) can change the SSP process. Therefore, all types of fillers might contribute to the rate of the SSP process. For example, by adding nanomaterials (montmorillonite), a positive but not exciting impact on the SSP process has been shown experimentally [57]. At 230°C and 25 h, pure PET achieves an intrinsic viscosity of 0.919 dL g^{-1} (starting value: 0.641 dL g^{-1}) and PET with 2.5% montmorillonite achieves a value of 0.996 dL g^{-1} (starting value: 0.651 dL g^{-1}). This acceleration is attributed to the nucleation of montmorillonite and interactions between montmorillonite and PET, resulting in less crystallinity and more amorphous regions. These data show as well that the crystallinity and heat of melting are reduced with the SSP time—but to a greater extent when montmorillonite is added.

Crystal structures and porosities might be modified as well by incorporating monomers with pendent side groups, such as 5-nitroisophthalic acid or 5-tertiary butylisophthalic acid [58]. The polycondensation takes place in the presence of Sb_2O_3 and cobalt acetate, phosphorous acid, and phosphoric acid. For example, by replacing 2 mol% of isophthalic acid by 5-nitroisophthalic acid, an increase in the intrinsic viscosity of the SSP process at 208°C from 0.017 dL g^{-1} per hour to 0.0898 dL g^{-1} per hour is reported.

5.6. CONCLUSIONS

Catalysts are essential components in manufacturing polycondensates, such as polyesters and polyamides via solid state processes, used to reduce the reaction time and improve the product quality. The polyester SSP process is based traditionally on metal catalysts, such as antimony oxide or titanates, which are used for the polycondensation step and show, on top, activity in the SSP stage. Catalysts to accelerate the SSP of polyamides are often based on phosphorus compounds. Despite the use of catalysts in SSP processes over many years, there are mechanistic aspects that are not yet fully evaluated, in particular when mixtures of different catalysts are used.

A number of improvements in SSP processes have been published that combine classical and nonclassical catalysts or that utilize salts of organic acids or metal-free organic compounds. However, it must be recognized that catalysis of the SSP process is only one part, and the resulting material properties will influence the potential selection of an efficient catalyst. Cost vis-á-vis performance is an additional criterion in industrial processes. Among the more sophisticated structures of SSP catalysts enhancing the SSP process and acting beneficially on the final properties, complex multifunctional phosphorus compounds such as hydroxyphenylphosphonic acid esters have to be mentioned.

An alternative approach to achieving high-molecular-weight polymers through SSP is the use of reactive additives (e.g., tetracarboxylic acid anhydrides), preferably in combination with catalysts. By selecting multifunctional reactive additives, modified polymer structures such as branched polymers are accessible with tailormade properties for new applications. The area of modified material properties through combining the catalyzed SSP process with chemical reactions is still a vast field for further research.

REFERENCES

1. Fakirov S. In: *Solid State Behavior of Linear Polyesters and Polyamides*, Schultz JM, Fakirov S, eds. Prentice Hall, London, 1990, pp. 37–40.

2. Vouyiouka SN, Karakatsani EK, Papaspyrides CD. Solid state polymerization. *Prog. Polym. Sci.* 2005;30:10–37.

3. Gantillon B, Spitz R, McKenna TF. The solid state postcondensation of PET. *Macromol. Mater. Eng.* 2004;289:88–105.

4. Burch RR. Addition of treatment agents to solid phase polymerization processes (E.I. du Pont de Nemours & Company). WIPO Patent WO 00/49065, 2000.

5. Nichols CS, Moore TC, Schiavone RJ, Edwards WL. Methods of post-polymerization extruder injection in polyethylene terephthalate production (Wellmann). U.S. Patent 6,569,991, 2003.

6. Hsu LC. Synthesis of ultrahigh-molecular-weight poly(ethylene terephthalate). *Macromol. Sci. Phys. B* 1967;1:801–813.

7. Hovenkamp SG. Kinetic aspects of catalyzed reactions in the formation of poly(ethylene terephthalate) *J. Polym. Sci. A-1.* 1971;9:3617–3625.

8. Duh B. Effect of antimony catalyst on solid-state polycondensation of poly(ethylene terephthalate). *Polymer.* 2002;43:3147–3154.

9. Kokkalas DE, Bikiaris DN, Karayannidis GP. Effect of the Sb_2O_3 catalyst on the solid-state postpolycondensation of poly(ethylene terephthalate). *J. Appl. Polym. Sci.* 1995;5:787–791.

10. Coover HW, Shearer NH. Solid phase process for linear superpolyesters (Eastman Kodak Company). U.S. Patent 3,075,952, 1963.

11. Wiegner JP, Voerckel V, Runkel D, Feix G, Staeuber H. Catalyst systems for polycondensation reactions (Dow Global Technolgies Inc.). WIPO Patent WO 2004/014982, 2004.

12. Massey FL, Callander DD. Polyester solid phase polymerization catalyst for low acetaldehyde generating resins. U.S. Patent 2007/0,191,582, 2007.

13. Bonte GIV, Weremus Buning GH, Dijkstra K, Warnier JMM. Preparation of a copolyether ester (DSM N.V.). WIPO Patent WO 00/24803, 2000.

14. James NR, Ramesh C, Sivaram S. Crystallization and solid-state polymerization of poly(aryl ester)s. *Polym. Int.* 2004;53:664–669.

15. Parthasarathy A. Method for increasing solid state polymerization rate of polyester polymers (E.I. du Pont de Nemours & Company). WIPO Patent WO 03/055931, 2003.

16. Waggoner MG. Process for producing a crystalline liquid polymer (E.I. du Pont de Nemours & Company). WIPO Patent WO 2004/058851, 2004.

17. Pfaendner R, Hoffmann K, Herbst H. Increasing the molecular weight of polycondensates (Ciba-Geigy AG). WIPO Patent WO 96/11978, 1996.

18. Steiner UB, Borer C, Oertli AG, Pfaendner R. The role of additives in solid state polycondensation of recycled PET. Presented at the SPE Recycling Division's 3rd Annual Recycling Conference, Chicago, *Proceedings*, Nov. 7–8, 1996.

19. Simon D. Attractive additives for polyesters to stimulate your business. Presentation at Polyester 2004, Zurich, Switzerland, Dec. 7–9, 2004.

20. Zu-Chun J. A manufacturing method for decreasing the cyclic oligomer content in polyester (Nan Ya Plastics Corp.). European Patent EP 1262514, 2002.

21. Schumann HD, Thiele U. Process for the acceleration of the polycondensation of polyester (Zimmer AG). U.S. Patent 5,744,572, 1998.

22. Horlbeck G, Heuer H. Condensation of poly(alkylene terephthalates) (Chemische Werke Hüls AG). European Patent EP 090915, 1983.

23. Browne AAB, McIntyre JE. Polyester manufacture (Imperial Chemical Industries Ltd.) U.S. Patent 4,088,635, 1978.

24. Volokhina AV, Kudryavtsev GI, Skurtaov SM, Bonetskaya AK. Polyamidation process in the solid state. *J. Polym. Sci.* 1961;53:289–294.

25. Bruck SD. New polyoxamidation catalysts. *Ind. Eng. Chem. Prod. Res. Dev.* 1963;2:119–121.

26. Papaspyrides CD, Kampouris EM. Influence of acid catalysts on the solid-state polyamidation of dodecamethylenediammonium adipate. *Polymer.* 1986;27:1433–1436.

27. Papaspyrides CD, Kampouris EM. Influence of metal catalysts on solid-state polyamidation of nylon salts. *Polymer.* 1986;27:1437–1440.

28. Katsikopoulos PV, Papaspyrides CD. Solid-state polyamidation of hexamethylenediammonium adipate: II. The influence of acid catalysts. *J. Polym. Sci. A.* 1994;32:451–456.

29. Papaspyrides CD. Solid-state polyamidation processes. *Polym. Int.* 1992;29:293–298.

30. Papaspyrides CD, Vouyiouka SN, Bletsos IV. New aspects on the mechanism of the solid state polyamidation of PA 6,6 salt. *Polymer.* 2006;47:1020–1027.

31. Vouyiouka SN, Koumantarakis GE, Papaspyrides CD. Preparation and solid-state polyamidation of hexamethylenediammonium adipate: the effect of sodium 5-sulfoisophthalic acid. *J. Appl. Polym. Sci.* 2007;104:1609–1619.

32. Gabler R, Giesen J, Zehnder W. Process for the production of polyamides (Inventa AG). U.S. Patent 2,993,879, 1961.

33. Process for solid polyamides. British Patent GB 1305246 (Inventa AG), 1973.

34. Gaymans RJ, Amirtharaj J, Kamp H. Nylon 6 polymerization in the solid state. *J. Appl. Polym. Sci.* 1982;27:2513–2526.

35. Leetsch N, Schumacher S, Mai HJ, Trittmacher K, Heilmann E, Dalcolmo HJ, Wilde G. Verfahren zur Herstellung von hochviskosen Polyamiden (VEB Leuna-Werke). German Patent DD 274823, 1990.

36. Leetsch N, Trittmacher K, Schumacher S, Mai HJ, Heilmann E, Nitzsche R. Verfahern zur Herstellung von Polyamid-6-formmassen (VEB Leuna-Werke). German Patent DD 248131, 1987.

37. Heinz HD, Schulte H, Buysch HJ. Process for preparing high molecular weight polyamides (Bayer AG). European Patent EP 410230, 1991.

38. Gaymans RJ, Bour EHJP. High-molecular-weight poly(tetramethyleneadipamide) (Stamicarbon B.V.). European Patent EP 038094, 1981.

39. Bonner WH. Process for spinning and drawing polyalkylene isophthalamides (E.I. du Pont de Nemours & Company). U.S. Patent 3,088,794, 1963.

40. Bub L, Baxmann F, Franke W. Verfahren zur Herstellung von hochmolekularen Polyamiden (Chemische Werke Huels AG). German Patent DE 1028779, 1958.

41. Brignac EP, Duke BH, Nunning WJ, Snooks RJ. Method for spinning polyamide yarn of increased relative viscosity. U.S. Patent 3,551,548, 1970.

42. Buzinkai JF, DeWitt MR, Wheland RC. Process for increasing the relative viscosity of polyamides with reduced thermal degradation (E.I. du Pont de Nemours & Company). U.S. Patent 5,116,919, 1992.

43. Dujari R, Cramer G, Marks DN. Method for solid phase polymerization (E.I. du Pont de Nemours & Company). WIPO Patent WO 98/23666, 1998.

44. Vouyiouka SN, Papaspyrides CD, Pfaendner R. Catalyzed solid state polyamidation. *Macromol. Mater. Eng.* 2006;291:1503–1512.

45. Fink J, Pfaendner R, Simon D, Papaspyrides CD, Vougiouka SN. Process for the preparation of polyamides in the presence of a phosphonate (Ciba Specialty Chemicals Holding Inc.). WIPO Patent WO 2007/006647, 2007.

46. Ghisolfi G. Process for the continuous production of high molecular weight polyester resin (Phobos N.V.). WIPO Patent WO 91/05815, 1991.

47. Ghisolfi G. Process for the production of high molecular weight polyester resin (M.&G. Ricerche S.P.A.). WIPO Patent WO 92/17520, 1992.

48. Ghisolfi G. Process for the production of high molecular weight polyester resin (M.&G. Ricerche S.P.A.). WIPO Patent WO 92/17522, 1992.

49. Stouffer JM, Blanchard EN, Leffew KW. Production of poly(ethylene terephthalate) (E.I. du Pont de Nemours & Company). U.S. Patent 5,830,982, 1998.

50. Sid-Ahmed AH, Tung WC. Solid state polymerization of polyesters in the presence of a diisocyanate (Goodyear Tire & Rubber Company). U.S. Patent 3,853,821, 1974.

51. Brink AE, Owens JT. Thermoplastic polyurethanes additives for enhancing solid state polymerization rates (Eastman Chemical Company). WIPO Patent WO 99/11711, 1999.

52. Bormann WFH. Process for the solid state polymerization of branched poly(alkylene terephthalates) using aromatic polycarbonates (General Electric Company). U.S. Patent 4,147,738, 1979.

53. Kao HC, Chen LH, Wu CL, Wong JJ, Chan SY, Yang ST. Composition and process for preparing high molecular weight polyester (Industrial Technology Research Institute). U.S. Patent 6,239,200, 2001.

54. Pfaendner R, Herbst H, Hoffmann K. Increasing the molecular weight of polyesters (Ciba Specialty Chemicals Corporation). U.S. Patent 5,693,681, 1997.

55. Simon D, Pfaendner R, Herbst H. Molecular weight increase and modification of poly-condensates (Ciba Specialty Chemicals Corporation). U.S. Patent 6,469,078, 2002.

56. Duh B. Effects of crystallinity on solid-state polymerization of poly(ethylene tereph-thalate). *Appl. Polym. Sci.* 2006;102:623–632.

57. Yu H, Han K, Yu M. The rate acceleration in solid-state polycondensation of PET by nanomaterials. *J. Appl. Polym. Sci.* 2004;94:971–976.

58. Gardner MW, Cunningham JB, Winter DJ, Jenkins SD. Method for producing enhanced solid state (Invista North America S.AR.L.). U.S. Patent 7,250,484, 2007.

6

HIGH-PRESSURE SOLID STATE POLYMERIZATION OF POLYAMIDE MONOMER CRYSTALS

TOKIMITSU IKAWA

2-6-36, San-yo, Akaiwa City 709-0827, Okayama Prefecture, Japan

6.1. INTRODUCTION

Thermally induced solid state polymerization (SSP), in which no solvents are used, is believed to be an environmentally sound procedure. In addition to SSP advantages, it is found to be an effective method of preparing perfect (i.e.,

Solid State Polymerization, Edited by Constantine D. Papaspyrides and Stamatina N. Vouyiouka
Copyright © 2009 John Wiley & Sons, Inc.

defect-free) polymer crystals. In this regard, in this chapter we describe a direct SSP process as a crystallization technique in parallel with the solid state polymerizability of linear polyamide (PA) monomer crystals under high pressure.

Generally, during direct SSP, the mobility of both reacting and formed molecules is restricted and the polymerization reaction is considered to occur across the molecular arrangement of monomer crystals, often resulting in controlled polymer structures. Accordingly, SSP was carried out on crystalline trioxane and on various vinyl monomers and diacetylene derivatives, and it was found that topochemical or topotactic reactions occurred in the solid phase. An example is that of Hayashi et al. [1], who performed trioxane radiation-induced SSP and got fibrillar crystals of polyoxymethylene (POM) irrespective of the molecular arrangement in the monomer crystals. The SSP of diacetylene derivatives [2–4] and diolefins [5] was also found to proceed topochemically. SSP topochemical reactions accompanied by crystallization have been reviewed regarding chain- [6,7] and step-growth polymerization [8–19], where perfect PA and polyester crystals have been prepared using the regular molecular arrangement in monomer crystals.

To date, the crystal structure of PAs has been controlled primarily by using drawn fibers [20,21]. However, the molecular chain packing in drawn fibers is not as regular, due to inclusion of the undistinguished packing of asymmetrical molecular chains and other factors. In addition, small crystallite size along molecular chains in PA single crystals complicates crystal structure analysis [22–24]. Therefore, thermally induced direct SSP, utilizing the molecular arrangement of monomer single crystals, was examined to prepare large PA single crystals. SSP in vacuo was first applied to 6-amino-n-caproic acid (6-ACA), 11-amino-n-undecanoic acid (11-AUA), and PA salt crystals from diamine–dicarboxylic acid. Although PA 6 and PA 11 single crystals were observed under an electron microscope, the resulting SSP PAs were composed macroscopically of randomly oriented crystallites [8–12]. It was concluded, however, that the SSP of PA monomers proceeded topotactically. On the other hand, Papaspyrides and Kampouris [17] reported that direct SSP (at atmospheric pressure) of PA 66 and PA 610 salt proceeded in a partially molten phase, and the relevant SSP activation energy was also estimated. According to this theory, although SSP was conducted at a low temperature (e.g., 10 to 20°C below the melting temperature of monomer crystals), partial melting of the reacting mass occurred.

The topochemical reactions occurring during direct SSP can be favored by high pressure, due to the more intensive restriction of molecular mobility, even in the aforementioned partially molten systems. Therefore, high-pressure SSP (HP-SSP) was investigated by the author [25–27], first considering 11-amino-n-undecanoic acid (11-AUA) and 12-amino-n-dodecanoic acid (12-ADA), resulting in well-oriented PA 11 and PA 12. Pertinent monomer molecules are linked by zwitterions between amine and carboxylic groups, and such a crystal structure is preferable for enhanced perfection of PA single crystals.

However, HP-SSP use has a significant drawback: the reduced condensate (H_2O) removal. The accumulated water destroys crystals and increases activated sites, which in turn prevent nuclei growth and significant increase in molecular weight. The latter is also supported through the relevant equilibrium nature of the reaction. The equilibrium constant (K_{eq}) changes with pressure (P) according to:

$$\frac{\delta \ln K_{eq}}{\delta P} = -\frac{\Delta V}{RT} \tag{6.1}$$

where P is the reaction pressure, T the reaction temperature, R the universal gas constant, and $\Delta V = V_p - V_m$, V_p and V_m being the specific volume of polymer and monomer crystals, respectively. ΔV is always positive, because the density of polymer crystals is lower than that of monomer crystals. Accordingly, $\delta \ln K_{eq}/\delta P$ is negative, suggesting that SSP is disadvantageous under high pressure.

HP-SSP is examined and discussed here in terms of the polymerizability and structure formation of a series of ω-amino acids with different numbers of carbon atoms (n) in a monomeric unit, and of various PA salts. Molecular arrangements in lateral packing of even–even and odd–odd PA crystals must be considered. For example, in lateral packing of the PA 66 salt crystal, diamine and dicarboxylic acid are arranged opposite to dicarboxylic acid and diamine, respectively [28], which is referred to as a *staggered arrangement* (*SA*). On the other hand, the same moieties (diamine and dicarboxylic acid) in PA 66 molecules are arranged opposite to each other in lateral packing [called a *parallel arrangements* (*PA*)], as shown in Figure 6.1(a). In the case of *SA* arrangement of PA 66 as shown in Figure 6.1(b), hydrogen-bond formation is possible by twisting amide groups (skew conformation) to form γ-crystal, according to the general view that all amide groups must take part in hydrogen-bond formation. However, any γ-form crystals could not actually be observed.

During HP-SSP of PA *mm* salt, growing polymer chains in which respective groups are arranged alternately in lateral packing (*SA*) must slide by half a monomeric unit to form an α-form PA 66 crystal with perfect hydrogen-bond formation [29]. In other words, crystallization to α-form crystal must accompany *SA* → *PA* sliding. On the other hand, there is no need to grow molecular chains to slide during HP-SSP of PA $m(m + 2)$ salt crystals such as PA 810 salt, since in both *SA* and *PA* arrangements it is possible to form polymer α-crystals with perfect hydrogen-bond formation [30,31]. Also, the length between hydrogen-bonding positions along the chains (in 010 planes) is nearly equal in PA $m(m + 2)$, different from PA *mm*. In PA $m(m - 2)$, $m(m - 4)$, and $m(m + 4)$ salt crystals, the same types of molecules must be arranged side by side to form parallel PA salt crystals.

Considering the effects that *SA* and *PA* monomer arrangements may have on PA (polyamide) structure formation, the HP-SSP of a series of PA $m(m \pm 2x)$ salt crystals was carried out and solid state polymerizability under high pressure was investigated. More specifically, the HP-SSP of PAs 810, 812, 1010, 128, and 108 salt crystals is discussed. The HP-SSP of odd–odd, odd–even, and

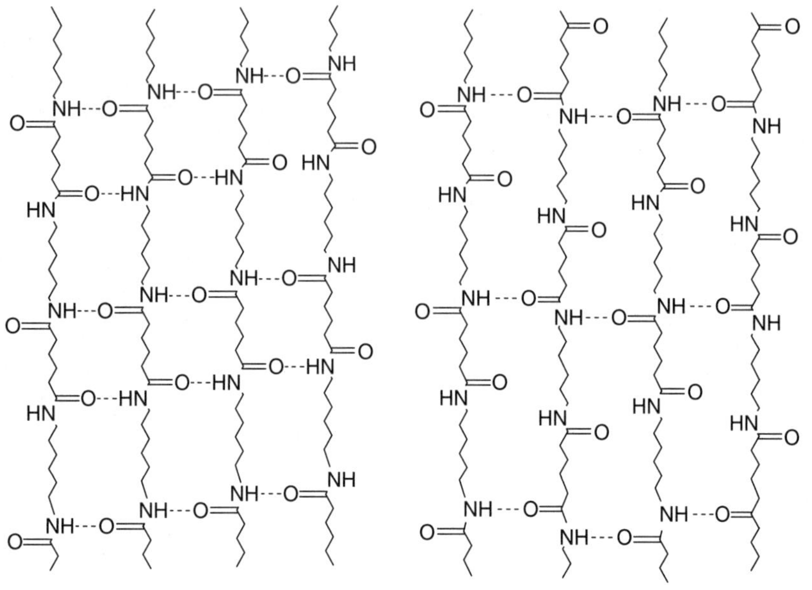

Fig. 6.1. (a) Parallel (*PA*) and (b) staggered arrangements (*SA*) of PA 66.

even–odd PAs was also conducted, and polymerizability and structure formation were investigated [32–36].

6.2. HIGH-PRESSURE SOLID STATE POLYMERIZATION

6.2.1. Crystals and Characteristics of Monomers

A series of ω-amino acids (PA n monomer) and even–even and odd–odd PA $m(m \pm 2x)$ salt crystals were prepared by recrystallization from a H_2O–CH_3OH mixture. In the case of ω-amino acids, $n = 2, 3, 4, 6, 8, 11, 12$, and in PA salts, $m = 6$ to 14 and $x = 0, 1, 2$. Even–odd and odd–even PA $m(m \pm x)$ salt crystals were also prepared for $m = 6$ to 12 and $x = 1$ and 3. Regarding the salt structure, it is well known that amine and carboxyl groups form a zwitterion structure, and the molecular axis of monomer chains is vertical to the basal plane. Zwitterion structure in PA $m(m \pm 2x)$ salt crystals was also confirmed by infrared spectroscopy (IR), even in mosaic PA salt crystals.

Generally, ω-amino acid crystals and PA salt crystals with fewer than six carbon atoms are large transparent single crystals of well-defined shape; they do not contain crystallization solvent (e.g., water) [28], and do not mosaic after enough drying. On the other hand, ω-amino acid and PA salt crystals with more than 10 carbon atoms (m and $n > 10$), are transparent long, narrow platelet single crystals containing crystallization solvents [37]. These single crystals are destructed due to solvent removal upon drying in vacuo after enough air drying at room temperature, and become brittle with cracks. These brittle crystals are composed of monocrystallites, called *mosaic crystals*, that keep the orientation of the original transparent crystals unchanged.

Melt-crystallized and crushed mosaic crystals of 12-ADA were prepared to investigate the relationship between polymerizability and crystalline state. The crystal shapes of PA mm, $m(m + 2)$, and $m(m + 4)$ salt crystals with $m \geq 8$ are almost the same, and PA salt crystals with $m = 4$ and 6 have the shape of thicker platelets or rodlike needles without water of crystallization. On the other hand, the crystal shape of the PA $m(m - 2)$ and PA $m(m - 4)$ series for $m \leq 10$ is that of a thick platelet, while for $m = 12$ the crystal is thin, without a well-defined shape.

The melting temperature (T_m) of ω-amino acid crystals rises with a decrease in the number of carbon atoms (n) in a monomeric unit at atmospheric pressure. At high pressure (500 MPa), T_m rises with an increase in n. The melting temperatures of PA $m(m \pm 2)$ and PA $m(m \pm 4)$ salt crystals are generally lower at atmospheric pressure by 10 to 20°C than those of PA mm crystals with the same numbers of carbon atoms. In the differential scanning calorimetry (DSC) melting curves of PA $m(m \pm 2)$ and PA $m(m \pm 4)$ salt crystals, a small endothermic peak analogous to the sample weight was observed at considerably lower temperature than T_m, while this was not the case for PA mm salt and ω-amino acid crystals. The endothermic peaks at lower temperature disappeared with weight loss by heat treatment at this temperature. It can be assumed that mosaic crystals of PA $m(m \pm 2)$ and $m(m \pm 4)$ may contain water or CH_3OH of crystallization, which was not, however, found to influence the polymerizability of PA $m(m \pm 2)$ and $m(m \pm 4)$ salts versus PA mm salts.

Finally, it should be mentioned that any crystal structure of PA salt crystals is unknown except for PA 66 salt crystals [28]. However, WAXD patterns of PA salt crystals show that monomer molecular chains are arranged vertically or tilting with some angles to the basal plane of platelet crystals. Further, IR spectra show that diamine and dicarboxylic acid molecules are linked by zwitterions to form monomer chains in a crystal.

6.2.2. HP-SSP Method for Polyamide Monomer Crystals

ω-Amino acid and PA salt crystals sealed in a Teflon tube with silica gel and silicone oil were put in a high-pressure autoclave and polymerized at various temperatures (190 to 240°C) under 500 MPa for given reaction time (25, 50, 100, and 200 h). High hydrostaticity was held in the oil-pressured system, so that the monomer crystals would not collapse during HP-SSP. Polymerization temperatures under high pressure, $T_p(P)$ were chosen to avoid monomer melting. Further, HP-SSP using a high-pressure vessel with a piston cylinder was also carried out to investigate the effect of the crystalline state on polymerizability in 11-AUA and 12-ADA. The resulting PA crystals were characterized by WAXD, IR, DSC, and viscosity measurements. Intrinsic viscosity (IV) was estimated from an m-cresol solution at 25°C.

In the present work, the term *polymerizability* does not always represent an increase in the degree of polymerization, mainly an increase in molecular weight. Thus, polymerizability reflecting the molecular weight and structure formation was estimated by IV and WAXD, respectively.

6.3. POLYMERIZABILITY AND STRUCTURE FORMATION

6.3.1. Polymerizability and Structure Formation of ω-Amino Acid Crystals to Polyamide Crystals

Starting with 11-AUA, the fracture surfaces of prepared PA 11 grown at the initial stage of HP-SSP are shown in Figure 6.2, depending on whether the polymer was washed with hot methanol or heated to remove residual monomers (170°C, in vacuo). The SEM pictures showed that polymerization proceeds one-dimensionally initially, and that activated sites are in interior surfaces parallel to the basal plane of monomer crystals on mosaicking. The fibrillar texture shown in Figure 6.2(a) is converted to a sheet structure by sintering among fibrils with increasing temperature and time.

The viscosity-average molecular weight ($\overline{M_v}$) estimated by the intrinsic viscosity of the polymers resulting from HP-SSP of 11-AUA and 12-ADA mosaicked single crystals increases with increasing polymerization temperature, pressure (to 500 MPa), and time and reaches 15,000 g mol^{-1}. However, $\overline{M_v}$ begins to decrease as the temperature rises above the $T_m(P)$ value of the monomer crystals. Further, polymerizability was reduced by half during the HP-SSP of crushed single crystals with small crystallite size along the molecular axis, although polymerization occurs at lower temperature than that of mosaic crystals and the polymerization rate is higher. Figure 6.3 shows the $T_m(P)$ of 12-ADA and PA 12, and the relationship between the polymerizable temperature range and the morphology of

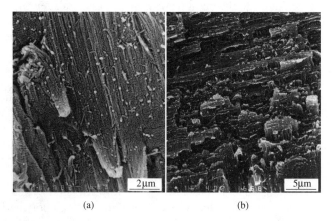

(a)   (b)

Fig. 6.2. SEM of the fracture surface of PA 11 in initial stage (a) washed by hot methanol (b) heated to 170°C in vacuo. (From Ikawa et al.[25] by permission of John Wiley & Sons, Inc.)

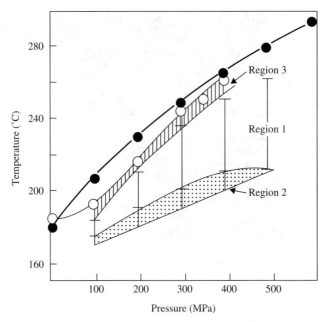

Fig. 6.3. Dependence of the morphology of PA 12 on the polymerization conditions. In region 1, well-oriented single-crystal-like PA 12 is obtained, but in region 3, sintering becomes predominant. In the condition in region 2, polymerization proceeds only with crushed monomer crystals. ● and ○, T_m values of 12-ADA and PA 12. (From Ikawa et al.[26] by permission of John Wiley & Sons, Inc.)

the resulting crystals under high pressure. This relationship was also observed in PA 11.

The shape and molecular axis of monomer crystals are kept in region 1 unchanged after HP-SSP, the polymerizable temperature increases, and its range expands with increasing $T_p(P)$. In region 3, sintering occurs between the resulting PA crystallites and may be related to the molten state [18]. In region 2, melt-crystallized and crushed 12-ADA crystals with small crystallite sizes are polymerizable, although mosaic 12-ADA crystals with larger crystallite sizes, whose excess surface free energy is smaller than those of melt-crystallized or crushed 12-ADA crystals with smaller crystallite sizes, do not polymerize at such a lower $T_p(P)$. In other words, the $T_m(P)$ values of crushed and melt-grown crystals with partial melting at lower temperatures are lower than those of mosaic crystals. As a result, the molecular weight of the resulting PAs obtained by HP-SSP depends on crystalline states of monomer crystals.

Changes in x-ray diffraction patterns of resulting PA 11 and 12 with $t_p(P)$ are shown in Figure 6.4. These patterns suggest that structure formation of PA 11 and PA 12 crystals during HP-SSP of 11-AUA and 12-ADA (i.e., ω-amino acids with long aliphatic chains) proceeds one-dimensionally along the chain axis of monomer crystals. The x-ray patterns of PA 11 (odd polyamide) shows that

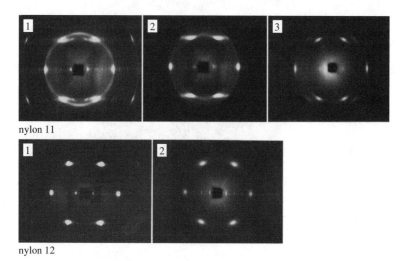

Fig. 6.4. Changes in x-ray diffraction patterns of resulting PA 11 and PA 12 with t_p (220°C at 500 MPa).

oligomer crystals with pseudohexagonal lattice are first formed and successively transformed to β-form crystals with increasing molecular weight, which are finally transformed into α-form. In PA 12 (even polyamide), γ-form crystals are formed via pseudohexagonal crystals of an oligomer which has lattice constants similar to those of monomer crystals. (0k0) reflections on crystallite size along the polymer chains are very sharp in these x-ray diffractions. Such a well-oriented structure is formed increasingly with an increase in the number of carbon atoms in a monomeric unit [25–27].

At this point it should be noted that the x-ray diffraction patterns of the resulting PA 11 and PA 12, 00l reflections in PA 11 and 0k0 reflections in PA 12, which are related to crystallite size along the chain axis, are very sharp. The melting temperature of these large crystallite crystals is higher by about 40°C than those of usual crystals of PA 11 and PA 12. However, long-range ordering in lateral packing cannot be seen from x-ray diffraction. This diffraction suggests that polymerization proceeds one-dimensionally and fibrillar crystals are formed.

On the other hand, molecular orientation of the resulting PA crystals decreases with decreasing n. In HP-SSP of the ω-amino acids above, adjacent growing polymer chains do not need to slide largely by each other to form hydrogen bonding in both parallel and antiparallel packing of polymer chains. Glycine, β-alanine, and 4-amino-n-butyric acid crystals, which give rise to large single crystals, did not polymerize. The degree of shrinkage accompanying polyamidation from 6-ACA and 12-ADA to PA 6 and PA 12 is 17 and 9.3%, respectively. The large shrinkage by polymerization may result in a molten state in the reaction field, so that the difference in shrinkage may control polymerizability and crystal orientation.

6.3.2. Polymerizability of Polyamide Salt Crystals to Polyamide Crystals

Both odd–odd and even–even PA $m(m \pm 2x)$ ($x = 0$, 1, and 2) salt crystals are easily converted into the corresponding PA crystals through HP-SSP with increasing T_p and t_p, as was the case in ω-amino acid crystals. The shapes of the resulting PA crystals are apparently kept unchanged after HP-SSP, showing that polycondensation occurs in a monocrystallite. Increases in IV of the resulting PA crystals, polymerized over 100 h, were low and must be compared under the same $t_p(P)$ and $\Delta T_p[= T_m(\text{monomer}) - T_p(P)]$.

The polymerizability of PAs prepared after 100 h under 500 MPa was compared, and Figure 6.5 shows the changes in IV of PA 810 and PA 108 with $T_p(P)$. Figure 6.6 also shows the changes in IV of PA 812, PA 1010, and PA 128, in which the total number of carbon atoms in PA mm and PA $m(m \pm 4)$ is 20. It can be seen that IV increases with increasing $T_p(P)$, and the HP-SSP of PA $m(m \pm 2)$ and PA $m(m \pm 4)$ salts may proceed over a wide range of temperature (190 to 240°C), which is narrower, however, in the case of PA 1010 (210 to 240°C). The IV achieved for PA $m(m \pm 2)$ and PA $m(m \pm 4)$ is not as high as in ω-amino acid and in PA mm salt mosaic crystals ($m > 10$). This can be attributed to the semimosaic structure of PA $m(m \pm 2)$ and PA $m(m \pm 4)$, which is intermediate between PA 66 and PA mm with $m > 10$. In the case of PA single crystals, such as PA 66, the destruction of monomer crystals by evolution of H_2O occurs more violently than in mosaic crystals where passages have been created during drying to remove water and to increase activated sites. In a sense, the IV of resulting PA $m(m \pm 2)$ and PA $m(m \pm 4)$ crystals is intermediate between PA 1010 and PA 66.

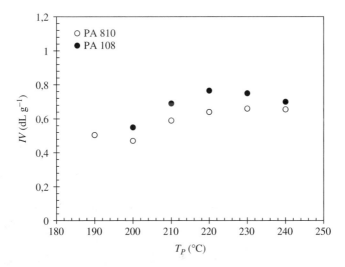

Fig. 6.5. Changes in intrinsic viscosity of PA 810 and PA 108 with T_p.

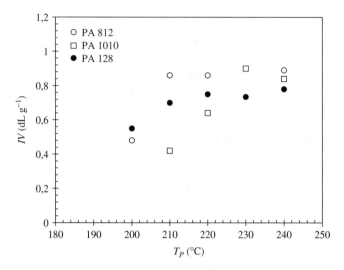

Fig. 6.6. Changes in intrinsic viscosity of PA 812, PA 1010, and PA 128 with T_p.

The melting temperatures (T_m) of the resulting PA $m(m \pm 2x)$ crystals pre-pared by HP-SSP rise with increasing t_p and T_p, and the maximum T_m reached is also higher by about 10 to 20°C than PA $m(m \pm 2x)$ crystallized from melt. Figure 6.7 shows the changes in T_m of the resulting PA 810 and PA 108 (total numbers of carbon atoms: 18), and Figure 6.8 shows the values for PA 812, PA

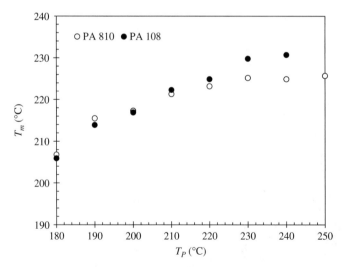

Fig. 6.7. Changes in melting temperature of the resulting PA 810 and PA 108 with T_p $(P = 500 \text{ MPa}, t_p = 100 \text{ h})$.

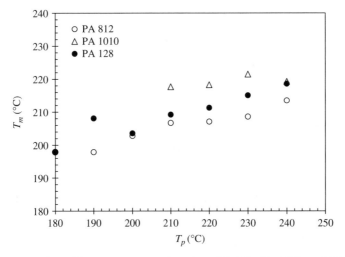

Fig. 6.8. Changes in melting temperature of resultant PA 812, PA 1010, and PA 128 with T_p ($P = 500$ MPa, $t_p = 100$ h).

1010, and PA 128 crystals (total numbers of carbon atoms: 20) by HP-SSP with T_p. These changes are similar to the changes in IV.

Significant differences between PA 108 and PA 810, and between PA 128 and PA 812, cannot be seen. Regarding PAs 810, 108, 812, 1010, and 128, HP-SSP always gives rise to α-form crystals. Only for PA 810 is hydrogen bonding to form α-crystals possible using all amide groups in both *PA* and *SA* arrangements. With the other polyamides, full hydrogen bonding is a *PA* arrangement, while in *SA* there is every other hydrogen formation, where the contribution of hydrogen-bond formation to the cohesive energy in crystallization is half. However, significant T_m differences in these PAs cannot appear. On the contrary, the T_m values of PA 1010 are higher than those of PA 128 and PA 812 with the same number of carbon atoms in the monomeric unit.

The T_m values of the above odd–odd and even–even PA $m(m \pm 2)$ and $m(m \pm 4)$ crystals are higher by 10 to 20°C than those of as-polymerized PA crystals prepared by the usual crystallization. The T_m values of the forego-ing PAs increase with increasing m and the above higher T_m of PAs is 20°C lower than those of ω-amino acid HP-SSP crystals. Such higher melting tem-peratures suggest that HP-SSP may result in a chain-extending effect on the polymerization–crystallization process. Different from the HP-SSP of ω-amino acids, melting of residual PA salt crystals could not be observed in the DSC melt-ing curves, in addition to the x-ray diffraction patterns, suggesting that PA salt crystals with smaller crystallite sizes contain many defects produced in mosaick-ing during HP-SSP, and that polymerization is initiated on these defects. As a result, lower-molecular-weight PAs of small crystallite size are produced. Crys-tallite sizes of the foregoing PAs estimated by WAXD are not large compared with those of PA n crystals. Figure 6.9 shows the changes in IV of even–even and

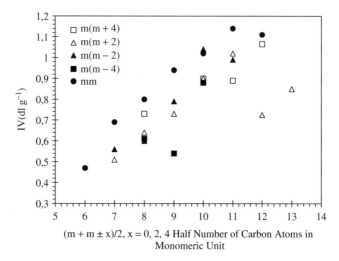

Fig. 6.9. Changes in intrinsic viscosity of even–even PA $m(m \pm 2x)$ with $m/2$.

odd–odd PA $m(m \pm 2x)$ prepared at 230°C for 100 h under 500 MPa with $m/2$ (half the number of carbon atoms in a monomeric unit). Generally, the molecular weight of PA $m(m \pm 2x)$ $(x \neq 0)$ is slightly lower than those of PA mm as shown in Figure 6.9. Therefore, polymerizability of PA $m(m \pm 2x)$ increases with increasing m except for the PA $m(m + 2)$ series. It should be noted here that the IV of the PA mm series is the highest, without regard to odd–odd or even–even PAs. Finally, significant differences cannot be seen among various PA $m(m \pm 2x)$.

Figure 6.10 shows IV changes of odd–even and even–odd PA $m(m \pm 1)$ and PA $m(m \pm 3)$ with $m/2$. In these series, the relation between IV and $m/2$ is

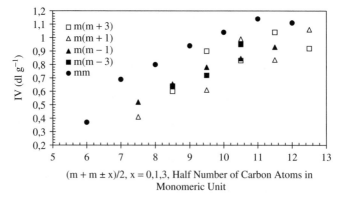

Fig. 6.10. Changes in initrinsic viscosity of even–odd and odd–even PA $m(m \pm 2x)$ with $m/2$.

similar to that of even–even and odd–odd PA $m(m \pm 2x)$ series. The changes in IV and T_m of the resulting odd–odd PAs with T_p are almost the same with even–even PAs.

6.3.3. Structure Formation of Polyamide Salt Crystals to Polyamide Crystals

Based on the above, PA salt crystals are easily transformed into PA crystals by polycondensation in intramonomer chains linked by zwitterion structure. Most crystal structures of even–even PAs prepared by HP-SSP are triclinic α-form, which is slightly different from the usual α-form observed in PA 66 and PA 610 [28,29]. Compared with PA mm and PA n with long aliphatic chains, the crystal orientation of PA $m(m \pm 2)$ and $m(m \pm 4)$ is not as good generally. One reason is because of the many defects arising from the thickness of crystals of PA salts, and the other reason may be because of differences in the mechanism of structure formation from monomer to polymer crystals. In the HP-SSP of PA 11 and PA 12, stable PA crystal structures, α- and γ-form, are formed through oligomer crystals similar to monomer crystal structures. In the HP-SSP of PA 1010 salt and 1212 salt crystals, growing PA crystals with small $\overline{M}_v$ are transformed into β-form at first and transformed successively into α-form crystals, although the existence of oligomer crystals with a monomer-like structure could not be observed in the initial stage due to the rapid HP-SSP rate of PA salts. Figure 6.11 shows the typical x-ray diffraction patterns of the resulting PA 108, 128 1010, 1212, 810, and 1012 crystals. The crystal structure of these even–even PAs is α-form, and no β-form was observed in the resulting PA 810 and PA 1012. The tilting angle of the molecular axis to the basal plane in α-form crystals by HP-SSP ranges from $70°$ to $60°$, which is smaller than that of the well-known α-form crystals with a $49°$ tilting angle (Bunn's α-form [29]). On the other hand, the 00l

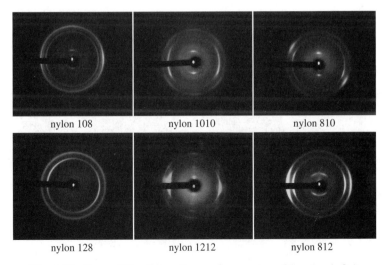

Fig. 6.11. X-ray diffraction patterns of even–even PA $m(m \pm 2x)$.

plane corresponds to the basal plane in α-form of PA 810 and PA 108 crystals (triclinic cell). On considering the existence of various α-forms with different 00*l* reflections, molecular arrangements in the lateral packing of PA salt and PA crystals must be taken into account. Therefore, the growing polymer chains during the HP-SSP of PA *mm* salts, such as PA 66 salt in the *SA* arrangement, must slide by half a monomeric unit to form α-crystal of PA 66 with perfect hydrogen-bond formation. Here, the *SA* in PA 66 and PA 812 has to take the γ-form with chain twisting. However, any γ-form crystals could not be observed.

This means that crystallization to α-form crystal must accompany *SA* → *PA* sliding. However, there remains great doubt regarding this *SA* → *PA* sliding because of keeping original molecular arrangements unchanged in PA *mm* with a large number of aliphatic groups, such as PA 1212. This suggests that every other hydrogen-bond formation of amide groups is possible to make PA crystals with *SA*. On the other hand, there is no need for growing molecular chains to slide in HP-SSP of PA $m(m+2)$ salt crystals, such as PA 810 salt, irrespective of the arrangement of diamine and dicarboxylic acid molecules in both monomer and polymer crystals. In both *SA* and *PA* (Fig. 6.12), it is possible to form α-crystal with perfect hydrogen-bond formation. Besides, the length between hydrogen-bonding positions along the chains (in 010 planes) is equal in PA $m(m+2)$ (Fig. 6.13), different from PA *mm*. In PA $m(m-2)$, $m(m-4)$, and $m(m+4)$ salt crystals such as PAs 108, 128, and 812, the same type of molecules must be arranged side by side to form PA salt crystals (PA) to form α-crystals of PA $m(m-2)$, $m(m-4)$, and $m(m+4)$.

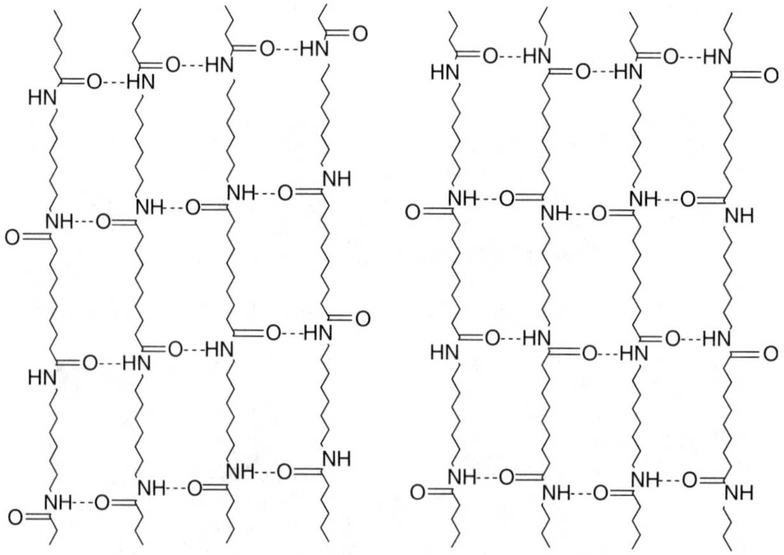

(a)　　　　　　　　　　　　　　　　(b)

Fig. 6.12. (a) Parallel and (b) staggered arrangements of PA 810.

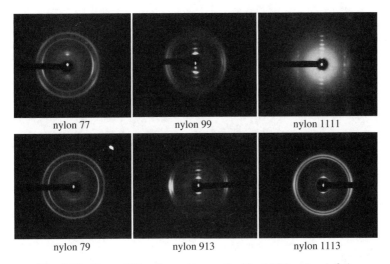

| nylon 77 | nylon 99 | nylon 1111 |
| nylon 79 | nylon 913 | nylon 1113 |

Fig. 6.13. X-ray diffraction patterns of odd-odd PA $m(m \pm 2x)$.

On the other hand, it is impossible to form α-crystals with perfect hydrogen-bond formation in *SA* and *PA*, considering the molecular arrangements of odd–odd PA *mm* and PA $m(m \pm 2x)$ (Fig. 6.14). In such a case, odd–odd PA *mm* and PA $m(m \pm 2x)$, such as PA 1111 and PA 913 with long aliphatic chains, crystallize to γ-form crystal, generally with chain twisting. However, odd–odd PA *mm* crystallized into α-form in the initial HP-SSP stage with lowering T_p and with decreasing number of carbon atoms. PA 77 and PA 79, which are expected to crystallize to γ-form, result in α-form initially and are transformed to γ structure with increasing T_p and t_p. Such crystallization to α-form suggests that hydrogen bonds are formed between every other amide groups. The existence of a free NH band was also confirmed by Fourier transform infrared spectroscopy (FTIR). Crystal structure in PA 1111 seems to be pseudohexagonal (δ-form: monoclinic) with similar lattice constant with γ-form. But chain shrinkage by twisting to form hydrogen-bond formation cannot be seen in this δ-form. δ-Form structure formation in PA 1111 is also observed in odd–odd PA $m(m \pm 2x)$ as PA 913. It suggests that such structure formation without chain shrinkage proceeds with every other hydrogen-bond formation. In HP-SSP of PA 77 and PA 79 salt with short methylene groups, α-form crystals are often observed in the initial stage at lower T_p and are transformed into γ-form with increasing T_p and t_p. Such a phenomenon was also observed in the α-form of even–even PAs, and this means that perfect hydrogen-bond formation is not necessarily a requirement for α structure formation.

As shown in Figure 6.15, x-ray diffraction patterns of odd–even PAs such as PA 910, and of even–odd PAs such as PA 109, show that orientation of molecular chains is kept unchanged but is a little short of long-range order in lateral packing. The crystal structure of odd–even and even–odd PA *mn* by HP-SSP is α-like crystals or δ-form, similar to γ-form. Equatorial reflection angles (2θ) in these

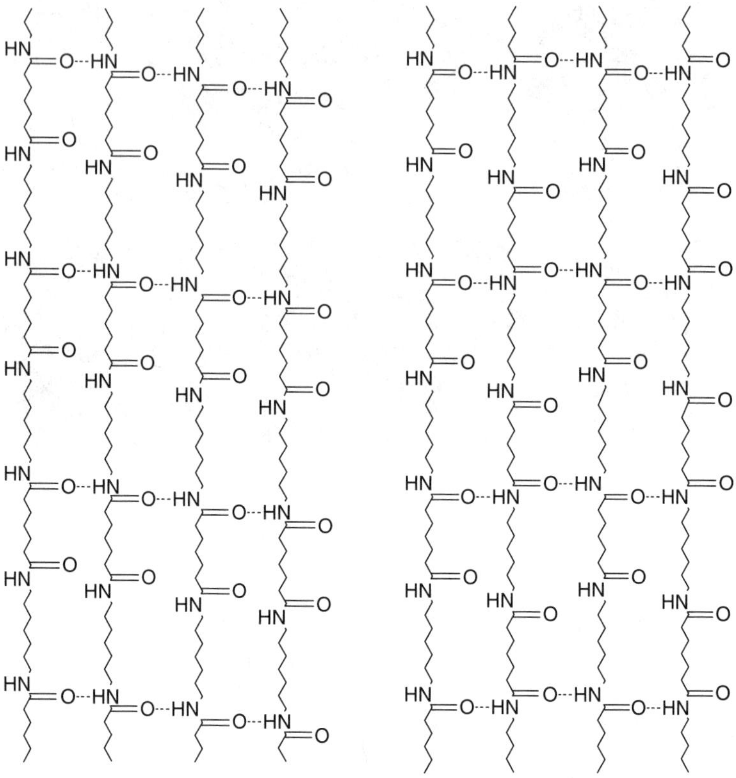

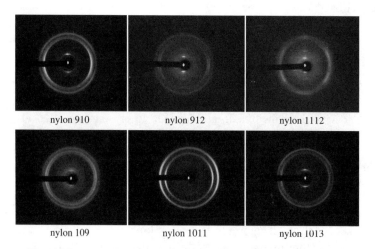

Fig. 6.14. Representation of every other hydrogen formation in PA 77 for *PA* (a) and *SA* (b) arrangements.

nylon 910	nylon 912	nylon 1112
nylon 109	nylon 1011	nylon 1013

Fig. 6.15. X-ray diffractions of odd–even and even–odd PA mn.

did not show the well-defined 200 and (020, 220) reflection angles characteristic of α- or γ-form crystals. This structure formation suggests that hydrogen-bond formation is incomplete in these series. FTIR also shows the existence of free NH based on incomplete hydrogen-bond formation. Systematic differences based on "$m - n = 1$ or 3, $m > n$ or $m < n$" have not been found.

6.4. CONCLUSIONS

6.4.1. Polymerizability

HP-SSP of a series of ω-amino acid and PA salt crystals was conducted from the following viewpoint: How does the number of carbon atoms, m or n, in a monomeric unit of PA n and PA $m(m \pm 2x)$ or PA mn salt crystals affect polymerizability and structure formation? Although ω-amino acids with short aliphatic chains ($n < 5$) do not polymerize under high pressure, ω-amino acids with $n = 6$, 7, 8 give rise to low-molecular-weight PAs, and 11-AUA and 12-ADA with long aliphatic chains result in high-molecular-weight PA 11 and PA 12. On the other hand, the polymerizability of PA salt crystals is almost the same with ω-amino acid crystals, IV increasing with increasing n. Further, the IV of the PA mm series is highest among the PA mm, $m(m \pm 2x)$, and mn series with the same m and n, and there are no significant relations between even–even and odd–odd PAs. HP-SSP of odd–even and even–odd PA mn salts without well-defined crystal shapes gives almost the same results with odd–odd and even–even PAs. Further, it was confirmed that by removing the H_2O evolved from the reaction field, molecular weight increases by successive HP-SSP. In addition, polymerizability of PA monomers in HP-SSP depends on crystalline states of starting monomer crystals. High-molecular-weight PA crystals with single-crystal orientation were obtained by HP-SSP of mosaic single crystals, but PA monomer single crystals with fewer defects do not polymerize in HP-SSP. Further, a decreasing number of carbon atoms in a monomer unit increases the destruction of zwitterion structure. It is assumed that mosaicking single crystals favor passages for H_2O to escape.

6.4.2. Structure Formation

a. ω-Amino Acids HP-SSP rates of 11-AUA and 12-ADA are generally slow, and HP-SSP gives rise to oligomer crystals of PA 11 and PA 12 with pseudohexagonal cell in the initial stage. The oligomer crystal of PA 12 is first transformed into metastable α-form and further transformed into stable γ-form crystal with increasing $T_p(P)$ and $t_p(P)$. In the resulting PA 11, oligomer crystals are further transformed into metastable βII-form (monoclinic) and then finally transformed into α-form crystals (triclinic). On the contrary, crystal structures of resulting PA 6 and 8 are only α-form. The structure formation in the initial stage of HP-SSP could not be observed, due to the rapid HP-SSP rate.

b. PA Salt Crystals   In HP-SSP of even−even PA mm and $m(m \pm 2x)$ (e.g., PA 810, 1010, and 108) salt, crystallization to α-form is generally observed. X-ray diffraction patterns of PA mm with long aliphatic chains show that PA mm with a β-form in the initial stage is transformed into a well-oriented α-form. Small disorientation during crystal transition suggests that chain sliding from *SA* to *PA* does not occur in crystals under HP-SSP. On the other hand, an α-form different from Bunn's α-form is often observed in most even−even PA $m(m \pm 2x)$ $(x \neq 0)$. In this case, we can assume that monomer molecules take PA to form monomer crystals.

As-polymerized odd−odd PA mm and $m(m \pm 2x)$ (e.g., PAs 77, 79, 913) generally crystallize to γ-form. On the contrary, during HP-SSP, α-form crystals are often observed in the resulting PA mm and PA $m(m + 2)$, such as PA 77 and 79 prepared at lower T_p, and α-form crystals are transformed into stable γ-form after enough HP-SSP time.

HP-SSP of even−odd and odd−even PA $m(m \pm x)$ salt crystals results in PAs with α-, α-like, or new crystalline δ-form, and no γ-form crystals can be obtained. Structure formation to α-form suggests that every other hydrogen-bond formation must be in parallel arrangement. Lattice spacing of δ-form, d002, is close to the repeating unit of molecular chain length. It should be noted that δ-form is not easily transformed into α- or γ-form crystals.

HP-SSP of PA monomer crystals gives well-oriented PA crystals that retain their shape. The molecular weight and degree of crystal orientation of the resulting PAs increase generally with increasing m or n. It is believed that the degree of orientation of the resulting crystals and successive SSP depend on the degree of retention of the zwitterion structure in the monomer−polymer crystal boundary by chain contraction brought about by every amidation during polymerization.

REFERENCES

1. Hayashi K, Nishii M, Okamura S. Structure of polymers formed by radiation-induced solid phase polymerization of cyclic monomers. *J. Polym. Sci. C*. 1964;4:839−847.

2. Wegner G. Topochemical reactions of monomers with conjugated triple bonds: I. Polymerization of derivatives of 2,4-hexadiyne-1,6-diols in the crystalline state. *Z. Naturforsch. B*. 1969;2:824−832.

3. Wegner G. Topochemical reactions of monomers with conjugated triple bonds: III. Solid-state reactivity of derivatives of diphenyldiacetylene. *J. Polym. Sci. B*. 1971;9:133−144.

4. Wegner G. Solid state polymerization and polymer single crystals, *Macromol. Chem. Phys. Suppl*. 1984;6:347−357.

5. Hasegawa M, Suzuki Y, Suzuki F, Nakanishi H. Four-center type photopolymerization in the solid state: I. Polymerization of 2,5-distyrylpyrazine and related compounds. *J. Polym. Sci. A-1*. 1969;7(2): 743−752.

6. Wunderlich B. Crystallization during polymerization. *Adv. Polym. Sci*. 1968;5:568−619.

7. Matsumoto S. *Top. Curr. Chem*. 2005; 254−263.

8. Morosoff M, Lim D, Morawetz H. Preparation of biaxially oriented polycapramide by the solid state polycondensation of a single crystal of aminocaproic acid. *J. Am. Chem. Soc.* 1964;86:3167.

9. Macchi E, Morosoff M, Morawetz H. Polymerization in the crystalline state. Solid-state conversion of 6-aminocaproic acid to oriented PA 6. *J. Polym. Sci. A-1.* 1968;6:2033–2049.

10. Macchi E. Crystallization and polymerization of 6-aminocaproic acid as simultaneous processes. *J. Polym. Sci.* 1972;6:45–56.

11. Macchi E. Preparation of crystalline PA 11 by solid state polymerization. *Macromol. Chem.* 1979;180:1603–1605.

12. Macchi E. Extended chain crystals of monoclinic PA 11. *Macromol. Chem. Rapid Commun.* 1980;1:563–567.

13. Oya S, Tomioka M, Araki T. Polyamides prepared in the solid state. Polymerization mechanism. *Koubunshi Kagaku.* 1966;23:415–421.

14. Kiyotukuri T, Otsuki F. Solid-phase polycondensation of aromatic polyamides. *Koubunshi Kagaku.* 1972;29:159–163.

15. Yamazaki T. Thesis, Kyoto University, 1983.

16. Kimura H, Sakabe H, Konishi T. Structure of polyglycine crystals formed by thermal solid state polymerization. *Sen'i Gakkaishi.* 1994;50:395–398.

17. Papaspyrides C, Kampouris E. Solid-state polyamidation of dodecamethylenediammonium adipate. *Polymer.* 1984;25:791–796.

18. Kampouris E, Papaspyrides C. Solid state polyamidation of PA salts: possible mechanism for the transition solid-melt. *Polymer.* 1985;26:413–417.

19. Vouyiouka S, Karakatsani E, Papaspyrides C. Solid state polymerization. *Prog. Polym. Sci.* 2005;30:10–37.

20. Kinoshita Y. An investigation of the structures of polyamide series. *Macromol. Chem.* 1959;33:1–20.

21. Kinoshita Y. The crystal structure of polyheptamethylene pimelamide (PA 77). *Macromol. Chem.* 1959;33:21–31.

22. Walner LG. *Monatsh. Chem.* 1948;79:86.

23. Walner LG. *Monatsh. Chem.* 1948;79:279.

24. Kaji K, Sakurada I. Determination of the elastic modulus of polyamide crystals along the chain axis by x-ray diffraction. *Macromol. Chem.* 1978;179:209–217.

25. Ikawa T, Shimamura K, Yokoyama F, Monobe K, Mori Y, Tanaka Y. Solid state polycondensation of 12-aminododecanoic acid to Nylon 12 under high pressure, *Sen'i Gakkaishi.* 1986;42: T403–T410.

26. Ikawa T, Maeda W, Date S, Shimamura K, Yokoyama F, Monobe K. Formation of Nylon 11 crystals by solid state polycondensation of 11-aminoundecanoic acid under high pressure. *Sen'i Gakkaishi.* 1988;44:385–393.

27. Ikawa T. Nylon (by high pressure solid state polycondensation). In: *Polymeric Materials Encyclopedia*, Vol. 6, Salamone JC, ed. CRC Press, Boca Raton, FL, 1996, pp. 4689–4694.

28. Brown C. Further refinement of the crystal structure of hexamethylenediammonium adipate. *Acta Crystallogr.* 1966;21:185–190.

29. Bunn C, Garner E. The crystal structure of two polyamides (PA). *Proc. R. Soc. A.* 1947;189:39–68.

30. Hill M, Atkins E. Morphology and structure of PA 68 single crystals. *Macromolecules*. 1995;28:604–609.

31. Jones N, Atkins E, Hill M, Cooper M, Franco L. Polyamides with a choice of structure and crystal surface chemistry: studies of chain-folded lamellae of nylons 8,10 and 10,12 and comparison with the other $2N,2(N+1)$ nylons 4,6 and 6,8. *Macromolecules*. 1997;30:3569–3578.

32. Ikawa T. Structure formation of Nylon 1010, 1212 and Nylon 1012 during high pressure-solid state polycondensation of Nylon salt crystals with even–even carbon number. Presented at the Polymer Processing Society, Europe/Africa Meeting, Greece, 2003.

33. Ikawa T. Structure formation of odd–odd Nylon $m \cdot m$ and PA $m \cdot m + 2$ during high pressure-solid state polycondensation of Nylon salt crystals. Presented at the World Polymer Congress, Macro2004, Paris, 2004.

34. Ikawa T, Oohara M. Structure formation of Nylon $m \cdot m \pm 2n$ during high pressure solid state polycondensation of Nylon salt crystals. Presented at the Polymer Processing Society Meeting, Leipzig, Germany, 2005.

35. Ikawa T. Direct conversion of Nylon salt crystals to Nylon crystals by high pressure solid-state polyamidation. Presented at the First European Chemistry Congress, M-Po-32, Budapest, Hungary, 2006.

36. Ikawa T. Polymerizability and structure formation to Nylon crystals by high pressure solid state polycondensation of even–odd and odd–even Nylon salt crystals. Presented at the 16th Fine Chemistry and Functional Polymers Conference, China, 2006.

37. Riggoti G, Rivero B, Macchi E. Crystal data for the dehydrate and anhydrate of 11-aminoundecanoic acid. *J. Appl. Crystallogr*. 1981;14:466–467.

7

FUNDAMENTAL PROCESS MODELING AND PRODUCT DESIGN FOR THE SOLID STATE POLYMERIZATION OF POLYAMIDE 6 AND POLY(ETHYLENE TEREPHTHALATE)

KEVIN C. SEAVEY AND Y. A. LIU

Virginia Polytechnic Institute and State University, Blacksburg, Virgimia

Solid State Polymerization, Edited by Constantine D. Papaspyrides and Stamatina N. Vouyiouka
Copyright © 2009 John Wiley & Sons, Inc.

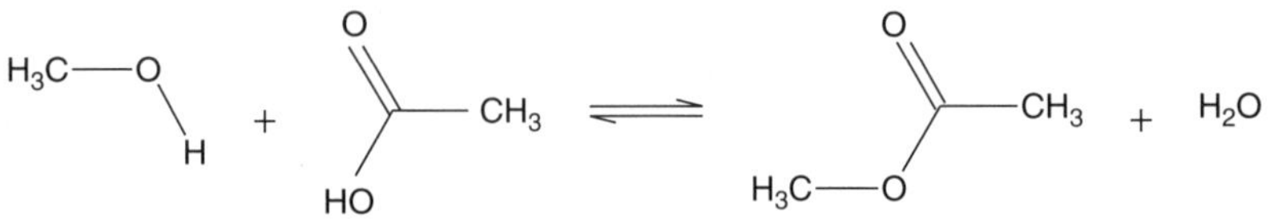

Fig. 7.1. Esterification reaction that joins two molecules of PET to form a single larger molecule and eliminates water.

7.1. INTRODUCTION

For most commercial step-growth polymers such as polyamides and polyesters, the molecular weight builds via an equilibrium reaction that eliminates a condensate molecule C:

$$P_n + P_m \rightarrow P_{n+m} + C \tag{7.1}$$

where P_i represents a polymer molecule of degree of polymerization i. Making a high-molecular-weight polymer depends on removing essentially all of the condensate from the polymer.

An example of such a reaction is esterification to form poly(ethylene terephthalate) (PET, Fig. 7.1). Here a PET molecule with a hydroxyl group —OH reacts with another PET molecule with a carboxylic acid group —COOH to form a single larger PET molecule with an ester functionality and the condensate water.

Beyond a certain molecular weight (ca. 15,000 to 25,000 g mol^{-1} [1]) the melt viscosity of the polymer becomes so high that we cannot effectively devolatilize it, even in specialized equipment such as wiped-film evaporators and rotating-disk finishers. Furthermore, increasing the reaction temperature frequently accelerates undesirable side reactions more than the polycondensation reaction. As a result, the polymer is not able to be devolatilized effectively, thus halting molecular-weight growth. To achieve higher molecular weights, upward of 30,000 g mol^{-1}, solid state polymerization (SSP) is used. Figure 7.2 shows an industrial process for making high-molecular-weight PA 6. In this process, the relatively low-molecular-weight polymer is made in a conventional VK-tube reactor. The molten polymer is then quenched and cut into small, uniform pellets. This dramatically increases the surface area available for mass and heat transfer. The pellets are then extracted using hot water in a leacher, thereby removing most of the unreacted monomer and by-products. Next, the wet pellets are conveyed to the top of the SSP reactor, where they are dried by hot gas. Further drying of the polymer removes the residual condensate within the polymer (water, in this case), allowing polycondensation to proceed.

7.2. SOLID STATE POLYMERIZATION MODELING GUIDE

Industrial practitioners of solid state polymerization have many options available in the literature for modeling SSP reactors. They only need to keep in mind the

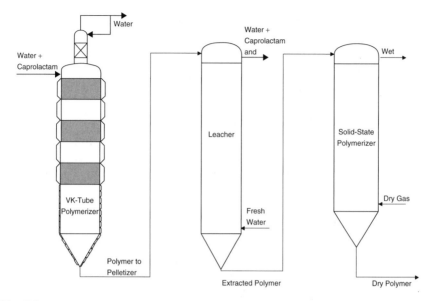

Fig. 7.2. High-molecular-weight PA 6 manufacturing process: conventional vertical tube polymerization reactor, followed by a hot-water leacher and a solid state polymerization reactor.

following guiding principle when choosing one of these modeling options: The needs of manufacturing should be met in the minimum time. Usually, this means the simplest model possible. In many industrial cases, simplicity is forced on the modeler, due to the lack of sufficient data and time to characterize free parameters.

Before embarking on the development of a particular SSP model, we recommend considering the following: model needs, data availability, and model context. Table 7.1 shows some of the questions we may consider before choosing a particular modeling option. The answers to these questions influence our choice of SSP model. For a particular SSP model, we first choose the material and energy balance (M&EB) type and then the details within the M&EB equations (Fig. 7.3).

The first consideration concerns the type of M&EB equations to use. While the SSP reactor is usually plug flow in reality, we have two choices for approximating its behavior. We can either use continuous-stirred-tank reactors (CSTRs) in series approximation, or discretize and integrate the plug flow reactor (PFR) equations directly. Importantly, using CSTRs in series is usually required for steady-state integrated process models with recycle loops. We may also be able to leverage a CSTR model for modeling similar unit operations with different flow patterns, such as a crystallizer. However, we have to use many CSTRs in series to get high fidelity in dynamic simulations. Usually, we can achieve higher accuracy with less computational cost using the PFR equations directly with high-order approximations for the derivatives.

After we choose the M&EB type, the specific model details that are needed are then chosen. Importantly, we should consider whether or not to include

TABLE 7.1. Questions to Consider Before Modeling SSP Reactors

Model Needs	Data Availability	Model Context
• Does crystallinity change significantly in the SSP reactor? If so, we may need to model crystal growth and estimate its effects on reaction and mass transfer. • Is temperature an input, or does it need to be predicted to optimize the reactor? This can be important for optimizing hotgas flow rates.	• Are laboratory data available for the reaction kinetics for the range of temperatures, condensate concentrations, molecular weights, and crystallinity of interest? • Are laboratory data available for diffusion behavior for the range of temperatures, condensate concentrations, molecular weights, and crystallinity of interest? • Are plant data available and complete?	• Is this a dynamic or a steady-state model? • Is this an integrated model; that is, does it include other unit operations and recycle loops?

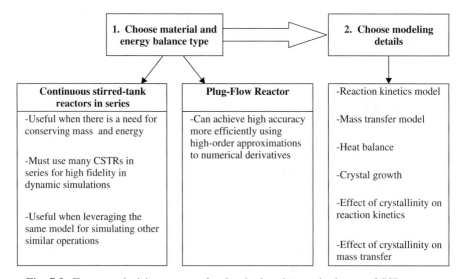

Fig. 7.3. Two-step decision process for developing the required type of SSP reactor.

models, and what types of models, for reaction kinetics, mass transfer, and crystal growth. Will we model only the main polycondensation reactions, or do we need to include reactions for important side products? Will we model rigorous diffusion through a polymer pellet or use a mass-transfer coefficient approach? Will we model crystal growth, and if so, will we model its impact on reaction and crystallization kinetics? Finally, do we need a heat balance? Honest answers to these questions, particularly with respect to actual manufacturing needs, strongly influence the type of SSP model that should be built.

Once we know what type of SSP model we would like to build, we can consult the literature to speed model development. Table 7.2 lists selected studies useful for building a model for an industrial SSP reactor.

7.3. FUNDAMENTALS OF SOLID STATE POLYMERIZATION REACTORS

In this section we describe one particular model for a solid state polymerization reactor, that of Yao et al. [12] and Rovaglio et al. [9]. For simplicity, we have omitted some finer details, such as modeling crystal growth, changes in the pellet bed level, and heat transfer through the wall of the reactor. Readers should consult the references cited in Table 7.2 for these and additional details.

7.3.1. Material and Energy Balances

As shown in Figure 7.2, an SSP reactor is typically a vertically oriented PFR containing a bed of polymer pellets. These polymer pellets, which collectively make up the polymer phase, move down the PFR while an inert gas is forced up through the bed. The material balance for each component in the polymer phase P is

$$
\underbrace{\frac{\partial C_{P,i}}{\partial t}}_{\substack{\text{time rate} \\ \text{of change}}} + \underbrace{u_P \frac{\partial C_{P,i}}{\partial z}}_{\text{convection}} = \underbrace{D_{P,i} \frac{\partial^2 C_{P,i}}{\partial z^2}}_{\text{axial diffusion}} + \underbrace{N_{P-G,i} a_P}_{\substack{\text{diffusion into} \\ \text{gas phase}}} + \underbrace{r_i}_{\text{reaction}} \tag{7.2}
$$

Here C_i is a concentration of species i (mol m^{-3}) in either the polymer (P) or gas (G) phase, u the velocity (m s^{-1}), z the axial position in the vessel (m), D_i the axial diffusion coefficient of species i (m^2 s^{-1}), $N_{P-G,i}$ the diffusion flux of species i between the polymer and gas phases (mol m^{-2} s^{-1}), a the specific area for diffusion (m^2 m^{-3}), and r_i the reaction rate (mol m^{-3} s^{-1}).

The convective term captures the bulk movement of each phase in the axial domain z; the diffusion term captures the molecular diffusion through a phase in the axial domain. We assume that the velocity of each phase is constant, calculable from feed conditions. This is a source of minor violations in the laws of material and energy conservation, as a small amount of material moves between the phases, changing the velocity of each phase. These violations can be avoided by using a CSTR-in-series approach. This approach is mandatory when simulating steady-state recycle loops where minor material balance errors are intolerable.

TABLE 7.2. Summary of Select References from the Solid State Polymerization Reactor Simulation Literature

Reference	PFR	Stirred-Tank Reactor	Heat Balance	Rigorous Diffusion Through Pellets	Mass-Transfer Coefficient Approach	Crystallization Kinetics	Effect of Crystallinity on Reaction Kinetics	Effect of Crystallinity on Mass Transfer
[2]	X		X	X		X		
[3]		X		X		X	X	X
[4]	X			X		X	X	X
[5]		X		X		X	X	X
[6]		X		X	X	X	X	X
[7]		X		X		X	X	X
[8]		X	X	X		X	X	X
[9]	X		X		X	X		
[10]		X		X		X	X	X
[11]	X		X	X		X		
[12]	X		X		X	X		

204

The last two terms of (7.2) represent the crux of modeling SSP processes. These are the diffusion term between the polymer and gas phases and the reaction term. The success or failure of any SSP model depends on developing expressions that accurately represent mass-transfer and reaction rates. The material balance for the gas phase G is similar, except that it does not contain a reaction term:

$$\underbrace{\frac{\partial C_{G,i}}{\partial t}}_{\substack{\text{time rate} \\ \text{of change}}} + \underbrace{u_G \frac{\partial C_{G,i}}{\partial z}}_{\text{convection}} = \underbrace{D_{G,i} \frac{\partial^2 C_{G,i}}{\partial z^2}}_{\text{axial diffusion}} - \underbrace{N_{P-G,i} a_G}_{\substack{\text{diffusion with} \\ \text{polymer phase}}} \qquad (7.3)$$

Note that the mass-transfer flux from polymer to gas is the same as that in (7.2), with the exception of sign reversal and multiplication by the gas-phase specific surface area as opposed to the polymer-phase specific surface area.

The gas not only serves to remove diffusing condensate, but also heats the polymer. To capture these effects, we develop energy balances for each phase, assuming that the wall of the SSP reactor is insulated (i.e., there is no heat transfer between the polymer or gas phase and the wall of the reactor). The resulting equations are

$$\underbrace{\frac{\partial T_P}{\partial t}}_{\substack{\text{time rate} \\ \text{of change}}} + \underbrace{u_P \frac{\partial T_P}{\partial z}}_{\text{convection}} = \underbrace{D_{T_P} \frac{\partial^2 T_P}{\partial z^2}}_{\text{axial diffusion}} + \underbrace{\frac{v_P}{c_{P,P}} \sum_i a_P N_{P-G,i} h_i^V}_{\text{heat of vaporization}}$$

$$+ \underbrace{\frac{v_P}{c_{P,P}} a_P h_0 (T_G - T_P)}_{\text{heat transfer between phases}} - \underbrace{\frac{v_P}{c_{P,P}} \sum_j R_j \Delta H(R_j)}_{\text{heat of reaction}} \quad (7.4)$$

$$\underbrace{\frac{\partial T_G}{\partial t}}_{\substack{\text{time rate} \\ \text{of change}}} + \underbrace{u_G \frac{\partial T_G}{\partial z}}_{\text{convection}} = \underbrace{D_{T_G} \frac{\partial^2 T_G}{\partial z^2}}_{\text{axial diffusion}} - \underbrace{\frac{v_G}{c_{P,G}} a_G (T_P - T_G) \sum_i c_{P,V,i} N_{P-G,i}}_{\substack{\text{heat required to raise temperature of volatilized} \\ \text{material from polymer temperature} \\ \text{to gas temperature}}}$$

$$- \underbrace{\frac{v_G}{c_{P,G}} a_G h_0 (T_G - T_P)}_{\substack{\text{heat transfer between} \\ \text{phases}}} \qquad (7.5)$$

Here T is the temperature of either the polymer (P) or gas (G) phase (K), u the velocity (m s^{-1}), z the axial position within the vessel (m), D the axial diffusion coefficient for heat (m^2 s^{-1}), c_P the constant-pressure heat capacity (J mol^{-1} K^{-1}), $N_{P-G,i}$ the diffusion flux of species i between the polymer and gas phases (mol m^{-2} s^{-1}), h_i^V the heat of vaporization of species i (J mol^{-1}), ΔH_j the heat of reaction for reaction j (J mol^{-1}), R_j the reaction rate of reaction j (mol m^{-3} s^{-1}), ν the molar volume of either the polymer (P) or gas (G) phase (m^3 mol^{-1}), a the specific surface area of either the polymer (P) or gas (G) phase (m^2 m^{-3}), and h_0 the heat transfer coefficient between the polymer and gas phases (J m^{-2}s^{-1} K^{-1}).

7.3.2. Mass and Heat Transfer

Rigorous modeling of mass transfer within the pellet requires the partial second derivative of concentration with respect to radial position within the pellet r (m), $D_{P,i} \nabla^2 C_{P,I}$ (see, e.g., Mallon and Ray [7]). However, most engineering analysis is less exact and the additional computational load of modeling this term is usually not justified. Yao et al. [12] approximated this term using a simpler mass-transfer coefficient approach, which we develop below. Figure 7.4 shows a composition gradient across a vapor–liquid interface. p_i^* is the partial pressure of component i in equilibrium with the polymer interface (Pa), $p_{G,i}$ the bulk partial pressure of component i in the gas phase (Pa), and p^* the partial pressure of component i in equilibrium with the bulk polymer phase (Pa). C_i^* is the interfacial concentration of component i (mol m^{-3}), $C_{P,i}$ the bulk polymer concentration (mol m^{-3}), and C^* the liquid concentration in equilibrium with the bulk vapor (mol m^{-3}).

We begin our analysis of the mass-transfer flux with a general equation for the mass-transfer flux in terms of phase-specific mass-transfer coefficients:

$$N_{P-G,i} = k_{G,i}(p_i^* - p_{G,i}) = k_{P,i}(C_{P,i} - C_i^*) \tag{7.6}$$

$N_{P-G,i}$ is the mass-transfer flux of component i (mol m^{-2} s^{-1}), $k_{G,i}$ the gas-side mass-transfer coefficient (mol m^{-2} s^{-1} Pa^{-1}), and $k_{P,i}$ the polymer-side mass-transfer coefficient (m s^{-1}). In terms of overall mass-transfer coefficients, the flux equation is

$$N_{P-G,i} = K_{G,i}(p^* - p_{G,i}) = K_{P,i}(C_{P,i} - C^*) \tag{7.7}$$

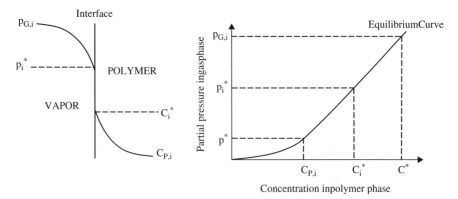

Fig. 7.4. Composition gradient across a vapor–liquid interface.

where $K_{G,i}$ is the overall mass-transfer coefficient in the gas phase (mol m^{-2} Pa^{-1} s^{-1}) and $K_{P,i}$ is the overall mass-transfer coefficient in the polymer phase (m s^{-1}).

We solve for the value of the overall mass-transfer coefficient in the gas phase by first substituting in the appropriate equation for the gas-phase mass-transfer coefficient:

$$\frac{1}{K_{G,i}} = \frac{p^* - p_{G,i}}{N_{P-G,i}}$$
$$= \frac{p^* - p_i^* + p_i^* - p_{G,i}}{N_{P-G,i}}$$
$$= \frac{1}{k_{G,i}} + \frac{p^* - p_i^*}{N_{P-G,i}} \tag{7.8}$$

Assuming that the activity coefficient of component i in the polymer phase, $\gamma_{P,i}$, the polymer-phase molar volume, $\upsilon_{L,P}$ (m^3 mol^{-1}), and the vapor pressure of component i, $P_{P,i}^{\text{sat}}$ (Pa), are all uniform throughout the polymer phase, we develop the remaining part of the overall gas-phase mass-transfer coefficient equation:

$$\frac{p^* - p_i^*}{N_{P-G,i}} = \frac{\upsilon_{L,P}\gamma_{P,i} P_{P,i}^{\text{sat}}(C_{P,i} - C_i^*)}{N_{P-G,i}}$$
$$= \frac{\upsilon_{L,P}\gamma_{P,i} P_{P,i}^{\text{sat}}}{k_{L,i}} \tag{7.9}$$

When we substitute equations (7.8) and (7.9) into (7.7), we arrive at the following equation for the mass-transfer flux:

$$N_{P-G,i} = K_{G,i}(p^* - p_{G,i})$$
$$= \left(\frac{1}{k_{G,i}} + \upsilon_{L,P}\gamma_{P,i} P_{P,i}^{\text{sat}}\frac{1}{k_{P,i}}\right)^{-1} (C_{P,i}\upsilon_{L,P}\gamma_{P,i} P_{P,i}^{\text{sat}} - y_i P) \tag{7.10}$$

We can use the polymer nonrandom, tow-liquid (PolyNRTL) activity coefficient model [13] or other activity coefficient model to estimate the activity coefficient.

We estimate the mass-transfer coefficient on the polymer side by [12]

$$k_{P,i} = 1.5\frac{D_{P,i}\pi^2}{3R} \tag{7.11}$$

where $k_{P,i}$ is the mass-transfer coefficient of species i on the polymer side (m s^{-1}), $D_{P,i}$ the diffusivity of species i in the polymer phase (m^2 s^{-1}), and R the sphere radius (m). Next, we estimate the transport coefficients on the gas side using the Chilton–Colburn analogy, which is used to quantify the effects of laminar and turbulent flow on transport phenomena. As explained in Yao et al. [12], we first compute the Reynolds number for a bed of packed spheres:

$$N_{Re} = \frac{\dot{m}R}{3(1-\varepsilon)a_c\mu} \qquad (7.12)$$

N_{Re} is the Reynolds number, $\dot{m}$ the mass flow rate of fluid (kg s^{-1}), R the sphere radius (m), ε the void fraction of the bed, a_c the cross-sectional area of the vessel (m^2), and μ the fluid viscosity (Pa · s).

The Colburn factor then follows:

$$j_H = j_D = \begin{cases} 0.91N_{Re}^{-0.51} & N_{Re} < 50 \\ 0.61N_{Re}^{-0.41} & N_{Re} \geq 50 \end{cases} \qquad (7.13)$$

The gas-side mass-transfer coefficient (mol m^{-2} s^{-1} Pa^{-1}) is

$$k_{G,i} = \frac{j_D\dot{n}_G}{p_{N_2m}a_c}\left(\frac{\mu_G}{\rho_G D_i}\right)_f^{-2/3} \qquad (7.14)$$

$$p_{N_2m} = \frac{p_{N_2R} - p_{N_2G}}{\ln(p_{N_2R}/p_{N_2G})} \qquad (7.15)$$

$$p_{N_2R} = P - \sum_{i=\text{volatiles}} p_{iR} \qquad (7.16)$$

$$p_{N_2G} = y_{N_2}P \qquad (7.17)$$

p_{N_2m} is the log mean of the partial pressures of nitrogen at the sphere surface and in the bulk gas phase, P the system pressure (Pa), p_{N_2G} the partial pressure of nitrogen in the gas phase (Pa), $k_{G,i}$ the mass-transfer coefficient of species i (mol m^{-2} Pa^{-1} s^{-1}), $\dot{n}_G$ the molar flow rate of gas (mol s^{-1}), ρ_G the density of the gas (kg m^{-3}), D_i the diffusivity of species i in the fluid (m^2 s^{-1}), y_{N_2} the mole fraction of nitrogen in the bulk gas phase, and the subscript f indicates that properties are to be evaluated at the film temperature. We can approximate the film temperature as the arithmetic average between the temperature of the bulk fluid and solid phases. We typically use (7.14) to approximate the mass-transfer coefficient on the vapor side together with (7.10) to estimate the mass-transfer flux.

We model the heat-transfer coefficient (h_0, J m^{-2} s^{-1} K^{-1}) similarly, as the sum of resistances in the gas and polymer phases:

$$\frac{1}{h_0} = \frac{1}{h_P} + \frac{1}{h_G} \qquad (7.18)$$

The heat-transfer coefficient due to gas-phase resistance is

$$h_G = \frac{j_H c_{P,G} F_G}{a_c}\left(\frac{\mu_G c_{P,G}}{\kappa_G}\right)^{-2/3} \qquad (7.19)$$

j_H is the Colburn factor from (7.13), $c_{P,G}$ the heat capacity of the gas (J mol^{-1} K^{-1}), F_G the mole flow rate of gas (mol s^{-1}), a_c the cross-sectional area of the reactor (m^2), μ_G the gas viscosity (Pa · s), and κ_G the thermal conductivity of the gas (W m^{-1} K^{-1}).

We estimate the polymer-side heat-transfer coefficient using [12]

$$h_P = 1.5 \frac{\kappa_P \pi^2}{3 R_P} \tag{7.20}$$

where κ_P is the solid thermal conductivity (W m^{-1} K^{-1}).

7.3.3. Reaction Kinetics

The primary reactions of interest are the polycondensation and side reactions that form by-products. It is usually necessary to model side reactions because these by-products usually make it difficult to meet product specifications, and thus their generation should be modeled. For example, food-grade PET should not contain too much of the by-product acetaldehyde.

When proprietary reaction kinetics are not available, we use corresponding melt-phase kinetics. We use a functional group approach instead of the method of moments to represent the polymerization kinetics. Advantages of the functional group approach include the ease of derivation as well as the ease of incorporating multiple comonomer or terminator types. However, a disadvantage of the functional group approach is that unlike the method of moments, it does not track the weight-average and higher average molecular weights. Next we summarize polymerization kinetics for PA 6 and PET.

a. Polyamide 6 For PA 6 polymerization, we consider the nonpolymeric components depicted in Table 7.3. The polymer contains the segments displayed in Table 7.4. The primary initiation reaction is the ring opening of caprolactam via water, which generates amine and carboxylic acid groups (Fig. 1.5). Polycondensation can then proceed via a condensation reaction between amine groups and carboxylic acid groups. Amine groups can also add caprolactam directly via polyaddition. Cyclic dimer undergoes similar reactions. Table 7.5 shows equilibrium reactions and the accompanying rates for the polymerization of PA 6 with these components. This particular form of the reaction rate equations is discussed in detail by Seavey et al. [14].

Arai et al. [15] give the rate constants displayed in Table 7.6. We have converted the numbers into SI units. The reaction kinetics allows us to write the time rate of change of concentration via reaction. Table 7.7 gives the time rate of change for all species and functional groups due to reaction. The rate constants are expressed in terms of mass concentration, not molarity. Therefore, we need to multiply the reaction rates by density (kg m^{-3}) before using them in (7.2).

TABLE 7.3. Chemical Formulas and Molecular Weights of Nonpolymeric Species (PA 6 Polymerization)

Species	Chemical Formula	Molecular Structure	MW (kg mol^{-1})
Aminocaproic acid (ACA)	$C_6H_{13}NO_2$		1.312E−01
Caprolactam (CL)	$C_6H_{11}NO$		1.132E−01
Cyclic dimer (CD)	$C_{12}H_{22}N_2O_2$		2.263E−01
Nitrogen (N_2)	N_2	$N \equiv N$	2.801E−02
Water (W)	H_2O	$H-O-H$	1.802E−02

TABLE 7.4. Segment Names, Formulas, and Molecular Weights for PA 6 Polymerization

Species	Chemical Formula	Molecular Structure	MW (kg mol^{-1})
B-ACA	$C_6H_{11}NO$		0.113
T-COOH	$C_6H_{12}NO_2$		0.130
T-NH$_2$	$C_6H_{12}NO$		0.114

b. Poly(ethylene terephthalate)   As we did for PA 6, we begin developing the kinetic model by identifying the components and polymer segments. Table 7.8 contains the nonpolymeric components that we consider in our reaction model for the SSP of PET. We also consider the segments that make up PET shown in Table 7.9. The primary reactions that we consider are the water formation and ester interchange reactions. We also include side reactions such as degradation of diester groups, diethylene glycol formation, ethylene glycol dehydration, and acetaldehyde formation. Table 7.10 shows all of our reactions and corresponding rates.

TABLE 7.5. Equilibrium Reactions and Accompanying Reaction Rates for PA 6 Polymerizations

Equilibrium Reaction	Reaction Rate
Ring Opening of Caprolactum	
1. $CL + W \underset{k_1' = k_1/K_1}{\overset{k_1}{\rightleftharpoons}} P_1$	$R_1 = k_1[CL][W] - k_1'[P_1]$
Polycondensation	
2. $P_1 + P_1 \underset{k_2' = k_2/K_2}{\overset{k_2}{\rightleftharpoons}}$	$R_2 = 2k_2[P_1]^2$
$\text{T-COOH} : \text{T-NH}_2 + W$	$-k_2'[W][\text{T-NH}_2]\dfrac{[\text{T-COOH}]}{[\text{B-ACA}] + [\text{T-COOH}]}$
3. $P_1 + \text{T-COOH} \underset{k_2' = k_2/K_2}{\overset{k_2}{\rightleftharpoons}}$	$R_3 = k_2[P_1][\text{T-COOH}]$
$\text{T-COOH} : \text{B-ACA} + W$	$-k_2'[W]\dfrac{[\text{B-ACA}]}{[\text{B-ACA}] + [\text{T-NH}_2]}$
4. $\text{T-NH}_2 + P_1 \underset{k_2' = k_2/K_2}{\overset{k_2}{\rightleftharpoons}}$	$R_4 = k_2[\text{T-NH}_2][P_1]$
$\text{T-NH}_2 : \text{B-ACA} + W$	$-k_2'[W][\text{T-NH}_2]\dfrac{[\text{B-ACA}]}{[\text{B-ACA}] + [\text{T-COOH}]}$
5. $\text{T-NH}_2 + \text{T-COOH} \underset{k_2' = k_2/K_2}{\overset{k_2}{\rightleftharpoons}}$	$R_5 = k_2[\text{T-NH}_2][\text{T-COOH}]$
$\text{B-ACA} : \text{B-ACA} + W$	$-k_2'[W][\text{B-ACA}]\dfrac{[\text{B-ACA}]}{[\text{B-ACA}] + [\text{T-NH}_2]}$
Polyaddition of Caprolactam	
6. $P_1 + CL \underset{k_3' = k_3/K_3}{\overset{k_3}{\rightleftharpoons}}$	$R_6 = k_3[P_1][CL]$
$\text{T-NH}_2 : \text{T-COOH}$	$-k_3'[\text{T-NH}_2]\dfrac{[\text{T-COOH}]}{[\text{B-ACA}] + [\text{T-COOH}]}$
7. $\text{T-NH}_2 + CL \underset{k_3' = k_3/K_3}{\overset{k_3}{\rightleftharpoons}}$	$R_7 = k_3[\text{T-NH}_2][CL]$
$\text{T-NH}_2 : \text{B-ACA}$	$-k_3'[\text{T-NH}_2]\dfrac{[\text{B-ACA}]}{[\text{B-ACA}] + [\text{T-COOH}]}$
Ring Opening of Cyclic Dimer	
8. $CD + W \underset{k_4' = k_4/K_4}{\overset{k_4}{\rightleftharpoons}}$	$R_8 = 2k_4[CD][W]$
$\text{T-COOH} : \text{T-NH}_2$	$-k_4'[\text{T-NH}_2]\dfrac{[\text{T-COOH}]}{[\text{B-ACA}] + [\text{T-COOH}]}$

(Continued overleaf)

TABLE 7.5. (*Continued*)

Equilibrium Reaction	Reaction Rate

Polyaddition of Cyclic Dimer

9. $P_1 + CD \underset{k_5' = k_5/K_5}{\overset{k_5}{\rightleftharpoons}}$ $R_9 = 2k_5[P_1][CD]$

T-NH$_2$: B-ACA : T-COOH

$$-k_5'[\text{T-NH}_2]\left(\frac{\dfrac{[\text{B-ACA}]}{[\text{B-ACA}] + [\text{T-COOH}]}}{\dfrac{[\text{T-COOH}]}{[\text{B-ACA}] + [\text{T-COOH}]}}\right)$$

10. $\text{T-NH}_2 + CD \underset{k_5' = k_5/K_5}{\overset{k_5}{\rightleftharpoons}}$ $R_{10} = 2k_5[\text{T-NH}_2][CD]$

B-ACA : B-ACA : T-NH$_2$

$$-k_5'[\text{T-NH}_2]\left(\frac{[\text{B-ACA}]}{[\text{B-ACA}] + [\text{T-COOH}]}\right)^2$$

TABLE 7.6. Reaction Rate Constants for Hydrolytic PA 6 Polymerization

Rate constant expression $k_i = A_i^0 \exp\left(-\dfrac{E_i^0}{RT}\right) + A_i^c \exp\left(-\dfrac{E_i^c}{RT}\right)$ ([ACA] + [T-COOH])

Equilibrium constant expression $K_i = \dfrac{k_i}{k_i'} = \exp\left(\dfrac{\Delta S_i - \Delta H_i/T}{R}\right)$

i	A_i^0 (kg mol^{-1}s^{-1})	E_i^0 (J mol^{-1})	A_i^c (kg^{2}mol^{-2}s^{-1})	E_i^c (J mol^{-1})	ΔH_i (J mol^{-1})	ΔS_i (J mol^{-1}K^{-1})
1	1.6632E+02	8.3234E+04	1.1965E+04	7.8722E+04	8.0287E+03	−3.3005E+01
2	5.2617E+06	9.7431E+04	3.3650E+06	8.6525E+04	−2.4889E+04	3.9505E+00
3	7.9328E+05	9.5647E+04	4.5492E+06	8.4168E+04	−1.6927E+04	−2.9075E+01
4	2.3827E+08	1.7585E+05	6.4742E+08	1.5656E+05	−4.0186E+04	−6.0781E+01
5	7.1392E+04	8.9179E+04	8.3639E+05	8.5394E+04	−1.3266E+04	2.4390E+00

TABLE 7.7. Reaction Rates for All Nonpolymeric Components and Segments in the Polymerization of PA 6

Functional Group	Time Rate of Change
W	$r_W = R_2 + R_3 + R_4 + R_5 + R_{11} + R_{12} + R_{14} + R_{15} - (R_1 + R_8)$
CL	$r_{CL} = -(R_1 + R_6 + R_7 + R_{13})$
CD	$r_{CD} = -(R_8 + R_9 + R_{10})$
ACA (P$_1$)	$r_{ACA} = R_1 - (2R_2 + R_3 + R_4 + R_6 + R_9 + R_{11} + R_{14})$
B-ACA	$r_{\text{B-ACA}} = R_3 + R_4 + 2R_5 + R_7 + R_9 + 2R_{10} + R_{12} + R_{15}$
T-NH$_2$	$r_{\text{T-NH}_2} = R_2 + R_6 + R_8 + R_9 + R_{13} + R_{14} - (R_5 + R_{12})$
T-COOH	$r_{\text{T-COOH}} = R_2 + R_6 + R_8 + R_9 + R_{11} - (R_5 + R_{15})$

TABLE 7.8. Chemical Formula and Molecular Weight of Nonpolymeric Species (PET Polymerization)

Species	Chemical Formula	Molecular Structure	MW (kg mol^{-1})
Acetaldehyde (AA)	C_2H_4O		4.405E−02
Antimony triacetate (SBOAC$_3$)	$C_6H_9O_6Sb$		2.988E−01
Diethylene glycol (DEG)	$C_4H_{10}O_3$		1.061E−01
Ethylene glycol (EG)	$C_2H_6O_2$		6.207E−02
Nitrogen (N$_2$)	N_2	$N \equiv N$	2.801E−02
Terephthalic acid (TPA)	$C_8H_6O_4$		1.661E−01
Water (W)	H_2O	H—O—H	1.802E−02

TABLE 7.9. Segment Names, Formulas, and Molecular Weights for PET Polymerization

Species	Chemical Formula	Molecular Structure	MW (kg mol^{-1})
B-DEG	$C_4H_8O_3$		0.104
B-EG	$C_2H_4O_2$		0.0601
B-TPA	$C_8H_4O_2$		0.132
T-DEG	$C_4H_9O_3$		0.105
T-EG	$C_2H_5O_2$		0.0611
T-TPA	$C_8H_5O_3$		0.149
T-VIN	C_2H_3O		0.0430

TABLE 7.10. Equilibrium Reactions and Accompanying Reaction Rates for PET Polymerization

Reaction Stoichiometry	Reaction Rate

Main Reactions

Water Formation

1. EG + TPA $\rightleftharpoons$ T-EG + T-TPA + W

$$R_1 = 4k_1[\text{EG}][\text{TPA}]$$
$$-k_1/K_1[\text{T-EG}]\frac{[\text{T-TPA}]}{[\text{T-TPA}]+[\text{B-TPA}]}[\text{W}]$$

2. EG + T-TPA $\rightleftharpoons$ T-EG + B-TPA + W

$$R_2 = 2k_1[\text{EG}][\text{T-TPA}]$$
$$-k_1/K_1[\text{T-EG}]\frac{[\text{B-TPA}]}{[\text{T-TPA}]+[\text{B-TPA}]}[\text{W}]$$

3. DEG + TPA $\rightleftharpoons$ T-DEG + T-TPA + W

$$R_3 = 4k_1[\text{DEG}][\text{TPA}]$$
$$-k_1/K_1[\text{T-DEG}]\frac{[\text{T-TPA}]}{[\text{T-TPA}]+[\text{B-TPA}]}[\text{W}]$$

4. DEG + T-TPA $\rightleftharpoons$ T-DEG + B-TPA + W

$$R_4 = 2k_1[\text{DEG}][\text{T-TPA}]$$
$$-k_1/K_1[\text{T-DEG}]\frac{[\text{B-TPA}]}{[\text{T-TPA}]+[\text{B-TPA}]}[\text{W}]$$

5. T-EG + TPA $\rightleftharpoons$ B-EG + T-TPA + W

$$R_5 = 2k_2[\text{T-EG}][\text{TPA}]$$
$$-k_1/K_1[\text{B-EG}]\frac{[\text{T-TPA}]}{[\text{T-TPA}]+[\text{B-TPA}]}[\text{W}]$$

6. T-EG + T-TPA $\rightleftharpoons$ B-EG + B-TPA + W

$$R_6 = k_2[\text{T-EG}][\text{T-TPA}]$$
$$-k_1/K_1[\text{B-EG}]\frac{[\text{B-TPA}]}{[\text{T-TPA}]+[\text{B-TPA}]}[\text{W}]$$

7. T-DEG + TPA $\rightleftharpoons$ B-DEG + T-TPA + W

$$R_7 = 2k_2[\text{T-DEG}][\text{TPA}]$$
$$-k_1/K_1[\text{B-DEG}]\frac{[\text{T-TPA}]}{[\text{T-TPA}]+[\text{B-TPA}]}[\text{W}]$$

8. T-DEG + T-TPA $\rightleftharpoons$
   B-DEG + B-TPA + W

$$R_8 = k_2[\text{T-DEG}][\text{T-TPA}]$$
$$-k_1/K_1[\text{B-DEG}]\frac{[\text{B-TPA}]}{[\text{T-TPA}]+[\text{B-TPA}]}[\text{W}]$$

9. T-VIN + T-TPA + W $\rightarrow$ AA + TPA

$$R_9 = k_1/K_1[\text{T-VIN}:\text{T-TPA}][\text{W}]$$

10. T-VIN + B-TPA + W $\rightarrow$ AA + T-TPA

$$R_{10} = k_1/K_1[\text{T-VIN}:\text{B-TPA}][\text{W}]$$

Ester Interchange

11. EG + B-EG $\rightleftharpoons$ T-EG + T-EG

$$R_{11} = 2k_3[\text{EG}][\text{B-EG}]\frac{[\text{T-TPA}]}{[\text{T-TPA}]+[\text{B-TPA}]}$$
$$-k_3/K_3[\text{T-EG}][\text{T-EG}]\frac{[\text{T-TPA}]}{[\text{T-TPA}]+[\text{B-TPA}]}$$

12. EG + T-DEG $\rightleftharpoons$ DEG + T-EG

$$R_{12} = 2k_3[\text{EG}][\text{T-DEG}]\frac{[\text{T-TPA}]}{[\text{T-TPA}]+[\text{B-TPA}]}$$
$$-2k_3/K_3[\text{DEG}][\text{T-EG}]\frac{[\text{T-TPA}]}{[\text{T-TPA}]+[\text{B-TPA}]}$$

13. EG + B-DEG $\rightleftharpoons$ T-DEG + T-EG

$$R_{13} = 2k_3[\text{EG}][\text{B-DEG}]\frac{[\text{T-TPA}]}{[\text{T-TPA}]+[\text{B-TPA}]}$$
$$-k_3/K_3[\text{T-DEG}][\text{T-EG}]\frac{[\text{T-TPA}]}{[\text{T-TPA}]+[\text{B-TPA}]}$$

14. EG + B-EG $\rightleftharpoons$ T-EG + T-EG

$$R_{14} = 2k_3[\text{EG}][\text{B-EG}]\frac{[\text{B-TPA}]}{[\text{T-TPA}]+[\text{B-TPA}]}$$
$$-k_3/K_3[\text{T-EG}][\text{T-EG}]\frac{[\text{B-TPA}]}{[\text{T-TPA}]+[\text{B-TPA}]}$$

15. EG + T-DEG $\rightleftharpoons$ DEG + T-EG

$$R_{15} = 2k_3[\text{EG}][\text{T-DEG}]\frac{[\text{B-TPA}]}{[\text{T-TPA}]+[\text{B-TPA}]}$$
$$-2k_3/K_3[\text{DEG}][\text{T-EG}]\frac{[\text{B-TPA}]}{[\text{T-TPA}]+[\text{B-TPA}]}$$

16. EG + B-DEG $\rightleftharpoons$ T-DEG + T-EG

$$R_{16} = 2k_3[\text{EG}][\text{B-DEG}]\frac{[\text{B-TPA}]}{[\text{T-TPA}]+[\text{B-TPA}]}$$
$$-k_3/K_3[\text{T-DEG}][\text{T-EG}]\frac{[\text{B-TPA}]}{[\text{T-TPA}]+[\text{B-TPA}]}$$

17. DEG + T-EG $\rightleftharpoons$ EG + T-DEG

$$R_{17} = 2k_3[\text{DEG}][\text{T-EG}]\frac{[\text{T-TPA}]}{[\text{T-TPA}]+[\text{B-TPA}]}$$
$$-2k_3/K_3[\text{EG}][\text{T-DEG}]\frac{[\text{T-TPA}]}{[\text{T-TPA}]+[\text{B-TPA}]}$$

18. DEG + B-EG $\rightleftharpoons$ T-EG + T-DEG

$$R_{18} = 2k_3[\text{DEG}][\text{B-EG}]\frac{[\text{T-TPA}]}{[\text{T-TPA}]+[\text{B-TPA}]}$$
$$-k_3/K_3[\text{T-EG}][\text{T-DEG}]\frac{[\text{T-TPA}]}{[\text{T-TPA}]+[\text{B-TPA}]}$$

19. DEG + B-DEG $\rightleftharpoons$ T-DEG + T-DEG

$$R_{19} = 2k_3[\text{DEG}][\text{B-DEG}]\frac{[\text{T-TPA}]}{[\text{T-TPA}]+[\text{B-TPA}]}$$
$$-k_3/K_3[\text{T-DEG}][\text{T-DEG}]\frac{[\text{T-TPA}]}{[\text{T-TPA}]+[\text{B-TPA}]}$$

(Continued overleaf)

TABLE 7.10. (*Continued*)

Reaction Stoichiometry	Reaction Rate
20. $DEG + T\text{-}EG \rightleftharpoons EG + T\text{-}DEG$	$R_{20} = 2k_3[DEG][T\text{-}EG]\frac{[B\text{-}TPA]}{[T\text{-}TPA]+[B\text{-}TPA]}$
	$\qquad -2k_3/K_3[EG][T\text{-}DEG]\frac{[B\text{-}TPA]}{[T\text{-}TPA]+[B\text{-}TPA]}$
21. $DEG + B\text{-}EG \rightleftharpoons T\text{-}EG + T\text{-}DEG$	$R_{21} = 2k_3[DEG][B\text{-}EG]\frac{[B\text{-}TPA]}{[T\text{-}TPA]+[B\text{-}TPA]}$
	$\qquad -k_3/K_3[T\text{-}EG][T\text{-}DEG]\frac{[B\text{-}TPA]}{[T\text{-}TPA]+[B\text{-}TPA]}$
22. $DEG + B\text{-}DEG \rightleftharpoons T\text{-}DEG + T\text{-}DEG$	$R_{22} = 2k_3[DEG][B\text{-}DEG]\frac{[B\text{-}TPA]}{[T\text{-}TPA]+[B\text{-}TPA]}$
	$\qquad -k_3/K_3[T\text{-}DEG][T\text{-}DEG]\frac{[B\text{-}TPA]}{[T\text{-}TPA]+[B\text{-}TPA]}$
23. $EG + T\text{-}VIN \rightarrow AA + T\text{-}EG$	$R_{23} = 2k_3[EG][T\text{-}VIN]\frac{[T\text{-}TPA]}{[T\text{-}TPA]+[B\text{-}TPA]}$
24. $EG + T\text{-}VIN \rightarrow AA + T\text{-}EG$	$R_{24} = 2k_3[EG][T\text{-}VIN]\frac{[B\text{-}TPA]}{[T\text{-}TPA]+[B\text{-}TPA]}$
25. $DEG + T\text{-}VIN \rightarrow AA + T\text{-}DEG$	$R_{25} = 2k_3[DEG][T\text{-}VIN]\frac{[T\text{-}TPA]}{[T\text{-}TPA]+[B\text{-}TPA]}$
26. $DEG + T\text{-}VIN \rightarrow AA + T\text{-}DEG$	$R_{26} = 2k_3[DEG][T\text{-}VIN]\frac{[B\text{-}TPA]}{[T\text{-}TPA]+[B\text{-}TPA]}$
27. $T\text{-}EG + T\text{-}VIN \rightarrow AA + B\text{-}EG$	$R_{27} = k_3[T\text{-}EG][T\text{-}VIN]\frac{[T\text{-}TPA]}{[T\text{-}TPA]+[B\text{-}TPA]}$
28. $T\text{-}EG + T\text{-}VIN \rightarrow AA + B\text{-}EG$	$R_{28} = k_3[T\text{-}EG][T\text{-}VIN]\frac{[B\text{-}TPA]}{[T\text{-}TPA]+[B\text{-}TPA]}$
29. $T\text{-}DEG + T\text{-}VIN \rightarrow AA + B\text{-}DEG$	$R_{29} = k_3[T\text{-}DEG][T\text{-}VIN]\frac{[T\text{-}TPA]}{[T\text{-}TPA]+[B\text{-}TPA]}$
30. $T\text{-}DEG + T\text{-}VIN \rightarrow AA + B\text{-}DEG$	$R_{30} = k_3[T\text{-}DEG][T\text{-}VIN]\frac{[B\text{-}TPA]}{[T\text{-}TPA]+[B\text{-}TPA]}$

Side Reactions
Degradation of Diester Group

31. $B\text{-}TPA + B\text{-}EG \rightarrow T\text{-}VIN + T\text{-}TPA$	$R_{31} = k_4[B\text{-}EG]\frac{[B\text{-}TPA]}{[T\text{-}TPA]+[B\text{-}TPA]}$

Diethylene Glycol Formation

32. $B\text{-}TPA + T\text{-}EG + T\text{-}EG \rightarrow T\text{-}TPA + T\text{-}DEG$	$R_{32} = k_5[T\text{-}EG]\frac{[B\text{-}TPA]}{[T\text{-}TPA]+[B\text{-}TPA]}[T\text{-}EG]$
33. $T\text{-}VIN + T\text{-}EG \rightarrow B\text{-}DEG$	$R_{33} = k_6[T\text{-}VIN][T\text{-}EG]$

Ethylene Glycol Dehydration

34. $2T\text{-}EG \rightarrow B\text{-}DEG + W$	$R_{34} = k_7[T\text{-}EG]^2$
35. $T\text{-}EG + EG \rightarrow T\text{-}DEG + W$	$R_{35} = 2k_7[T\text{-}EG][EG]$
36. $2EG \rightarrow DEG + W$	$R_{36} = 4k_7[EG]^2$

Acetaldehyde Formation

37. $B\text{-}TPA + T\text{-}EG \rightarrow AA + T\text{-}TPA$	$R_{37} = k_8[T\text{-}EG]\frac{[B\text{-}TPA]}{[T\text{-}TPA]+[B\text{-}TPA]}$

We use the rate constants from Bhaskar et al. [16]. Table 7.11 summarizes these values. The rate constants k_1 through k_8, along with the equilibrium constants K_1 and K_3, are used to compute the reaction rates for reactions 1 to 37 in Table 7.10. Note that the activation energies in Table 7.11 are the intended values; Bhaskar et al. [16] convert the units incorrectly from Saint Martin and Choi [17].

We compute the rate constant using the equation

$$k_i = k_{0,i}\frac{w_{\text{SBOAC3}}}{0.0004}\exp\left(-\frac{E_{a,i}}{RT}\right) \tag{7.21}$$

where w_{SBOAC3} represents the mass fraction of SBOAC3. When no SBOAC3 is present, the term $w_{\text{SBOAC3}}/0.0004$ should be disregarded. This creates an inconsistency in that reaction rates with no SBOAC3 will be higher than when

TABLE 7.11. Rate Constants for the Polycondensation Stage of PET Melt Polymerization[a]

Reaction Rate Constant k_i and Equilibrium Constant K_i	$k_{0,i}$ (concentration in mol m^{-3}, time in min)	$E_{a,i}$ (J mol^{-1})	K_i
k_1, K_1	2.08E+03	7.36E+04	2.50
k_2	2.08E+03	7.36E+04	
k_3, K_3	1.76E+02	7.74E+04	0.161
k_4	2.22E+08	1.61E+05	
k_5	8.32E+04	1.25E+05	
k_6	2.50E+05	1.25E+05	
k_7	1.14E+05	1.25E+05	
k_8	4.77E+07	1.25E+05	

[a] $SBOAC_3 = 0.04\%$ by mass.

TABLE 7.12. Reaction Rates for all Nonpolymeric Components and Segments in the Polymerization of PET

Functional Group	Time Rate of Change
AA	$r_{AA} = R_9 + R_{10} + \sum\limits_{i=23}^{30} R_i + R_{37}$
B-DEG	$r_{\text{B-DEG}} = R_7 + R_8 - R_{13} - R_{16} - R_{19} - R_{22} + R_{29} + R_{30} + R_{33} + R_{34}$
B-EG	$r_{\text{B-EG}} = R_5 + R_6 - R_{11} - R_{14} - R_{18} - R_{21} + R_{27} + R_{28} - R_{31}$
B-TPA	$r_{\text{B-TPA}} = R_2 + R_4 + R_6 + R_8 - R_{10} - R_{31} - R_{32} - R_{37}$
DEG	$r_{\text{DEG}} = -R_3 - R_4 + R_{12} + R_{15} - \sum\limits_{i=17}^{22} R_i - R_{25} - R_{26} + R_{36}$
EG	$r_{\text{EG}} = -R_1 - R_2 - \sum\limits_{i=11}^{16} R_i + R_{17} + R_{20} - R_{23} - R_{24} - R_{35} - 2R_{36}$
T-DEG	$r_{\text{T-DEG}} = R_3 + R_4 - R_7 - R_8 - R_{12} + R_{13} - R_{15} + R_{16} + \sum\limits_{i=17}^{22} R_i + R_{25} +$ $R_{26} - R_{29} - R_{30} + R_{32} + R_{35}$
T-EG	$t_{\text{T-EG}} = R_1 + R_2 - R_5 - R_6 + 2(R_{11} + R_{14}) + R_{12} + R_{13} + R_{15} + R_{16} -$ $R_{17} + R_{18} - R_{20} + R_{21} + R_{23} + R_{24} - R_{27} - R_{28} - 2R_{32} - R_{33} - 2R_{34} -$ $R_{35} - R_{37}$
TPA	$r_{\text{TPA}} = -R_1 - R_3 - R_5 - R_7 + R_9$
T-TPA	$r_{\text{T-TPA}} =$ $R_1 - R_2 + R_3 - R_4 + R_5 - R_6 + R_7 - R_8 - R_9 + R_{10} + R_{31} + R_{32} + R_{37}$
T-VIN	$r_{\text{T-VIN}} = -R_9 - R_{10} - \sum\limits_{i=23}^{30} R_i + R_{31} - R_{33}$
W	$r_W = \sum\limits_{i=1}^{8} R_i - R_9 - R_{10} + \sum\limits_{i=34}^{36} R_i$

SBOAC3 is present below 0.04 wt%. This is because catalyzed reactions are not distinguished from uncatalyzed reactions as they are in the PA 6 polymerization kinetics set. Therefore, comparisons between polymerizations with and without SBOAC3 using this kinetics set should be avoided.

TABLE 7.13. Vapor-Pressure Parameters for Conventional Species in PA 6 Polymerization[a]

Species	A_i	B_i	C_i	D_i	E_i	F_i	G_i
Caprolactam (CL)	7.42E+01	−1.05E + 04	−6.89E + 00	1.21E − 18	6.00E + 00	342.36	806.00
Nitrogen (N_2)	5.98E+01	−1.10E + 03	−8.67E + 00	4.63E − 02	1.00E+00	63.15	126.1
Water (W)	7.36E+01	−7.26E + 03	−7.30E + 00	4.17E − 06	2.00E+00	273.16	647.13

Source: Daubert and Danner [18].
[a]Temperature in K, vapor pressure in Pa.

TABLE 7.14. Vapor-Pressure Parameters for Conventional Species in PET Polymerization[a]

Species	A_i	B_i	C_i	D_i	E_i	F_i	G_i
Acetaldehyde (AA)	2.06E+02	−8.48E + 03	−3.15E + 01	4.63E−02	1.00E+00	150.15	461.00
Diethylene glycol (DEG)	7.46E+01	−1.06E + 04	−6.82E + 00	9.10E−18	6.00E+00	262.70	680.00
Ethylene glycol (EG)	1.95E+02	−1.46E + 04	−2.54E + 01	2.01E−05	2.00E+00	260.15	645.00
Terephthalic acid (TPA)	−4.01E + 03	−1.77E + 05	−6.30E + 02	−5.12E − 01	1.00E+00	523.00	700.15

Source: Daubert and Danner [18].
[a]Temperature in K, vapor pressure in Pa. See Table 7.13 for the vapor-pressure parameters for nitrogen and water.

Table 7.12 shows all species balances due to reaction.

7.3.4. Physical Properties

We need a variety of physical properties to solve the material and energy balances for a SSP reactor. These include the vapor pressure, liquid molar volume, diffusivity, thermal conductivity, and polymer properties such as intrinsic viscosity.

a. Vapor Pressure We use the modified Antoine equation to model the vapor pressure for a pure component i, P_i^{sat} (Pa):

$$P_i^{sat} = \exp\left(A_i + \frac{B_i}{T} + C_i \ln T + D_i T^{E_i} \right) \qquad F_i \leq T \leq G_i \qquad (7.22)$$

Daubert and Danner [18] give known parameters for the volatiles (Tables 7.13 and 7.14).

We do not have vapor-pressure parameters for aminocaproic acid (ACA), antimony triacetate (SBOAC3), and cyclic dimer (CD); we treat these species as nonvolatile, like PA 6 and PET. To compute the vapor pressure of nonvolatile species (i.e., vapor pressure ca. 0), we set the first parameter to −40 and the rest to zero [19].

TABLE 7.15. DIPPR Parameters for Liquid Molar Volume for Conventional Species in PA 6 Polymerization (kmol m^{-3} K^{-1})

Species	A_i	B_i	C_i	D_i	E_i	F_i
Caprolactam (CL)	7.12E−01	2.54E−01	8.06E+02	2.86E−01	342.36	806.00
Nitrogen (N$_2$)	3.17E+00	2.85E−01	1.26E+02	2.93E−01	63.15	126.10
Water (W)	5.46E+00	3.05E−01	6.47E+02	8.10E−02	273.16	333.15

Source: Daubert and Danner [18].

TABLE 7.16. DIPPR Parameters for Liquid Molar Volume for Conventional Species in PET Polymerization^a

Species	A_i	B_i	C_i	D_i	E_i	F_i
Acetaldehyde (AA)	1.67E+00	2.60E−01	4.61E+02	2.78E−01	150.15	461.00
Diethylene glycol (DEG)	8.48E−01	2.64E−01	6.80E+02	1.97E−01	262.70	680.00
Ethylene glycol (EG)	1.34E+00	2.55E−01	6.45E+02	1.72E−01	260.15	645.00

Source: Daubert and Danner [18].
^{a}Values in (kmol m^{-3} K^{-1}; see Table 7.15 for the liquid molar volume parameters for water and nitrogen.

b. Liquid Molar Volume   We approximate the liquid molar volume of a mixture v_L (m^3 mol^{-1}) using the mole-fraction average of pure-component molar volumes $v_{L,i}$; this is *Amagat's law:*

$$v_L = \sum_i x_i v_{L,i} \tag{7.23}$$

For conventional species, we estimate the pure-component liquid density (kmol m^{-3}) using the *Design Institute for Physical Properties* (DIPPR) *correlation*:

$$\frac{1}{v_{L,i}} = \frac{A_i}{B_i^{1+(1-T/C_i)^{D_i}}} \qquad E_i \le T \le F_i \tag{7.24}$$

Tables 7.15 and 7.16 show known values for these four parameters.

We do not have parameters for aminocaproic acid (ACA), antimony triacetate (SBOAC3), terephthalic acid (TPA), and cyclic dimer (CD). For aminocaproic acid and cyclic dimer, we use the caprolactam parameters. For SBOAC3, we use the parameters for water. These approximations are valid as long as ACA, CD, and SBOAC3 do not occur in high quantities in liquid mixtures. For terephthalic acid, we assume that the pure component liquid–solid density is 1000 kg m^{-3}. The molecular weight is 166.1 kg kmol^{-1}. Dividing the density by the molecular weight gives the parameter A_i, which is 6.02 kmol m^{-3}. We set B_i and C_i equal

**TABLE 7.17. Parameters Needed to Estimate Polymer Density Using Equations
(7.26) and (7.27) for PA 6 and PET**

	PA 6	PET
$v_{g,298K}$ (cm^3 mol^{-1} seg^{-1})	104.4	144.5
$v_{c,298K}$ (cm^3 mol^{-1} seg^{-1})	92.0	130
E_g (cm^3 mol^{-1} seg^{-1} K^{-1})	4.45E−02[a]	4.42E−02
E_l (cm^3 mol^{-1} seg^{-1} K^{-1})	6.34E−02	1.29E−01
T_g (K)	323	340
T_m (K)	504	553

Source: Data from Van Krevelen [22], Mehta [23], and Rule [24].
[a]Assumed identical to the same parameter for PA 7.

to 1 and D_i equal to zero. This gives the desired approximation: for temperature
greater than 1 K, the liquid density of pure TPA is always 1000 kg m^{-3}.

For polymers, we compute the molar volume using a method outlined by Van
Krevelen [20]. One of the most practical methods treats the specific volume as a
linear function of temperature (Fig. 7.5) [21]. Semicrystalline polymers such as
polyamides and polyesters are composed of crystalline and amorphous domains
at temperatures below the crystalline melting point T_m. Let x_c represent the mole
fraction of crystalline domains. We write the average molar volume $\overline{v_i}$ (m^3 mol^{-1}

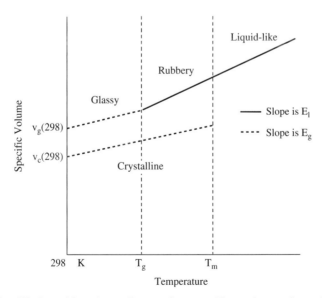

Fig. 7.5. Simplified specific volume diagram for crystalline and amorphous domains in a
polymer over a range of temperatures.

seg^{-1}) for a semicrystalline polymer as

$$\overline{v}_i = x_c v_c + (1 - x_c) v_a \tag{7.25}$$

where v_c is the molar volume for purely crystalline material and v_a is the molar volume of purely amorphous material. From Figure 7.5 we see that we calculate these molar volumes from the temperature T using the following relationships:

$$
\begin{aligned}
v_a &= f(T) \\
&= v_{g,298K} + E_g(T_g - 298) & 298\ \text{K} < T < T_g \\
&= v_{g,298K} + E_g(T_g - 298) + E_l(T - T_g) & T_g < T \tag{7.26} \\
v_c &= f(T) \\
&= 0 & T_m < T \\
&= v_{c,298K} + E_g(T - 298) & 298\ \text{K} < T < T_m \tag{7.27}
\end{aligned}
$$

where T_g is the polymer glass-transition temperature (K), T_m the crystalline melting temperature (K), E_g the amorphous glass thermal expansivity (m^3 mol^{-1} seg^{-1} K^{-1}), and E_l the amorphous liquid thermal expansivity (m^3 mol^{-1} seg^{-1} K^{-1}). Van Krevelen [22] tabulates values for the parameters $v_{g,298K}$, $v_{c,298K}$, E_g, E_l, T_g, and T_m for common polymers. Table 7.17 shows these parameters for PA 6 and PET. To convert "mol seg^{-1}" into mass, we need the molecular weight. For PA 6, we use the molecular weight of the aminocaproic acid repeat unit (B-ACA):

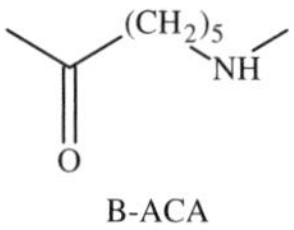

B-ACA

The monomer segment has a molecular weight of 1.132×10^{-1} kg mol^{-1}. For PET, we use

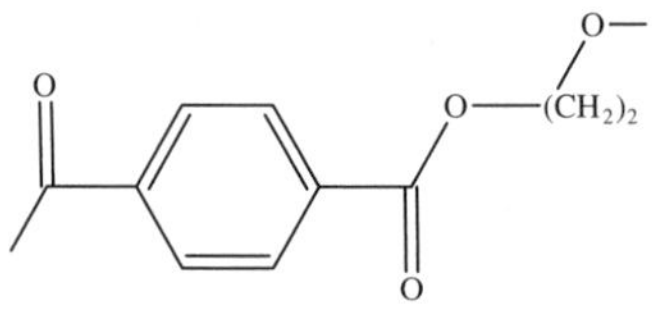

B-TPA:B-EG

where ":" represents a covalent bond. This monomer segment has a molecular weight of 1.922×10^{-1} kg mol^{-1}.

c. Diffusivity in Polymers   No values have been reported for the diffusivity of water and caprolactam in solid state PA 6 at SSP conditions. Therefore, we approximate the diffusivity using the values at melt conditions and adjust the parameter values empirically to fit reactor performance data. Seavey et al. [14] give approximations for the diffusivities of water and caprolactam in PA 6 melts. The equation for the diffusion coefficient is

$$D_i = D_{0,i} \exp\left(-\frac{E_i}{RT}\right) \tag{7.28}$$

where temperature is in kelvin. The parameters $D_{0,i}$ and E_i are, respectively, the preexponential factor for diffusion ($m^2 \ s^{-1}$) and the activation energy (J mol^{-1}). We use values of $2.21 \times 10^{-8} \ m^2 \ s^{-1}$ and $3010 \ J \ mol^{-1}$ for water, and $1.14 \times 10^{-8} \ m^2 \ s^{-1}$ and $33,457 \ J \ mol^{-1}$ for caprolactam.

For PET, Algeri and Rovaglio [2] collect and report values from Mallon and Ray [7,8] for the diffusivity of water, ethylene glycol, and acetaldehyde in solid state PET. For water and ethylene glycol, the diffusivity equation has the form

$$D_i = D_{0,i} \exp\left[-\frac{E_i}{R}\left(\frac{1}{T} - \frac{1}{493}\right)\right] \tag{7.29}$$

where temperature is in kelvin. The activation energy E_i is $1.24 \times 10^5 \ J \ mol^{-1}$ for both water and ethylene glycol. The preexponential factor $D_{0,i}$ is 1.93×10^{-10} $m^2 \ s^{-1}$ for ethylene glycol and $1.29 \times 10^{-10} \ m^2 \ s^{-1}$ for water.

For acetaldehyde in solid state PET, the diffusion coefficient is

$$D_{AA} = 1.15E + 04 \exp\left(-\frac{15,300}{T}\right) \tag{7.30}$$

The diffusion coefficient has units of $m^2 \ s^{-1}$ and the temperature is in kelvin.

d. Diffusivity in Nitrogen   Bird et al. outline a method [25] to predict the diffusivity of molecules in a gas:

$$D_i = \underbrace{a(T_{c,A}T_{c,B})^{5/12 - b/2}(p_{c,A}p_{c,B})^{1/3}\left(\frac{1}{M_A} + \frac{1}{M_B}\right)^{1/2}\frac{T^b}{p}}_{\text{lumped constant}} \tag{7.31}$$

The diffusion coefficient D_i has units of $cm^2 \ s^{-1}$, T and T_c are temperature and critical temperature (K), p and p_c are pressure and critical pressure (atm), and M is the molecular weight (g mol^{-1}). For nonpolar gas pairs, the constants a and b are 2.745×10^{-4} and 1.823, respectively. For water with a nonpolar gas, the constants are 3.640×10^{-4} and 2.334, respectively.

We use the water/nonpolar gas constants to describe all diffusing species in nitrogen. The lumped constants [as defined in (7.31), units are $m^2 \ Pa \ s^{-1} \ K^{-2.334}$]

for water, caprolactam, ethylene glycol, diethylene glycol, and acetaldehyde are 2.69×10^{-6}, 1.04×10^{-6}, 1.51×10^{-6}, 1.18×10^{-6}, and 1.84×10^{-6}, respectively.

e. Thermal Conductivity of Nitrogen   The thermal conductivity of the gas phase is easier to estimate. We use the thermal conductivity of nitrogen (κ, W m^{-1} K^{-1}) [18]:

$$\kappa_{nitrogen} = \frac{(3.51E - 4)T^{0.765}}{1 + (25.8/T)} \qquad 78K \leq T \leq 1500 \text{ K} \qquad (7.32)$$

f. Solution Viscosity   Polymer plants do not typically measure the molecular weight of the polymer. Instead, they measure a viscosity that correlates directly with molecular weight. However, our models predict the number-average molecular weight of the polymer MW_n (g mol^{-1}):

$$MW_n = 2\frac{\sum_{i=\text{all segments}} [i]}{\sum_{j=\text{end segments}} [j]} \qquad (7.33)$$

where $[i]$ is the concentration (mol m^{-3}) of a particular polymer segment i. To relate our models to plant measurements, we need a correlation to convert MW_n to the solution viscosity of interest. For extractables-free PA 6, a relevant solution viscosity is the sulfuric acid relative viscosity RV [26]:

$$RV = \frac{MW_n}{11,500} + 1 \qquad (7.34)$$

For PET, a corresponding measure of molecular weight is the intrinsic viscosity. The intrinsic viscosity, IV (dL g^{-1}), can be estimated as [27]

$$IV = 2.1E - 4M_n^{0.82} \qquad (7.35)$$

The intrinsic viscosity is determined in a solution of 1,1,2,2-tetrachloroethane and phenol solvent 1 : 1 by volume at 25°C.

7.4. NUMERICAL SOLUTION

We use the method of lines to discretize and solve our partial differential equation (PDE) set. We start by rewriting one of the PDEs that we wish to solve: the polymer-phase material balance equation (7.2):

$$\frac{\partial C_{P,i}}{\partial t} + v_P\frac{\partial C_{P,i}}{\partial z} = D_{P,i}\frac{\partial^2 C_{P,i}}{\partial z^2} + N_{P-G,i}a_P + r_i$$

This equation applies to the continuous spatial domain z (the axial domain in the SSP reactor). We discretize this domain using uniformly distributed points

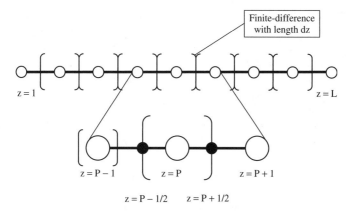

Fig. 7.6. One-dimensional axial domain.

(Fig. 7.6). A given interior point $z = \mathrm{P}$ lies at the center of a finite difference of length dz (i.e., we are using a cell-centered finite-difference method).

Since our PDEs are second order with respect to z, we need two boundary conditions. For our SSP equations, we prescribe a Dirichlet condition at the entrance $z = 1$ and a Neumann condition at the exit $z = \mathrm{L}$:

$$C_{\mathrm{P},i}|_{z=1} = C_{\mathrm{P},i}^{\mathrm{feed}} \tag{7.36}$$

$$\left.\frac{\partial C_{\mathrm{P},i}}{\partial z}\right|_{z=\mathrm{L}} = 0 \tag{7.37}$$

We can view the Neumann condition as an extrapolation boundary condition. This will be clear when we discretize that boundary condition.

We introduce spatial discretization of the convective derivative using a first-order upwind finite difference, used frequently in computational fluid dynamics [28]:

$$
\begin{aligned}
u_{\mathrm{P}}\frac{\partial C_{\mathrm{P},i}|_{\mathrm{P}}}{\partial z} &= u_{\mathrm{P}}\frac{C_{\mathrm{P},i}|_{\mathrm{P}+1/2} - C_{\mathrm{P},i}|_{\mathrm{P}-1/2}}{\Delta z} \\
&= \begin{cases} u_{\mathrm{P}}\dfrac{C_{\mathrm{P},i}|_{\mathrm{P}} - C_{\mathrm{P},i}|_{\mathrm{P}-1}}{\Delta z} & \text{for } u_{\mathrm{P}} > 0 \\[2ex] u_{\mathrm{P}}\dfrac{C_{\mathrm{P},i}|_{\mathrm{P}+1} - C_{\mathrm{P},i}|_{\mathrm{P}}}{\Delta z} & \text{for } u_{\mathrm{P}} < 0 \end{cases}
\end{aligned} \tag{7.38}
$$

We have assumed a constant value for the fluid velocity u (m s^{-1}). In the upwind finite difference, we use a backward finite difference (from point $z = \mathrm{P}$ to $z = \mathrm{P} - 1$) to approximate the partial derivative associated with convection when the fluid velocity is positive. If the fluid velocity is negative, we use a forward finite difference (from point $z = \mathrm{P} + 1$ to $z = \mathrm{P}$) to approximate the convective term. The term *upwind* comes from the fact that we are approximating values at particular points using upwind values (i.e., those values that lie upstream in relation to the direction of the fluid velocity). So, for example, if the fluid velocity is positive, the concentration value at point $P - 1$ is upwind from the point P. Similarly, if the fluid velocity is negative, the concentration value at point $P + 1$ is upwind from the point P.

The upwind finite-difference method is numerically stable because it mimics the underlying physics of the problem, specifically by using information from points upstream in the flow to approximate information at a downstream point. However, it achieves this by being numerically diffuse. Numerical diffusion means that in addition to material convecting across the domain, material also diffuses both up and down the domain (in addition to diffusion from the second derivative in the original PDE). We can reduce this numerical diffusion using more sophisticated techniques for representing the convective derivative. We shall see one such technique and its impact on reducing numerical diffusion at the end of this section.

Moving on to the second derivative, we use a second-order, centered difference approximation:

$$
\begin{aligned}
\frac{\partial^2 C_{P,i}|_{P}}{\partial z^2} &= \frac{\left.\frac{\partial C_{P,i}}{\partial z}\right|_{P+1/2} - \left.\frac{\partial C_{P,i}}{\partial z}\right|_{P-1/2}}{\Delta z} \\
&= \frac{(C_{P,i}|_{P+1} - C_{P,i}|_{P})/\Delta z - (C_{P,i}|_{P} - C_{P,i}|_{P-1})/\Delta z}{\Delta z} \\
&= \frac{C_{P,i}|_{P-1} - 2C_{P,i}|_{P} + C_{P,i}|_{P+1}}{\Delta z^2}
\end{aligned}
\tag{7.39}
$$

In sum, we can write a complete, discretized PDE at point P in the domain:

$$
\frac{\partial C_{P,i}}{\partial t} + u_{P}\frac{\partial C_{P,i}}{\partial z} = D_{P,i}\frac{\partial^2 C_{P,i}}{\partial z^2} + N_{P-G,i}a_{P} + r_i
$$

$$\downarrow \tag{7.40}$$

$$
\left.\frac{\partial C_{P,i}}{\partial t}\right|_{P} = -u_{P}\frac{C_{P,i}|_{P} - C_{P,i}|_{P-1}}{\Delta z} + D_{P,i}\frac{C_{P,i}|_{P-1} - 2C_{P,i}|_{P} + C_{P,i}|_{P+1}}{\Delta z^2}
$$

$$
+ N_{P-G,i}a_{P}|_{P} + r_i|_{P}
$$

Velocity is constant and positive in this case (i.e., flow is from left to right). Upon inspection of (7.40), we see that we cannot use the equation when $P = 1$ or N, as the values at $P - 1$ and $P + 1$ are not defined in these cases. Instead, at these points, we apply the boundary conditions

$$
C_{P,i}|_{z=1} = C_{P,i}^{\text{feed}} \tag{7.41}
$$

$$
\left.\frac{\partial C_{P,i}}{\partial z}\right|_{z=L} = 0 \rightarrow \frac{C_{P,i}|_{z=L} - C_{P,i}|_{z=L-1}}{\Delta z} = 0 \rightarrow C_{P,i}|_{z=L} = C_{P,i}|_{z=L-1} \tag{7.42}
$$

The coupled set of ordinary differential equations for each interior point (7.40), together with the gas-phase material balance and the polymer and gas-phase energy balances, can be solved using an ordinary differential equation (ODE) solver such as ODEPACK [29]. Since these are ODEs, we must provide initial conditions, dictated by the scenario we wish to simulate.

7.5. EXAMPLE SIMULATION AND APPLICATION

Consider a PET SSP reactor that is 30 m long and has a diameter of 5 m. The radius of the polymer pellets is 1 mm, and the void fraction is 0.35. The polymer crystallinity is 0.1, and the pressure is atmospheric. Table 7.18 shows the feed mole flow rates (mol s^{-1}) of both the polymer and gas phases. We wish to determine how sensitive the steady-state intrinsic viscosity is to changes in:

- Gas flow rate
- Radius of the polymer pellets
- Gas feed temperature

We assume that all axial diffusion coefficients are 1×10^{-10} m^2 s^{-1}.

We use 10 points to discretize the domain. Instead of using the first-order upwind approximation for the convective derivative (7.38), we use a more accurate third-order upwind approximation, the quadratic upwind interpolation for convective kinematics (QUICK) [30]:

$$\left.\frac{\partial C_i}{\partial z}\right|_P = \frac{C_i|_{P+1} - C_i|_{P-1}}{2dz} - \frac{C_i|_{P+1} - 3C_i|_P + 3C_i|_{P-1} - C_i|_{P-2}}{8dz} \qquad (7.43)$$

TABLE 7.18. Polymer and Gas Feeds for the PET Solid State Polymerizer

Species	Polymer Feed (mol s^{-1}) 230°C	Gas Feed (mol s^{-1}) 250°C
Acetaldehyde	1.89E−2	0
Diethylene glycol	2.60E−5	0
Ethylene glycol	3.10E−1	0
Nitrogen	0	185
Terephthalic acid	2.50E−4	0
Water	2.26E+0	0
PET	6.96E−1	0
B-EG	5.72E+1	0
B-TPA	5.78E+1	0
T-EG	1.14E−1	0
T-TPA	2.26E−1	0
T-VIN	9.42E−3	0
T-DEG	1.01E+0	0

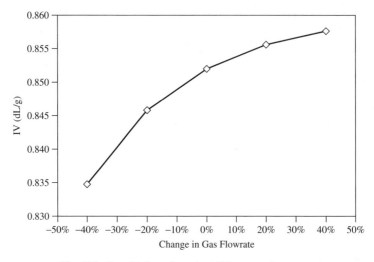

Fig. 7.7. Sensitivity of product IV on gas flow rate.

We can use this approximation for points P = 3 to $N - 1$, where N is the number of grid points, starting at 1. For the interior point P = 2, we use the first-order upwind approximation.

To represent steady state, we integrate the dynamic material and energy balance equations for 80 hours. The incoming IV is 0.603 dL g^{-1}. The base-case outgoing IV is 0.852 dL g^{-1}. In Figure 7.7 we show the sensitivity of product IV on gas flow rate. The results are intuitive—a higher gas flow rate removes more condensate, thus increasing the intrinsic viscosity. Figure 7.8 shows the

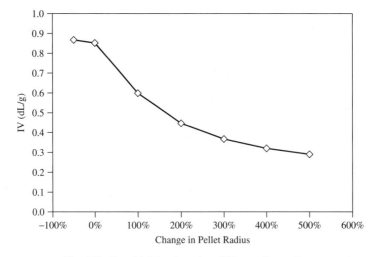

Fig. 7.8. Sensitivity of product IV on pellet radius.

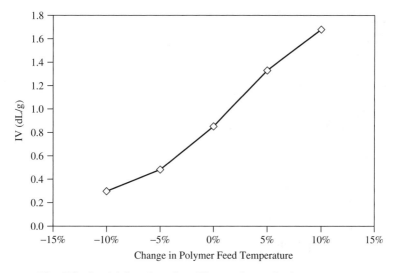

Fig. 7.9. Sensitivity of product IV on polymer feed temperature.

sensitivity of IV on pellet radius. Increasing the pellet radius decreases the product IV. Finally, Figure 7.9 shows the sensitivity of product IV on polymer feed temperature. Increasing the polymer feed temperature has a very strong effect on the final product molecular weight. The product IV increases dramatically with a relatively small temperature increase of 10%.

7.6. MODIFICATIONS TO ACCOUNT FOR CRYSTALLIZATION

A more detailed simulation for a SSP reactor may contain a model that tracks the growth of crystallinity in the polymer phase as well as the impact of crystallinity on reaction rates and diffusion.

We can derive a convection equation for the degree of crystallinity ϕ_c (m^3 m^{-3}, volume basis):

$$\frac{\partial \phi_c}{\partial t} + u_P \frac{\partial \phi_c}{\partial z} = G_c \tag{7.44}$$

The degree of crystallinity is ϕ_c (m^3 m^{-3}), u_P is the velocity of the polymer phase (m s^{-1}), z is the axial position in the reactor (m), and G_c is the growth rate of crystals (s^{-1}).

TABLE 7.19. Crystallization Kinetic Parameters

Parameter	PA 6	PET
E_x (kJ mol^{-1})	35.1	40.24
T_m^0 (K)	501	533
ψ (K)	225.8	201
K_1 (min^{-1})	1.03×10^6	4.33×10^6
K_2 (min^{-1})	4.326×10^7	6.6×10^4

Source: Malkin et al. [31,33].

Lucas et al. [6] model the growth rate of crystals G_c (s^{-1}) using an equation developed by Malkin et al. [31–33]:

$$G_c = \left[K_1 \exp \left(-\frac{E_x}{RT} - \frac{\psi T_m^0}{T(T_m^0 - T)} \right) \right.$$
$$\left. + K_2 \exp \left(-\frac{E_x}{RT} - \frac{\psi T_m^0}{T(T_m^0 - T)} \right) \phi_c \right] (\phi_c^0 - \phi_c) \qquad (7.45)$$

Here K_1, K_2, and ψ are constants, E_x the activation energy of the segment transfer across the nucleus–melt boundary, R the gas constant, T_m^0 the equilibrium melt temperature, and T the current temperature. The equilibrium melting temperature is not the same as the melting temperature observed experimentally. Malkin et al.'s parameter values for PA 6 and PET are given in Table 7.19.

An increase in crystallinity generally decreases diffusivity. Yoon et al. [34] model a linear dependence of diffusivity on crystallinity:

$$D_i = D_{0,i} \exp \left(-\frac{E_i}{RT} \right) (1 - z_c) \qquad (7.46)$$

where D_i is the diffusivity of species i in the semicrystalline polymer phase (m^2 s^{-1}), $D_{0,i}$ the preexponential factor (m^2 s^{-1}), E_i the activation energy (J mol^{-1}), T the temperature (K), R the ideal gas law constant (8.314 J mol^{-1}K^{-1}), and z_c the mass fraction of crystals in the polymer.

Regarding the effect of crystallinity on reactions, it is common to assume that nonpolymers and polymer end groups exist only in the amorphous phase [35]. Based on this assumption, we would adjust the concentration of these species in the reaction rate equations: consider, for example, the concentration of water in the reaction rate equations in Table 7.5:

$$[W]_{\text{amorphous}} = \frac{[W]_{\text{overall}}}{1 - z_c} \qquad (7.47)$$

This modification, required because reactions take place only in the amorphous phase, applies to all nonpolymers and end segments (i.e., those whose name is prefixed by a "T-"). We also need to adjust the concentration of repeat segments, which comprise the crystalline phase entirely. Their concentration adjustment is

$$[\text{B-ACA}]_{\text{amorphous}} = \frac{[\text{B-ACA}]_{\text{overall}} - \frac{z_c}{\text{MW}_{\text{B-ACA}}}}{1 - z_c} \tag{7.48}$$

Note that all of these concentrations as well as the crystal fraction are on a mass basis.

Since the reactions take place only in the amorphous phase, we multiply the reaction rate by the amorphous fraction to get the overall reaction rate:

$$r_i = r_{i,\text{amorphous}}(1 - z_c) \tag{7.49}$$

To demonstrate these modifications of the reaction kinetics, consider the polycondensation reaction between a terminal amine and carboxylic acid group in PA 6. The original reaction rate from Table 7.5 is

$$R_5 = k_2[\text{T-NH}_2][\text{T-COOH}]$$

$$-k_2'[\text{W}][\text{B-ACA}]\frac{[\text{B-ACA}]}{[\text{B-ACA}] + [\text{T-NH2}]}$$

When modifying this to account for crystallinity, we write

$$R_5 = (1 - z_c)k_2\frac{[\text{T-NH}_2]}{1 - z_c}\frac{[\text{T-COOH}]}{1 - z_c}$$

$$- (1 - z_c)k_2'\frac{[\text{W}]}{1 - z_c}\frac{[\text{B-ACA}] - \left(\frac{z_c}{\text{MW}_{\text{B-ACA}}}\right)}{1 - z_c}$$

$$\times \frac{[\text{B-ACA}] - \left(\frac{z_c}{\text{MW}_{\text{B-ACA}}}\right)}{[\text{B-ACA}] - \left(\frac{z_c}{\text{MW}_{\text{B-ACA}}}\right) + [\text{T-NH}_2]/(1 - z_c)}$$

$$= \frac{k_2}{1 - z_c}[\text{T-NH}_2][\text{T-COOH}]$$

$$- \frac{k_2'}{1 - z_c}[\text{W}]\frac{(1 - z_c)\left([\text{B-ACA}] - \left(\frac{z_c}{\text{MW}_{\text{B-ACA}}}\right)\right)^2}{(1 - z_c)\left([\text{B-ACA}] - \left(\frac{z_c}{\text{MW}_{\text{B-ACA}}}\right)\right) + [\text{T-NH}_2]} \tag{7.50}$$

There are many other modifications that are available to a SSP reactor modeler; see the studies cited in Table 7.2.

7.7. CONCLUSIONS

Scientists and engineers have many choices to make when building models for solid state polymerization (SSP) reactors. These choices are often dictated by the availability of laboratory and plant data as well as the needs of manufacturing.

We have outlined some key questions to consider for helping guide these choices, along with pertinent literature references documenting specific models.

To demonstrate the key concepts in SSP reactor modeling, we have outlined the basic material and energy balances. In particular, we consider the solid state polymerization of PA 6 and PET. These material and energy balance equations, accounting for convection and diffusion of heat and mass between the polymer and gas phases, along with polymerization, result in a system of PDEs that we discretize using the finite-difference method and integrate using the method of lines. We solve these PDEs for the SSP of PET and show model predictions for the dependence of the intrinsic viscosity of the polymer product on gas flow rate, pellet radius, and polymer feed temperature.

The next logical step in the development of these models is to increase the resolution further and include explicit modeling of the gas flow field within the SSP reactor. This will shed light on how to design gas distributors properly to avoid dead zones within the reactor. Another important aspect of SSP that has not been addressed in the literature is pellet agglomeration. Above a certain temperature, pellets tend to begin clumping; obviously, this should be avoided in an SSP reactor. Having a model that predicts this behavior allows us to avoid what are essentially infeasible operating conditions. Today's state-of-the-art SSP models do not have this feature. However, making these advances depends critically on the availability of complete data for industrial SSP reactors, which is usually difficult or impossible to collect.

Acknowledgments

We thank Alliant Techsystems (particularly Mark Matusevick, Vice President), Aspen Technology (particularly Larry Evans, Founder, and Mark Fusco, President), China Petroleum and Chemical Corporation (particularly Tianpu Wang, President, and Xianghong Cao, Chief Technology Officer), Formosa Petrochemical Corporation (particular Wilfred Wang, Chairman) and Milliken Chemical (particularly John Rekers, President; Jean Hall, Business Manager; Jack Miley, Director of R&D and Technology) for supporting our educational programs in computer-aided design and process systems engineering at Virginia Tech.

REFERENCES

1. Vouyiouka SN, Karakatsani EK, Papaspyrides CD. Solid-state polymerization. *Prog. Polym. Sci.* 2005;30:10–37.

2. Algeri C, Rovaglio M. Dynamic modeling of a poly(ethylene terephthalate) solid-state polymerization reactor: I. Detailed model development. *Ind. Eng. Chem. Res.* 2004;43:4243–4266.

3. Devotta I, Mashelkar RA. Modeling of polyethylene terephthalate reactors: X. A comprehensive model for the solid-state polycondensation process. *Chem. Eng. Sci.* 1993;48:1859–1867.

4. Lee EH, Yeo YK, Choi KY, Kim HY. Modeling of a solid-state polycondensation process for the production of PET. *J. Chem. Eng. Jpn.* 2003;36:912–925.

5. Li LF, Huang NX, Tang ZL, Hagen R. Reaction kinetics and simulation for the solid-state polycondensation of nylon 6. *Macromol. Theory Simul.* 2001;10:507–517.

6. Lucas B, Seavey KC, Liu YA. Steady-state and dynamic modeling for new product design for the solid-state polymerization of poly(ethylene terephthalate). *Ind. Eng. Chem. Res.* 2007;46:190–202.

7. Mallon FK, Ray WH. Modeling of solid-state polycondensation: I. Particle models. *J. Appl. Polym. Sci.* 1998;69:1233–1250.

8. Mallon FK, Ray WH. Modeling of solid-state polycondensation: II. Reactor design issues. *J. Appl. Polym. Sci.* 1998;69:1775–1788.

9. Rovaglio M, Algeri C, Manca D. Dynamic modeling of a poly(ethylene terephthalate) solid-state polymerization reactor: II. Model predictive control. *Ind. Eng. Chem. Res.* 2004;43:4267–4277.

10. Wang XQ, Deng DC. A comprehensive model for solid-state polycondensation of poly(ethylene terephthalate): combining kinetics with crystallization and diffusion of acetaldehyde. *J. Appl. Polym. Sci.* 2002;83:3133–3144.

11. Yao KZ, McAuley KB, Berg D, Marchildon EK. A dynamic mathematical model for continuous solid-phase polymerization of nylon 6,6. *Chem. Eng. Sci.* 2001;56:4801–4814.

12. Yao KZ, McAuley KB, Marchildon EK. Simulation of continuous solid-phase polymerization of nylon 6,6: III. Simplified model. *J. Appl. Polym. Sci.* 2003;89:3701–3712.

13. Chen CC. A segment-based local composition model for the Gibbs energy of polymer solutions. *Fluid Phase Equilibria*. 1993;83:301–312.

14. Seavey KC, Liu YA, Lucas B, Khare NP, Lee T, Pettrey J, Williams TN, Mattson J, Schoenborn E, Larkin C, Hu H, Chen CC. New mass-transfer model for simulating industrial PA 6 production trains. *Ind. Eng. Chem. Res.* 2004;43:5063–5076.

15. Arai Y, Tai K, Teranishi H, Tagawa T. The kinetics of hydrolytic polymerization of ε-caprolactam: III. Formation of cyclic dimer. *J. Polym.* 1981;22:273–277.

16. Bhaskar V, Gupta SK, Ray AK. Modeling of an industrial wiped-film poly(ethylene terephthalate) reactor. *Polym. React. Eng.* 2001;9:71–99.

17. Saint Martin HC, Choi KY. Two-phase model for continuous final-stage melt polycondensation of poly(ethylene terephthalate): 2. Analysis of dynamic behavior. *Ind. Eng. Chem. Res.* 1991;30:1712–1718.

18. Daubert TE, Danner RP. *Physical and Thermodynamic Properties of Pure Chemicals*. Data Compilation, Vols. 1 to 3. Hemisphere, New York, 1989.

19. Bokis CP, Orbey H, Chen CC. Properly model polymer processes. *Chem. Eng. Prog.* 1999;95(4): 39–52.

20. Van Krevelen DW. *Properties of Polymers*. Elsevier, New York, 1990, pp. 71–107.

21. Simha R, Boyer RF. General relation involving the glass temperature and coefficients of expansion of polymers. *J. Chem. Phys.* 1962;37:1003–1007.

22. Van Krevelen DW. *Properties of Polymers*. Elsevier, New York, 1990, pp. 82–85, 92–95, 144, 145, 164, 165, 792, 793, 803.

23. Mehta RH. Physical constants of various polyamides. In: *Polymer Handbook*, 4th ed., Vol. 1, Brandrup J, Immergut EH, Grulke EA, eds. Wiley, New York, 1999, pp. V/121–V/133.

24. Rule M. Physical constants of poly(oxyethylene-oxyterephthaloyl) (poly(ethylene terephthalate)). In: *Polymer Handbook*, 4th ed., Vol. 1, Brandrup J, Immergut EH, Grulke EA, eds. Wiley, New York, 1999, pp. V/113–V/118.

25. Bird RB, Stewart WE, Lightfoot EN. *Transport Phenomena*. Wiley, New York, 1960, p. 505.

26. Xiao W, Huang N, Tang Z, Filippini-Fantoni R. Simulation of polymerization in an industrial two-step VK tubular reactor. *Macromol. Mater. Eng.* 2003;288:235–244.

27. Zhi-Lian T, Gao Q, Nan-Xun H, Sironi C. Solid-state polycondensation of poly(ethylene terephthalate): kinetics and mechanism. *J. Appl. Polym. Sci.* 1995;57: 473–485.

28. Anderson JD. *Computation Fluid Dynamics: The Basics with Applications*. McGraw-Hill, New York, 1995, pp. 497–508.

29. Hindmarsh AC. ODEPACK, a systematized collection of ODE solvers. In: *Scientific Computing*, Stepleman RS, et al., eds. North-Holland, Amsterdam, 1983, pp. 55–64.

30. Leonard BP. A stable and accurate convective modeling procedure based on quadratic upstream interpolation. *Comput. Methods Appl. Mech. Eng.* 1979;19:59–98.

31. Malkin AY, Beghishev VP, Keapin IA. Macrokinetics of polymer crystallization. *Polymer*. 1983;24:81.

32. Malkin AY, Beghishev VP, Keapin IA, Bolgov SA. General treatment of polymer crystallization kinetics: 1. A new macrokinetic equation and its experimental verification. *Polym. Eng. Sci.* 1984;24:1396.

33. Malkin AY, Beghishev VP, Keapin IA, Andrianova ZS. General treatment of polymer crystallization kinetics: 2. The kinetics of nonisothermal crystallization. *Polym. Eng. Sci.* 1984;24:1402.

34. Yoon KH, Kwon MH, Jeon MH, Park OO. Diffusion of ethylene glycol in solid state poly(ethylene terephthalate). *Polym. J.* 1993;25:219.

35. Zimmerman J. Equilibria in solid phase polyamidation. *J. Polym. Sci. Polym. Lett. Ed.* 1964;2:955.

8

RECENT DEVELOPMENTS IN SOLID STATE POLYMERIZATION OF POLY(ETHYLENE TEREPHTHALATE)

S. A. WADEKAR

204, Mohandeep Society, Almeida Road, Chandanwadi, Panchpakhadi, Thane (West), State-Maharashtra, India

U. S. AGARWAL, W. H. BOON, AND V. M. NADKARNI

Reliance Industries Limited, Maharashtra, India

Solid State Polymerization, Edited by Constantine D. Papaspyrides and Stamatina N. Vouyiouka
Copyright © 2009 John Wiley & Sons, Inc.

8.1. INTRODUCTION

Step-growth polymerization processes involve buildup of chain lengths by reaction between end groups of participating molecules (i.e., monomers or oligomers). For example, the polycondensation during polymerization to poly(ethylene terephthalate) (PET), the most common thermoplastic polymer, involves reaction of glycol (or hydroxyl) end groups either with another glycol (or hydroxyl) end group [reaction (8.1)], or with an acid end group [reaction (8.2)]:

Transesterification/polycondensation:

$$(8.1)$$

Esterification:

$$(8.2)$$

These reversible reactions are accompanied by the evolution of a condensate molecule such as the ethylene glycol (EG) and water, and the desired forward reaction is facilitated by removal of these condensate molecules.

PET possesses very good mechanical properties, such as strength, stiffness, ductility, and good resistance to chemicals, and has better oxygen and carbon dioxide barrier properties. Therefore, PET is widely used for fiber, films, and container applications. For fiber and film applications, the intrinsic viscosity (IV) requirement is about 0.6 dL g^{-1}, whereas for container applications, PET resin that has an IV of 0.7 to 0.84 dL g^{-1} is required. For some specialized applications, such as technical yarns and tire cord, the IV requirement is in excess of 0.9 dL g^{-1}.

To produce PET of about 0.6 dL g^{-1}, melt polymerization technology is employed for reaction of EG with dimethyl terephthalate (DMT) or purified terephthalic acid (PTA), and the reaction mass flows through a series of reactors (Fig. 8.1) while increasing temperature and vacuum gradually to allow a sufficient rate of condensate removal despite the increasing melt viscosity [1].

When starting with PTA and EG as raw materials for melt polymerization, the condensate formed in the first step (esterification) [2–10] at 240 to 260°C is water. A low-molecular-weight oligomer, bishydroxyethyl terephthalate (BHET), is the main product. In the subsequent step, a prepolymer having a degree of polymerization (DP) of approximately 25 to 30 is produced. In the polycondensation step, the DP is increased further, up to 80 to 100, by carrying out a reaction in vacuum and at a temperature between 260 and 300°C. The residence time in the melt polycondensation reaction is on the order of 3 to 5 h. Due to the long thermal history of the reaction at higher temperature in the polycondensation stage, degradation reactions also take place. The degradation reactions generate certain undesirable volatiles, such as acetaldehyde (AA) and diethylene glycol (DEG), as well as ether groups and vinyl end groups in polymer chains [11,12].

When the IV requirement of PET is higher, the melt polymerization process becomes somewhat complex. This is due to difficulties in handling high-viscosity melt, continuous surface renewal required to maintain forward reaction, degradation due to very high temperature, and the longer reaction times and limited

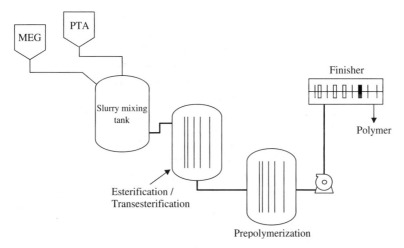

Fig. 8.1. Three typical vessels for a PET melt polymerization process.

efficiency of the catalyst system. These problems in producing higher-IV (IV >0.7 dL g^{-1}) PET resin are generally overcome by extrusion of polymer melt from the finisher reactor as strand, freezing with chilled water, cutting into particles (pellets) of dimensions a few millimeters, followed by crystallization, drying, and solid state polymerization. The SSP processes essentially involve heating the particles to 200 to 240°C (thus below the melting point) for 10 to 30 h under inert gas flow (or vacuum) to remove the condensates formed by reactions (8.1) and (8.2) [4,13–17]. The degradation reactions are reduced due to the SSP reaction temperature being 60 to 80°C lower than the melt polymerization temperature. Despite the lower temperature during SSP, the small diffusion length for the condensates to be removed from the pellets allows a reasonable rate of reaction.

Transport of polymer during industrial SSP process and subsequent polymer processing demands flowability of the solid particles. This flowability can be hindered by lump formation, due to sticking or sintering of the polymer particles above the glass-transition temperature ($T_g = 78$°C) of PET. Therefore, a crystallization process preceding and during the SSP is required to reduce sintering of the solid particle. Crystallization to the semicrystalline state also increases the concentration of reactive chain-end groups in the amorphous regions, thereby assisting the SSP kinetics. On the other hand, crystallization also reduces the molecular mobility of polymer chains, thereby reducing the frequency of interaction of the end groups and hence the SSP rate. Further, some of the end groups are trapped in the crystalline domains and become inactive and hence affect the overall reaction rate [14,15]. As the crystallization proceeds during SSP and the crystallinity reaches a critical value, end-group interaction and diffusion of by-products becomes extremely difficult, resulting in termination of polymerization [18–20].

It has been found that during solid state polymerization, transesterification is favored by higher concentration of the hydroxyl end groups [reaction (8.1)], while an intrinsically higher rate constant, faster diffusion, and removal of water as condensate favor esterification [reaction (8.2)]. Because of this difference in the reaction rate and the rate of diffusion of by-products, the ratio of end groups (OH/COOH) is very important for optimizing SSP reaction rates [21,22].

Solid state polymerization demands relatively simple equipment: commercial continuous SSP processes usually carried out at atmospheric conditions (i.e., low energy consumption, less degradation reactions) since SSP temperatures are much lower than that required for melt polymerization reactions, ensuring better product quality for food and beverage packing, for example (e.g., fewer oligomers and fewer acetaldehyde, acid end groups, and vinyl end groups).

When using an SSP process, there are certain limitations in producing PET resin with a high comonomer content, such as isophthalic acid (IPA) and cyclohexanedimethanol (CHDM). The comonomers often reduce the crystallizability of the base prepolymer. For a trouble-free SSP process, prepolymer must possess a certain crystallinity and melting point to avoid sintering or lump formation in reactor. A higher comonomer content increases sintering propensity in the reactor. To overcome this problem, residence time in the crystallizer section preceding the reactor during the SSP needs to be increased further and SSP temperatures need to be kept at lower levels. The net effect of these changes reduces plant throughput, since SSP reactivity is directly proportional to temperature [12,23].

Molecular-weight increase in the melt polymerization up to an IV value of about 0.6 dL g^{-1} occurs at a faster rate than with SSP. To achieve a higher-IV PET (>0.7 dL g^{-1}) in melt polymerization, a substantially higher residence time may be required. Due to the very high residence time at higher temperatures (>280°C), and to the presence of the catalyst, severe degradation reactions can take place. Degradation reactions occur at all stages: during transesterification, esterification, and polycondensation during the melt polymerization process as well as during SSP. The extent of degradation varies in different stages. For example, most DEG formation (ca. 70%) occurs during transesterification during preheating and during early stages in the polycondensation reaction [24].

The extent of DEG formation during PET synthesis is higher when produced by a direct esterification route between PTA and EG compared to the trans-esterification route of DMT and EG [15]. Higher DEG content was found to reduce the light stability of PET and the melting point by around 5°C for each 1% of DEG [26]. Other degradation reactions, such as vinyl and acetaldehyde end groups, are formed mainly at later stages of polycondensation reaction. The rate of formation of all degradation reaction by-products depends on process parameters such as residence time and temperature and on the type and concentration of catalyst used [27]. Acetaldehyde formation at even in a few parts per million causes off-flavor in mineral water and soft drinks packaged in PET bottles [28]. Degradation reactions and by-product generation become serious issues when higher-IV polymer is produced using melt-phase polymerization. Various functional additives [29–38] can be used to control degradation reactions and to control acetaldehyde in the resin as well as finished product. Alternatively, volatilization in an inert gas stream can be used to remove acetaldehyde from a polymer prior to processing. However, this may increase the production cost of resin. One of the advantages of making high-IV PET by all melt processes is the lower downstream processing temperature, since resin will have much lower crystallinity, due to solidification from melt of high-molecular-weight polymer. Additionally, resin with a high comonomer content, such as IPA and CHDM, can be produced favorably, since reduction of crystallization tendency is not limiting for all melt processes.

In a typical process, a base polymer, generally termed a *precursor*, that has a IV value of 0.50 dL g^{-1} or higher is subjected to SSP. These precursors are amorphous in nature, since they are produced by a strand-cutting method under water at the exit of a melt polymerization reactor. In other SSP patents, reaction may also begin with a base polymer of IV about 0.15 to 0.40 dL g^{-1} [39–44]. The main reasons for developing a SSP process for a low-IV precursor is to reduce both capital and operation expenses (CAPEX and OPEX) and better product attributes. Due to the low viscosity of these low-IV precursors, a standard strand-cutting method cannot be used, and hence they are produced by different techniques and with different shapes, such as hemispherical [41–44] or spherical

[45,46]. Due to the unique particle formation technologies, their physical properties are entirely different compared to standard base polymer. Conventional base polymer is amorphous in nature and has to be crystallized to the required level of crystallinity before it enters an SSP reactor to prevent sintering or lump formation. Low-IV prepolymers of new inventions are already crystalline in nature or crystallized in-line during the particle-former process [41–44].

8.2. CONVENTIONAL SOLID STATE POLYMERIZATION PROCESSES

Conventionally, high-IV PET resin is made by solid state polymerization of base polymer of IV about 0.6 dL g^{-1}. Conventional SSP process can be used for producing bottle-grade PET resin (IV about 0.7 to 0.84 dL g^{-1}). PET for technical yarn application with IV about 1 to 1.2 dL g^{-1} is also achieved through SSP, but this case is not addressed here. Due to the increased global demand for high-IV PET resin for beverage and food and packaging applications, several new continuous SSP plants have been installed around the world. Solid state polymerization during commercial production of high-IV PET is preceded by drying for moisture removal and crystallization to eliminate sintering of the amorphous chips.

8.2.1. Moisture Removal

It is well known that polyesters containing higher moisture contents are susceptible for hydrolytic degradation [the reverse of reaction of (8.2)] if exposed to higher temperature [12,47,48]. This may result in a decrease in IV during the initial stages at the higher temperatures involved during solid state polymerization, as hydrolytic degradation becomes faster at those temperatures. Therefore, it is necessary to dry the polyester chips before exposure to the SSP temperatures greater than 200°C. The effect of moisture content at various temperatures on SSP behavior of PET was studied by Duh [49]. He subjected undried crystalline prepolymer to SSP at various temperatures ranging from 170 to 220°C. He found that there was a net depolymerization during drying at starting temperatures above 190°C and some net polymerization at starting temperatures below 180°C. A starting temperature of 180°C gave the shortest process time and most satisfactory product quality. Drying during industrial SSP processes is achieved in the preheater and crystallizer sections, which therefore also act as "dryers" and result in a drop in the moisture content of the solid chips to below 50 ppm, thereby avoiding any hydrolytic degradation during IV buildup in an SSP reactor.

8.2.2. Crystallization

Most industrial PET processes produce amorphous prepolymer via melt polymerization, which is then subjected to SSP for further increasing the molecular weight. Before entering to the final SSP reactor operating at temperatures greater

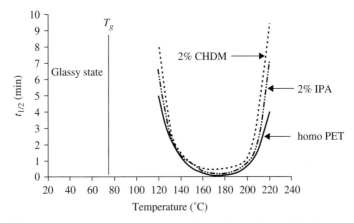

Fig. 8.2. Effect of comonomers on temperature dependence of crystallization rate of PET. (From Fakirov [50] by permission of John Wiley & Sons, Inc.)

than 200°C, the prepolymer crystallinity must be achieved as high as 40% for avoiding sintering or lump formation. A typical crystallization half-time curve for PET is given in Figure 8.2 [50]. The crystallization rate of PET is highest at a temperature 170 to 180°C. This information is very useful during the precrystallization step of the SSP process. If crystallization is carried out in the range 170 to 180°C in the crystallizer vessel, the required crystallinity can be achieved in a shorter duration.

Crystallinity in the prepolymer is also considered to increase the SSP reactivity [51–53]. The hypothesis is that during crystallization the reactive end groups are rejected from the crystalline unit cell, and thus their concentration in the amorphous phase and hence the SSP rate increase. While increasing crystallinity will continue to help prevent sintering, the SSP reactivity may begin to decrease beyond some level of crystallinity as the diffusivities of the reactive end groups and the condensate get reduced. For example, Duh found that if particle size increases beyond a critical value of about 7 mm, the SSP reactivity decreases with increasing crystallinity because of the excessively increased by-product-diffusion resistance within the PET particles. Thus, PET crystallization during the SSP process needs to be controlled by adjusting the process temperature such that the IV rate rise should dominate the rate of crystallization, yet the sintering should be avoided.

If the degree of crystallinity is very high (e.g., in SSP processes where the base polymer IV is low) [41–44], higher SSP temperatures are required for maintaining sufficient SSP reactivity. However, the higher SSP temperatures also increase the melting point of the product polymer, due to the formation of more perfect crystals at higher SSP temperatures. Higher melting may demand higher temperatures during processing, such as injection molding or extrusion, and thus increases the energy consumption.

8.2.3. Industrial Solid State Polymerization Processes

In the early stage, most SSP processes were operated in batch mode, usually in vacuo, using apparatus such as jacketed double-cone dryers or tumble dryers. In modern times, the batch mode of operation is used for small-volume specialty products, such as poly(ethylene naphthalate) (PEN), poly(trimethylene terephthalate) (PTT), or for recycling applications. Rotary or agitated paddle dryers are also sometimes used. Later development of continuous processes in tall reactors brought in increased severity of sintering of solid particles under load of the particles above. For example, for PET, the sintering is controlled largely by crystallization at lower temperature (ca. 140 to 180°C) prior to SSP at higher temperature (200 to 220°C). Although higher temperatures are desirable to reduce the SSP time and hence SSP reactor size, the increasing sintering tendency limits the SSP temperature in practical processes where uninterrupted continuous runs are expected for periods exceeding one year. In addition to preventing sticking, the other parameters that govern the technological advancement in SSP processes are moisture removal, efficient heat and mass transfer, and inert gas purification–recycle circuit. In the following paragraphs, two conventional industrial SSP processes, UOP-Sinco and Buhler SSP, are described.

a. UOP-Sinco Process   Today, more than 60 UOP-Sinco SSP units for producing PET resin for beverage applications are in operation worldwide. UOP-Sinco is a four-step SSP process, as shown in Figure 8.3. In this process, amorphous base chips of IV around 0.6 dL g^{-1} are fed to the fluidized-bed precrystallizer, where these chips are heated and crystallized in a multizone fluidized-bed heat exchanger. In this section the polymer dust is carried away with the fluidizing gas and thus removed from the chips. These partially crystallized chips are then fed to the crystallizer to achieve target crystallinity, crystal perfection, and high onset temperature of melting. Onset temperature of melting of chips is extremely important for avoiding lump formation and sintering in the main solid state polymerization reactor, and thus carrying out trouble-free SSP process. A key feature in this operation is that chips may be crystallized in a moist nitrogen environment to reduce acetaldehyde in the final product [54].

Crystallized chips are then fed to the main SSP reactor by gravity. The SSP reactor is a moving-bed reactor where IV is raised to the desired value. Reaction by-products, such as oligomers, AA, ethylene glycol (EG), and water are removed by carrier gas, which is nitrogen. An optimum gas/solid ratio is required in the reactor to ensure the best process performance. The temperature of the final product is brought down in the cooling section, where dust is also removed and is finally sent to the bagging section. Depending on the desired end application, an IV value as high as 1.2 dL g^{-1} can be obtained.

Use of mechanical agitation to keep pellets from aggregating in the heating vessels and reactors is generally avoided, to avoid dust generation. However, mechanical agitation is sometimes used to break up agglomerates at reactor outlet. Separators may also be used to remove the dust and oligomer at the vessel exit, to

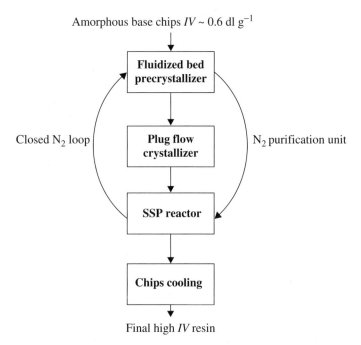

Amorphous base chips $IV \sim 0.6$ dl g^{-1}

Fig. 8.3. UOP-Sinco four-step SSP process. (From the UOP website [54].)

reduce subsequent contamination and sintering. In the UOP-Sinco process [55], the SSP is conducted in an inert gas (e.g., nitrogen) environment, where gas flows in a countercurrent direction. The gas is purified in the gas purification system, where hydrocarbon by-products are removed by a specially designed N_2 loop to convert organic by-products into carbon dioxide and water.

b. Buhler Process   Buhler AG is among the world's leading SSP technology suppliers. Buhler has commissioned over 100 SSP units worldwide for bottle-grade PET, technical fiber applications, and bottle-to-bottle recycle PET SSP. The four critical stages of the process are essentially similar to UOP-Sinco:

1. Primary crystallization in fluidized beds using hot air ($<185°$C). The fluidization provides a high heating rate, and the associated gentle agitation helps prevent agglomeration without significant dust formation.
2. Annealing or secondary crystallization with heated nitrogen (210 to 220°C) in a preheater, where roof-type internals reduce the pressure and help resist sintering.
3. SSP in a reactor column providing a residence time of 10 to 20 h at $210 \pm 5°$C with countercurrent N_2 flow, with gas flow/solid flow mass ratio below 1.
4. Cooling below 60°C in a fluidized-bed cooler with fresh air.

8.2.4. Nitrogen Purification Units in a Solid State Polymerization Process

SSP is carried out in an inert-gas atmosphere, such as nitrogen. Reaction by-products such as EG, water, AA, and low-molecular-weight oligomers are stripped from the system by carrier gas. These impurities are present in the carrier gas in quantities, defined as methane equivalent, up to 2000 to 3000 ppm or more. For economical considerations, carrier gas has to be reused in the system by maintaining its purity. If the impurity level of recirculated carrier gas is higher, it affects adversely the polymerization reaction and also the resulting polymer quality. For example, if the oxygen content in the carrier gas is higher than the maximum permissible limit (preferably less than 5 ppm), it can cause yellowing of the polymer.

In both the Buhler and the UOP-Sinco processes, the inert-gas purification is carried out in a nitrogen purification–cleaning unit [56–62] by oxidation to convert organic impurities to carbon dioxide. A schematic of the nitrogen purification unit (NPU) system of the UOP-Sinco [61] process is shown in Figure 8.4. The steps involved are oxidation for converting the organic impurities to CO_2, a deoxidation with hydrogen to eliminate any excess oxygen used in the first step and a drying step to eliminate the water formed during polymerization process. The oxidation step is carried out with the oxygen or with a gas containing oxygen, such as air, by using an oxygen concentration that is no more than in slight excess of stoichiometric quantity as regards the organic impurities. The oxidation step is usually carried out at a temperature between 250 and 600°C by circulating the inert gaseous stream over a catalyst bed formed of a support coated with platinum or with platinum and palladium. The gas-purifying step is carried out by circulating the gas over silica gel, molecular sieves, or other beds of drying materials. The water generated both by the polymerization process and during the oxidation step is thus eliminated. After drying, the inert gas is recirculated to the SSP process.

In particular, the inert gas from the SSP reactor containing reaction by-products is first passed through a filter, where solid particulates are removed. An airstream containing oxygen is injected into the inert gas before entering the NPU, which is essentially a catalyst bed and a drier. The ratio of oxygen to impurities is of great importance in this process. Control of the oxidation step is highly critical to ensure minimum oxygen content in the purified gas stream at the exit of the NPU section. Sometimes, a hydrogen gas stream is connected at the exit of the NPU system for deoxidation from the purified inert-gas stream.

The selection of oxidation catalysts is very important for ensuring complete oxidation of organic volatiles in the catalyst bed. Two patents [58,62] disclose the use of platinum or platinum/palladium-based catalysts for nitrogen purification in the continuous SSP process of PET. Buhler AG has developed a method [57] for purifying a gas stream by using rhodium or a rhodium alloy on an inert porous support at a temperature between 280 and 350°C. This process is based on a lambda probe–like measuring unit, which is based on a noble-metal film such as rhodium or its alloy disposed in the catalyzer. A lambda probe measures quantitatively the purity in the contaminated nitrogen gas stream entering the

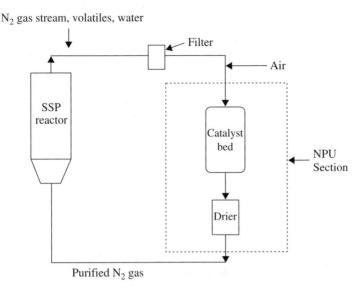

N₂ gas stream, volatiles, water

Fig. 8.4. Nitrogen gas purification system. (From Canadian Patent CA 2143099 [61].)

NPU and the nitrogen gas stream at the exit of the NPU. It is claimed that the process has excellent control of the purification of the inert gas stream. In another invention of UOP [56], a reduced platinum or platinum–palladium catalyst is used for the oxidation process. Support for the catalyst systems may be an alumina type [63,64]. According to the process, gas purification is achieved at low temperature (250°C) compared to the NPU systems described earlier. It is proposed that due to the reduced state of the metal catalyst (i.e., a valance state of zero compared to 2 or 4 in the case of metal oxides), significantly greater oxidative efficiency was achieved. Catalysts that are being used in NPU systems can be regenerated by carbon removal using known methods.

8.3. NEW SOLID STATE POLYMERIZATION PROCESSES

Due to the very high demands for high-IV PET resin for packaging applications, such as food, beverages, and technical yarns, the SSP technology suppliers have developed several novel processes, which claim to offer several advantages:

- Lower capital cost (CAPEX) and operating cost (OPEX)
- Large production capabilities with a single unit
- Easier product changeover
- Superior and consistent product quality (i.e., fewer by-products, improved color, low-melting PET, less IV variation)

8.3.1. Invista NG3 Process

Invista (formerly DuPont) Technologies has developed and commercialized a unique polymerization process, NG3 [41–44,65–70], for producing PET resin for bottle-grade applications. Unique features of this technology are:

1. The melt process is terminated at low-IV (0.2 to 0.3 dL g^{-1}) prepolymer. Prepolymerization is carried out in a column reactor operated at atmospheric pressure, and no finisher reactor or subsequent melt polymerization is involved. The vacuum processes (and associated high equipment costs) are thus eliminated.
2. Prepolymer is solidified into a hemispherical shape on a particle former.
3. Further crystallization and postpolymerization to IV = 0.74 to 0.84 dL g^{-1} is carried out in the solid state.

Since a larger part of the polymerization is in solid state, the time spent at the high temperatures of melt polymerization is reduced. This results in reduced degradation and reduced color formation. Melt polymerization sections consist of esterification and a prepolymerizer column–reactor. The IV value of prepolymer at the exit of a column reactor is in the range 0.2 to 0.3 dLg^{-1}. The low-IV prepolymer is solidified [67] into semicrystalline particles through a particle formation (Rotoformer) process developed by Sandvik (Fig. 8.5).

The prepolymer melt drops from a Rotoformer fall on the moving heated belt to solidify and achieve desired crystallinity. The hemispherical particles so-formed, called *pastilles*, are then used as a precursor for the solid state polymerization process. More specifically, hemispherical prepolymer particles are formed by dropping melt droplets on a preheated (100 to 200°C) moving metal belt. Due

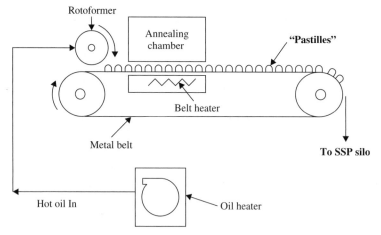

Fig. 8.5. Rotoformer particle-formation technology of NG3 process. (From U.S. Patent 5,730,913 [67].)

to the very high temperature (260 to 300°C) of melt droplet, a quenching effect occurs when it falls on the low-temperature metal belt. As a result, the bottom surface of hemispherical prepolymer particles exhibits different crystalline morphology from that of the top dome surface. It is observed that for every hemispherical crystalline prepolymer particle, the bottom surface remains weaker than the top dome surface in terms of the mechanical strength. Hence, during conveying to SSP silos or to the bagging section, larger amounts of dust and broken particles may be generated. This results in increased waste generation compared to the conventional melt polymerization process. We have demonstrated the use of crystallization enhancers for a reduction in waste generation [71,72].

In the solid state polymerization process, prepolymer is first heated to the temperature desired in a preheater section and then fed to a solid state polymerization reactor to achieve the target IV values (0.74 to 0.84 dL g^{-1}), depending on end-use application. In the conventional solid state polymerization process, base PET prepolymer (IV ca. 0.6 dL g^{-1}) is amorphous and thus needs to be crystallized (ca. 25 to 30%) in the crystallizer section before it is fed to the SSP reactor (Fig. 8.6). The solidified prepolymer (IV ca. 0.2 to 0.3 dL g^{-1}) in the NG3 process is already semicrystalline [41–44,65–67] (ca. 35%), and is crystallized to a higher degree before feeding to the SSP reactor. This higher level of crystallinity is responsible for the slow rate of SSP despite the higher

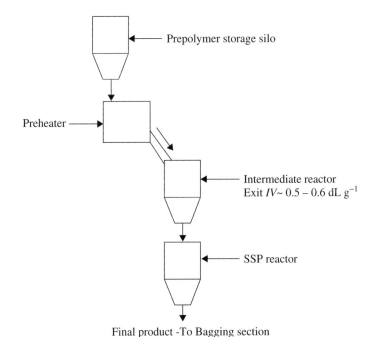

Final product -To Bagging section

Fig. 8.6. NG3 SSP process flowsheet. (From U.S. Patent 6,703,479 [68] and the UOP website [69].)

SSP temperature in the NG3 process. The large delta value between the final polymer and the base prepolymer IV in the NG3 process also adds to the high SSP time. The longer residence time and higher-temperature operation in SSP, as well as the higher crystal size and perfection, result in high melting temperatures [70]. The higher melting temperatures may demand a higher increased processing temperature during injection molding, thereby increasing energy consumption.

Due to the lower prepolymer IV, more oligomers get liberated. These oligomers are mainly low-melting substances such as monohydroxyethyl terephthalate (MHET) and bishydroxyethyl terephthalate (BHET), which are linear oligomers. Due to the very high nitrogen gas flow in the preheater, low-molecular-weight oligomers get carried away along with dust particles into nitrogen circulation lines. It is also observed that broken chips, dust, and low-molecular-weight oligomers stick to the preheater gas distribution plates, plenum, and sidewalls, all made of stainless steel 316. Over a period of time, bridging of prepolymer, dust, broken particles, or oligomers occurs in the preheater nitrogen gas distribution plates, plenum, and sidewalls of preheater and nitrogen gas circulation lines. This results in the blocking of the holes in the preheater nitrogen gas distribution plate, thereby interrupting the nitrogen gas flow and velocity and also the fluidization of prepolymer in the preheater. The pressure drop across the preheater nitrogen distribution plates increases due to this bridging at a faster rate, due to which frequent shutdown of the plant becomes necessary to clean these deposits.

Similarly, oligomer liberation is also seen in the conditioning vessel, which is downstream of the preheater in the solid state polymerization process. In the conditioning vessel, the IV value of the prepolymer gets built up, but at the same time the process also generates dust and liberates oligomers such as MHET and BHET together with other by-products, such as ethylene glycol and water, which causes deposit formation. To a large extent, oligomers get liberated in the conditioning vessel, where IV increases up to 0.6 dL g^{-1}. Oligomers get carried away along with nitrogen gas and tend to deposit on the surfaces of the conditioning vessel. Over a period of time, this increases the nitrogen flow pressure drop across the conditioning vessel and nitrogen line, and nitrogen gas flow becomes inconsistent. Deposits are also formed on the nitrogen gas circulation lines, filters, and heaters, thereby reducing their life, too. These deposits get degraded over a period of time, due to very high temperature and residence time. When these degraded deposits get dislodged during the process, they contaminate the final product in the form of black specs. We found that the use of low-energy surfaces such as fluoropolymer coatings on reactor internals and gas line reduces the formation of deposits and thus reduces product contamination [73].

8.3.2. Bepex Process

In the Bepex process [74–78], amorphous prepolymer chips are fed to the first-stage horizontal crystallizer (Bepex SolidAire). Material is heated through direct contact with the interior surface of the horizontal vessel. To prevent sticking during the heating stage, a single shaft with paddles rotates inside the crystallizer. Subsequently, chips are fed to the dryer to remove moisture from chips. Dried chips are then fed to the preheater and postcrystallizer to increase the temperature up to SSP reaction temperature. During this stage, PET achieves the desired crystallinity and reaction temperature. To stabilize the crystal structure to prevent sticking and lump formation in the SSP reactor, chips are fed to the annealer, which contains a screw to convey the material in a plug-flow manner. At this stage, no additional heating is provided to the chips, but the chip temperature is maintained with slight agitation. Additional crystallization, crystal perfection, and subsequent heat dissipation are achieved during this stage. The level of crystallinity at the annealer exit is greater than 45%. This makes possible SSP at high temperatures without processing problems (e.g., sintering, lump formation), even for PET containing larger amounts of comonomers, such as IPA and naphthalate.

Crystallized prepolymer chips are then fed to the SSP reactor to achieve the desired IV values. The SSP reactor is a plug flow continuous-moving-bed reactor. The IV value is increased by passing nitrogen to the chips in a countercurrent direction. To ensure a uniform plug flow and to minimize IV variation during SSP, a mechanical discharger is employed at the bottom of the reactor. This mechanical discharger also breaks the agglomerates formed in the SSP reactor. A product cooling unit is provided after the SSP reactor for cooling hot chips before these go to the bagging section. Chips exiting the SSP reactor are cooled by circulating air in a fluidized-bed cooler. This also helps in removing dust and fines generated during the SSP process.

8.3.3. Aquafil-Buhler Process

Aquafil Engineering has developed a unique reactor design named UPR (i.e., Universal Polymerization Reactor). It can be used to produce polyester up to an IV value of 0.98 dL g^{-1}. It can be a one- or two-reactor process. In the one-reactor process using UPR, a base prepolymer IV value of about 0.20 to 0.40 dL g^{-1} is achieved in the melt state (without finisher), and then subjected to SSP to increase IV suitable for bottle-grade applications. Thus, advantages associated with reduced residence time in melt are applicable. In contrast to the NG3 process, which utilizes a particle former (Fig. 8.5) for prepolymer solidification, Aquafil process utilizes underwater pelletizer. In this case, particle shape is spherical, due to this unique underwater pelletizing system [79].

Recently, Buhler has developed and commercialized a new SSP technology, S-HIP (i.e., Solid High IV Polycondensation) for the production of bottle-grade resin from a base polymer IV value of about 0.4 dL g^{-1}. S-HIP technology involves direct crystallization of resin with uniform control over crystallinity. Advantages claimed for S-HIP are reduced CAPEX, OPEX, and space requirements due to a reduced number of vessels, lower SSP process temperatures and shorter residence times, and better product properties, such as reduced AA concentrations.

8.3.4. M&G Process

Gruppo Mossi & Ghisolfi (M&G) has developed EasyUp, a new proprietary SSP technology claimed to have lower capital (around 40% lower than standard SSP) and operating costs than those of standard SSP technology. Daily plant capacity can be extended to 2000 tons. The main difference in the EasyUp and conventional SSP processes is that EasyUp uses a horizontal reactor, whereas conventional SSP reactors are vertical. Especially for applications such as tire cords and conveyer belts, where IV requirements are 0.9 to 1.1 dL g^{-1}, it is difficult to achieve high IV values in conventional vertical single-vessel SSP reactors. There are usually three ways of increasing IV with conventional vertical SSP reactors:

1. By increasing temperature
2. By increasing bed height
3. By using a second SSP vessel

In both cases, when increasing temperature and bed height in vertical SSP reactors beyond a certain limit, the tendency of agglomeration, sticking, or lump formation of chips is also increased, due to very high compressive forces at the bottom of the reactor. In commercial-scale vertical reactors that have a daily capacity of 500 to 700 tons, an IV value of 0.72 to 0.86 dL g^{-1} can be very well achieved for bottle-grade applications. However, there are challenges for vertical reactors in terms of capacity beyond 700 tons day^{-1}.

The M&G process avoids this by using a rotary kiln type of horizontal reactor (Fig. 8.7) that has a particular inclination angle with respect to the horizontal plane

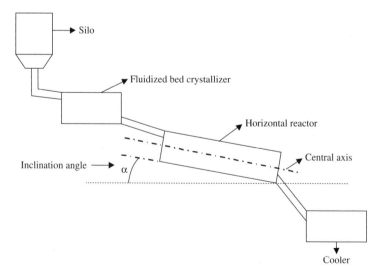

Fig. 8.7. Rotary kiln–type horizontal SSP reactor. (From WIPO Patent WO 2004/018541 [80].)

[80], thereby eliminating the requirement of high vertical SSP reactors for higher productivity, and hence reducing the tendency to agglomerate. This process is claimed to have lower production costs, due primarily to the decreased pressure drop required for the purge gas, ease of operation, and quicker product change. Additionally, this makes possible a high degree of plug flow and, consequently, a high degree of uniformity and consistency in the final product. Due to the reduced bed height, the crystallinity requirement in the base polymer is also greatly reduced, to avoid sticking and agglomeration during the process.

8.4. POLY(ETHYLENE TEREPHTHALATE) FLAKE RECYCLING USING SOLID STATE POLYMERIZATION

At the present time, post-consumer recycling (PCR) of PET bottles is being conducted under the 3R policy (reduce, reuse, recycle). It is driven by legislation encouraging collection, and to some extent by the economics of cheap bottle flakes as the raw material for polyester. PET is recycled primarily into textiles, carpets, strappings, and sheets. Recycling high-IV flakes into low-IV products can be done by direct processing after washing and drying, or through conversion to a monomer or oligomer through depolymerization. However, the process of recycling used high-IV PET bottles back to high-IV resin for beverage applications is particularly demanding, as it requires SSP to upgrade IV, in addition to achieving acceptable quality in terms of color, lower acetaldehyde content, and freedom from contaminants. In particular, the processing of virgin pellets into PET bottles, and the remelting of used bottle flakes for conversion to pellets, are both accompanied by reductions in IV, coloration, and formation of such by-products as AA, DEG, vinyl end groups, and excess acid end groups. The drop in IV demands IV upgrade to a level similar to that of virgin resin. This can be achieved by subjecting PET bottle flakes to SSP after removal of contaminants and aromatics. Following the collection of used PET bottles and their crushing, the flakes undergo sorting and washing. The flakes may then be subjected directly to crystallization and SSP, or be melted and pelletized prior to crystallization and SSP. Whereas the former process offers the advantage of a fast SSP rate, eliminates melting and pelletizing and the associated costs and polymer degradation, the latter process employs melt filtration to reduce contamination, reduces the SSP equipment size (lower bulk of pellets), and reduces the product variability associated with crystallization and SSP rate dependence on the flakes with a range of thickness and orientation (e.g. bottle neck vs. wall).

There are several PET flakes recycling technologies available worldwide. Among them, the processes employing SSP are: UOP-Sinco R-PET and Buhler's bottle–to–bottle recycling, and are discussed here. The Buhler process involves remelting of flakes into pellets. Other R-PET technologies available are VacuRema (Erema), Viscostar and Recostar (Starlinger), Closed Loop Process (OHL), etc.

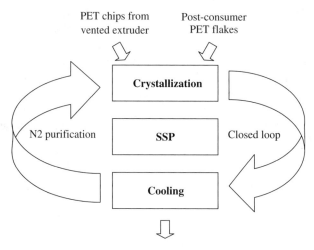

Fig. 8.8. Three-step SSP process of UOP-Sinco. (From the UOP website [81].)

a. UOP-Sinco R-PET Technology   UOP-Sinco has developed a PET recycling process named R-PET [81]. In this process, direct PET flakes, or PET chips produced from flakes by using vented extruder, can be subjected to SSP. Feed material IV can be 0.5 to 0.8 dL g^{-1}, and the three-step process (Fig. 8.8) can lead to IV up to 1 dL g^{-1} (SSP-R process for R-PET, www.uop.com). The high IV resin produced using R-PET process is suitable for various applications, including beverage containers, industrial yarns, sheets, and PET strapping.

b. Buhler's Bottle-to-Bottle Recycling   This process [82] has developed by keeping environmental aspects in view. A unique ring extruder was designed for melting and pelletizing PET flakes free of contaminants. It provides high surface/volume ratio, as well as a high surface renewal rate. Thus, aromas or migrated or diffused solvents are removed very effectively under reduced pressure. Thus, the predrying requirement can be eliminated. A melt filter is also provided to remove unwanted solid particles. These pellets are then fed to the continuous solid state polymerization reactor [83]. PET resin produced by this process is claimed to meet all FDA requirements for beverage applications.

8.5. PARTICLE FORMATION TECHNOLOGIES

One critical aspect of continuous commercial SSP processes is the need for uniform flow and avoiding sticking of the solid particles. These are, in turn, influenced by the shape and crystallinity and morphology of the particles. The shape, and to some extent the crystalline morphology, are decided by the process of forming particles at the end of the melt polymerization process. Conventionally, strand pelletizers have been used to quench the melt extrudate into nearly amorphous solid strands, using water slides for cooling and for forwarding the strands, followed by cutting into particles of somewhat cylindrical shape. Although some degree of crystallization can be induced during this process by pulling the strands at higher speed [84], most crystallization is achieved in the solid state step during pellet reheating to an intermediate temperature of 120 to 180°C for several hours (Section 8.2.3) prior to heating to SSP temperature.

In recent years, underwater pelletizers offered by companies such as Reiter, KBG, and Gala are drawing attention as alternative pelletizing systems. Here, the melt extruded from the die, while tending to form strands, is immediately cut by rotating cutter blades in a water box. The still-soft polymer has an opportunity to transform shape toward spherical. Water is flowing continuously through the water box to cool and solidify the polymer strands and pellets and carry the pellets out of the water box through transport piping to a dryer, such as a centrifugal dryer, where the water is removed from the pellets. A spherical shape is desirable for easy flow to avoid sticking during SSP at high temperatures.

Since some of the foregoing new SSP processes are driving toward lower-IV prepolymers, there is significant work focused on particle formation of these materials. Since low IV results in materials with lower cohesive strength and where the strand-cutting method cannot be used, particle shapes that minimize the impact of this is critical. The underwater pelletizers have been claimed to be used in Aquafil's process. Potential also exists for use of underwater pelletizers to obtain spherical pellets in the NG3 processes, as the edges of the hemispherical semicrystalline pastilles from the particle former of the NG3 process are a disadvantage, due to the associated dust generation during handling of the low-IV solid.

Gala Industries [85,86] offers an underwater pelletizer for producing particles from low-molecular-weight prepolymer. In this process, prepolymer melt is extruded and cut by specially designed die plate and knives under water. These prepolymer particles have a nearly spherical shape. Reiter-GMBH also offers similar particle formation technology, called Sphero [87]. Particle formation takes place under water via an extrusion die-face cutting process. Particles are then sent to a dryer chamber via process water itself, where water is removed and chips are dried and crystallized for further processing (i.e., SSP). Alternatively, free-flowing crystallized particles can be achieved directly (e.g., in the process CrystallCut of BKG). The underwater die-face-cut hot PET pellets are transported rapidly in hot water (of up to 95°C) to a pellet dryer in the closed conveying pipes, where pellet cooling and solidifying takes place. The PET pellets exit the pellet dryer at a temperature of about 150 to 160°C onto a vibrating conveyor, where the pellet motion enables them to prevent sticking and pass on their energy to other pellets. The exiting hot pellets with crystallinity of about 40% can be transported directly into the SSP.

Rieter also offers a drop-pelletizing system called Droppo for potential use with low-IV oligomer. A frequency is superimposed through a mechanical vibrator on the oligomer at a rigid diehead with a die plate. Frequency and amplitude are harmonized with the physical characteristics of the product such that the liquid emerging from the holes in the die plate breaks up into uniform droplets, and spherical particles of a predefined size are formed. Cooling and/or consolidation of the pellets takes place in air, inert gas, or a cooling liquid.

8.6. ALTERNATIVES TO SOLID STATE POLYMERIZATION

Over the years, solid state polymerization process is being practiced around the world for producing higher-IV PET suitable for food and beverages and technical yarn applications such as tire cord. This is due to the flexibility and ease of operation, lower production cost, better product quality, and other factors. Thanks to the extensive research, attempts to produce high-IV polyester suitable for packaging and other applications at reduced costs have lead to several new technologies. These new technologies are claimed to have lower investment and production cost and product quality similar or better than that of conventional SSP. One essential idea is to eliminate the following intermediate steps: melt cooling and solidification to pellets, reheating the pellets for SSP, cooling the heated high-IV product for packing and transportation, and remelting in extruders for processing into articles. Thus, costs can be reduced at least to the extent of energy costs involved in repeated heating and melting, as well as the equipment and labor costs. In brief, the steps in conventional technologies are:

- Esterification and prepolymerization in melt to IV about 0.3 dL g^{-1}
- Melt polycondensation in finisher to IV about 0.6 dL g^{-1}
- Cooling, solidification, and particle formation
- Heating and SSP to IV about 0.74 dL g^{-1} or more
- Melting and processing into articles

The steps in new technologies are:

- Esterification and prepolymerization in melt to IV about 0.3 dL g^{-1}
- Continue melt polycondensation (finisher) to IV about 0.74 dL g^{-1} or more
- Direct processing into articles

8.6.1. Additives for Direct Polymerization to High Intrinsic Viscosity

As discussed, process to direct high-IV polyester in the melt state is somewhat restricted by reduced diffusion in the high-viscosity melt, as well as the level of high generation of impurities such as aldehydes. The former can be addressed by use of chain extenders, and the latter can be addressed by use of aldehyde scavengers.

a. Chain Extenders   Although SSP is conventionally used to make high-IV PET, the disadvantages are the slow reaction rate and the need for additional equipment. An alternative is to reach high IV values in the melt state. However, as the polycondensation proceeds in melt polymerization, the reaction rate decreases and the degradation of terminal groups occurs, leading to an increase in carboxyl and vinyl chain ends as well as DEG and aldehydes. One of the promising ways to achieve a very high molecular weight of PET without substantially increasing the melt polycondensation time is to use chain extenders. Chain extenders are basically bis- or higher functional compounds, which can react with the polymer end groups very rapidly when simply mixed in molten polymer.

As shown in the Figure 8.9, the use of chain extender results in joining polymer chains and in a faster reaction rate to achieve high-molecular-weight PET, often in minutes as compared to several hours required for SSP. Chain extenders are also useful for achieving high-molecular-weight polyesters with low carboxyl content, which in turn improves mechanical properties and thermal stability during melt processing. Chain extender molecular design is the key to their effectiveness. In the past, various chemicals, such as diphenyl carbonate, diphenyl terephthalate, and diphenyl oxalate and phenyl orthocarbonates, were reported as effective chain extenders [88]. However, these chain extenders tend to give phenol and ethylene carbonate as reaction by-products. These by-products have high boiling points, and their removal from the resulting polymer is rather difficult. They can contaminate recovered ethylene glycol and therefore increase the cost of a commercial PET production processes.

Chain extenders generating no by-products are much preferable, as they eliminate the need for removal of by-product from the viscous polymer. These are additive-type chain extenders. Bis-epoxy compounds, bis(cyclic carboxylic anhydride) bisoxazines, bisoxazolines, and diisocyanates are reported as additive-type chain extenders [89–93]. Karayannidis et al. [94,95] studied diepoxides extensively for increasing the molecular weight of PET. Apart from the rapid rise in molecular weight, a significant increase was observed in the mechanical properties. Although the reactivity of other bisoxazolines is low, 2,2′–bisoxazoline is an exception.

Particularly interesting additive chain extenders are biscaprolactam terephthalates [96] and, more recently, carbonyl biscaprolactam [97], since they do not generate branching. On the other hand, some chain extenders can increase melt strength by additionally introducing branching: for example, pyromellitic dianhydride (PMDA) and trimellitic anhydride (TMA). PMDA is a tetrafunctional molecule, commercially available at low cost. Al Ghatta et al. [98] have studied

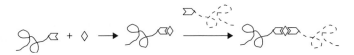

Fig. 8.9. Scheme showing role of a chain extender (◇) in increasing molecular weight by joining chain-end groups.

PMDA extensively for an increasing solid state polymerization rate of PET by addition during polymerization or by reactive extrusion. The reaction mechanism was explained by Khemani [99].

Chain extenders can be classified according to their reactivity with functional end groups (i.e., alcohol and acid). Carboxyl-reactive chain extenders are preferred since in addition to the molecular-weight increase, they reduce carboxyl content in the final polymer. Examples of carboxyl-reactive chain extenders are given in the literature [90]. Hydroxyl-reactive chain extenders are more effective for low-molecular-weight PET produced by melt polymerization processes, where hydroxyl-end-group content predominates over carboxyl end groups. An example of hydroxyl-reactive chain extenders are 2,2′-bis(3,1-benzoxnin-4-one) [92]. Diisocyanates are also reported to be effective chain extenders for PET but with the drawback of product coloration [100]. The resulting urethane linkages are typically not thermally stable at PET processing temperatures.

The issue with the use of chain extenders during polymerization is capital investment versus operating costs. Typically, chain extenders result in higher costs when all factors are considered, and this is responsible for their limited use (if at all) in large-scale polymerization from virgin raw material. The use of chain extenders is more common during small-scale recycling operations. For example, in the context of recycling of polyester (e.g., to polyester strappings), use of chain extenders in a reactive extrusion technique is being practiced to increase the molecular weight of recycled PET (R-PET). Here, chain joining by chain extenders can reverse the loss of molecular weight caused by chain scission. Several studies were conducted to explore chain extenders for this application [101–107]. To achieve the properties desired, the type and amount of chain extender need to be adjusted according to the chain extension reaction stoichiometry and reaction conditions. During R-PET processing, the chain extension reaction should dominate the degradation reaction to achieve PET of a desired molecular weight within a short period of time. Shorter reaction times and a low moisture content during the extrusion process can lead to an effective chain extension and restrain thermal and hydrolytic degradation.

b. Acetaldehyde Scavengers   Acetaldehyde level is one of the most important criteria for selecting PET resin for beverage applications, especially mineral water. Acetaldehyde is a highly volatile chemical having boiling point of 20°C. During the lifetime of a typical PET container, acetaldehyde can migrate from the container sidewall into the beverage, causing an off-flavor. Acetaldehyde gets generated due to the polymer degradation during melt polymerization process and also during polymer processing, such as injection molding, where the temperatures are in the range 270 to 300°C. Thus, the problem of high aldehydes is expected to be more severe in processes involving melt polymerization to high IV.

Various mechanisms have been proposed for the thermal degradation of PET. Marshall and Todd [108] postulated that thermal degradation is initiated at chain ends, whereas Goodings [109] believed that thermal degradation occurs by random chain scission at the ester linkages. Data suggest that both of these mechanisms are active. During production of PET, thermal degradation of polymer chains by random thermal cleavage of ester links generates carboxyl and vinyl end groups. Additionally, acetaldehyde is formed from a glycol or hydroxyl end group through ethylene oxide intermediate [110]:

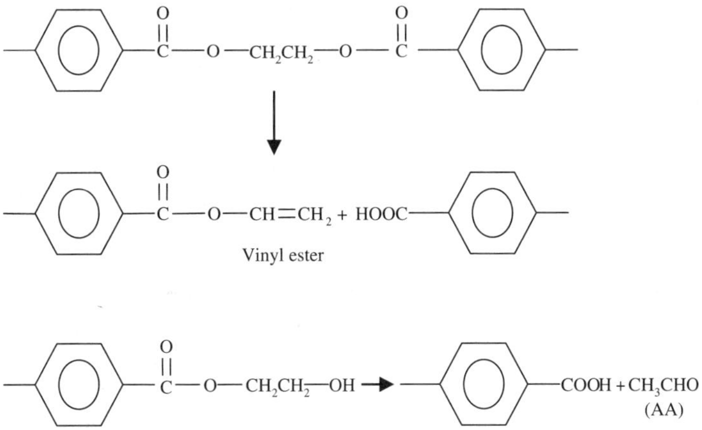

Kim and Jabarin [111] studied the vinyl end group formation and subsequent acetaldehyde generation during melt as well as a solid state polymerization process. Data indicated increased vinyl end-group formation (and subsequent generation of AA) with increasing base prepolymer IV with different thermal histories in the melt polymerization process. During a solid state polymerization process, AA content decreased with the increasing final IV of PET. It should be noted that these reactions do not have a significant impact on the melt or SSP reaction kinetics.

Acetaldehyde concentration decreases during the solid state polymerization process. Usually, a base polymer of IV of around 0.60 dL g^{-1} when prepared in the melt state contains acetaldehyde in the range 40 to 50 ppm. After a solid state polymerization process, acetaldehyde drops to less than 1 ppm in the resin. The concentration of AA decreases with increasing SSP reaction temperature and residence time. Invista's new polymerization technology (NG3) [41–44] claims to have reduced acetaldehyde in the PET resin due to a very short residence time in the melt-phase polymerization and high residence time at a low SSP temperature. On the other hand, a solid state devolatilization step [112,113] or use of acetaldehyde scavengers during a polymerization process can limit the amount of AA in the high-IV (0.74 to 0.84 dL g^{-1}) PET produced by a melt polymerization process where polymer stays longer in the melt phase at very high temperatures.

Apart from the melt polymerization process, acetaldehyde gets generated during polymer processing (injection molding or extrusion) as a result of thermal degradation [38]. The generation of acetaldehyde is directly correlated with the processing temperature and residence time. One of the ways of controlling AA regeneration is to use milder processing conditions, for example, low barrel temperatures, minimum screw speed, injection rate, back pressure, and screw design—all contributing to high shear heating and shorter cycle time to minimize thermal decomposition.

There are numerous patents [29–31,34–37] describing the use of acetaldehyde scavengers for PET and the relevant scavenging mechanisms. For example, metal phosphates [29] react with the aldehyde by the acid- or base-catalyzed addition of the P—H moiety across the carbonyl group of the aldehyde to form an α-hydroxyphosphonate. Incorporation of polyamide to remove acetaldehyde is believed to act by the nucleophilic addition of the free amino group on the polyamide to aldehydes or ketones to form imines (presumedly, via a Schiff-base reaction with the aldehyde).

AA scavengers are also commercially available (e.g., from ColorMatrix under the trade name Triple A) [32]. About a 75% reduction in AA is claimed with these additives. These chemicals are to be added during the injection molding process to produce performs. PolyOne Corporation [36,37] has developed anthranilamide and DL-α-tocopherol for reducing acetaldehyde in the finished polyester products, such as films, containers, and sheets for food packaging applications. Some of the heterocyclic compounds, such as 6-aminouracil, 6-aminocytosine, 6-amino-1-methyluracil, and 6-amino-1-3-dimethyluracil, were reported [30] to be effective for reducing AA generation during PET melt processing.

Franck et al. [38] evaluated various additives for reducing acetaldehyde during the injection molding process and found that p-aminobenzoic acid, 3,5-dihydroxybenzoic acid, and 4-hydroxybenzoic acids were effective under laboratory conditions. These additives were effective both for trapping free radicals and blocking hydroxyl chain ends and therefore more suitable for commercial applications.

8.6.2. Direct Melt Polymerization to High-Intrinsic-Viscosity Polymer Resin or Preform by UDHE-Inventa and Zimmer AG

Direct to preform process announced by UDHE-Inventa claims use of standard catalyst systems and better product characteristics. In their current two-reactor-melt-to-resin (2-R MTR) technology [114], the first reactor is a tower reactor, the Espree reactor, which replaces the first three reactors of a typical conventional polycondensation plant. The Espree reactor comprises several reaction sections and has a total height of up to 30 m. In addition to other new features, it uses a liquid film flow on the inner surfaces of vertical tubes to produce the prepolymer. Since there is intense mixing caused by physical phenomena within the various reaction sections, agitators and the associated seal systems are avoided. The advantages are lower energy and maintenance costs, higher raw material yield, and superior quality. The high-molecular-weight PET material required for bottle production up to an intrinsic viscosity of 0.85 dL g^{-1} is obtained directly by melt polycondensation in a special high-viscosity finisher, the Discage MV reactor, which has a lot of benefits, due to its unique design. The high mechanical stability of the horizontal cage-type agitator allows polycondensation at low temperature and forced transport of the polyester melt from inlet to outlet. Their invention [115] discloses the continuous inline polymerization and injection molding process, i.e., in-line production of preforms from polymer from the continuous melt polymerization to high IV. The polymer can be poly(ethylene terephthalate) and its copolyester. The acetaldehyde level in the preform can be restricted within an acceptable range (i.e., <10 ppm). An inert gas such as nitrogen or carbon dioxide is introduced into the flow of molten polyester having an IV value between 0.5 and 0.75 dL g^{-1} from the polycondensation, and then the acetaldehyde content was reduced below 10 ppm under vacuum at a temperature between 285 to 260°C (i.e., the temperature starts near 285°C and becomes less during the reaction and ends near 260°C) in a melt post-polycondensation reactor having vacuum devolatilization. The melt post-polycondensation reactor is a screw reactor with at least two screw shafts rotating in the same direction. IV up to 0.95 dL g^{-1} is achieved in less than 60 min and the melt is guided immediately to an injection-molding machine to produce preforms.

Zimmer AG has also developed a similar process for direct production of finished articles suitable for food packaging (preforms, films) and yarn applications. Their new process, direct high IV (DHI), claims to reduce entire plant expenditure by around 15 to 20% with better product performance. The same catalyst system as that being used for conventional polyester polymerization processes can be used in this process. Their process uses conventional catalyst as well as cobalt or cobalt plus manganese compound (in the molar ratio 1 : 1 to 3 : 1) and phosphorus compound [112]. It is claimed that polymer quality can be maintained by limiting the dwell time of the polyester melt between the outlet of the polycondensation reactor and the entrance of processing equipment such as injection molding or yarn spinning. This will minimize IV drop and color properties of finished product. To avoid higher acetaldehyde generation in the process, an inert gas is introduced in the polymer melt immediately after the outlet from the last melt phase condensation reactor and the subsequent vacuum degassing. Further acetaldehyde reduction was achieved by introducing amide compounds such as MXD6 (Mitsubishi) or Selar PA 3426 (Du Pont) next to the gas inlet.

Direct melt-phase high-IV processes may be limited by the requirement of matching capacities of continuous high-speed preform production and yarn spinning. Existing processing equipment is not capable of handling such a huge flow of polymer mass from the commercial plant. Disturbance in one of the processing machines would cause significant melt flow differences and would require a facility for high-IV melt recycling. One invention [116] suggests a complicated sidestream cutting and melt recycling system to handle such circumstances.

8.6.3. Du Pont Atmospheric-Pressure All-Melt Process

As an alternative to SSP, Du Pont developed a new technology [117–123] to produce polyesters while avoiding a high-vacuum process and SSP. This technology was developed to produce fiber-grade (IV = 0.6 dL g^{-1}) polyester, PET for beverage applications (IV = 0.74 to 0.84 dL g^{-1}) and very high IV technical-grade polyesters for tire cord applications (IV >0.9 dL g^{-1}).

The main objective to develop this process is to avoid costly high-vacuum equipment, to avoid volatile organic emissions and the wastewater discharge. In a well-practiced conventional melt polymerization process, reaction by-products such as ethylene glycol have to be removed to shift the reaction equilibrium toward chain-growth polymerization (Fig. 8.1). This is achieved by applying a vacuum of 1 to 3 mmHg for which costly and complicated hardware is required. To minimize the air leakages in the process, nitrogen purge seals and flanges are required. Reaction condensates and organic by-products from the process create wastewater stream from which volatile organic substances emission takes place in air, polluting the environment. This new process (Fig. 8.10) claims to eliminate the use of costly and complicated hardware requirements for vacuum, the emission of volatile organic by-products, and wastewater discharge, thereby becoming a more eco-friendly process.

According to the process, melt polymerization is carried out at atmospheric pressure. In this process the key element is a specially designed horizontal

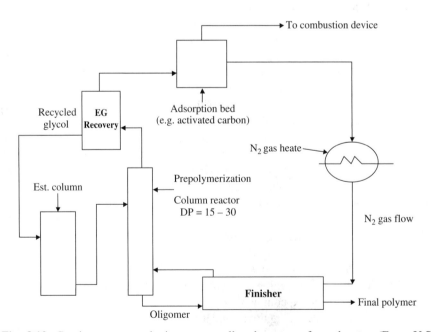

Fig. 8.10. Continuous atmospheric-pressure all-melt process for polyester. (From U.S. Patent 5,434,239 [117].)

finisher, where final IV is achieved from low-molecular-weight oligomers. The DP of oligomer entering the finisher is in the range 15 to 30 and that of the final polymer is between 60 (suitable for fiber grade) and 150 (suitable for bottle-grade application). An inert gas such as nitrogen is introduced in the finisher, and the flow of nitrogen is countercurrent to the direction of polymer melt in order to remove polymerization by-products such as EG, water, and AA. The quantity of an inert gas is adjusted so that the partial pressure of by-products is lower than the equilibrium pressure of by-products with the melt for their efficient removal with the gas stream. Glycol from the volatilized by-products is recovered and reused again in the system, whereas other by-products are adsorbed in an adsorption bed, such as activated carbon, and sent to a combustion device, where they are converted to CO_2 and water as done with conventional processes.

The residence time in the finisher varies from 3 to 5 h depending on the final target IV, and the temperature is in the range 270 to 300°C. No special catalyst system (apart from conventional antimony trioxide) is used to achieve higher-IV polyester. For obtaining high-IV polyester in molten-state polymerization, efficient and quick removal of reaction by-products is necessary. It can be facilitated by providing large interfacial area between the polymer melt and gas phase. In this process, a special type of agitator is designed to meet these objectives. The following criteria should be considered when designing an agitator:

- Large interfacial area
- Wiping of the vessel walls
- Frequent renewal of surface area
- Good mixing of polymer melt

Polyester produced using this process can be used to produce fibers, bottle-grade resin, and technical yarns such as tire cord. Since this is only a melt-phase polymerization process, it is possible to use a higher comonomer content, such as IPA or CHDM, for polyester modification.

8.6.4. Eastman Chemical's IntegRex Technology

In September 2004, Eastman announced the development of IntegRex for producing bottle-grade PET resin. High-IV PET (IV >0.70 dL g^{-1}) is achieved by melt-phase polymerization. Although detailed information is not available on this proprietary development, it is believed to link paraxylene–PTA–PET production. Eastman claims a 15 to 20% reduction in CAPEX and a substantial reduction in OPEX, possibly due to some or all of the following components:

- Eastman has filed some patents [124–126] describing the optimized liquid-phase oxidation of *p*-xylene to terephthalic acid. Such liquid-phase oxidation is carried out in a bubble column reactor that permits a highly efficient reaction at relatively low temperatures.

- Use of aqueous TA solution with EG in the polyester plant eliminates PTA drying and solidification equipment and the slurry mix equipment in the polyester plant.
- Direct coupling of the water distillation in the CTA oxidation section to the reactor eliminates the cost for removal of water produced in the CTA reaction and also lowers waste treatment cost (for the remaining acetic acid contained in the water).
- Eastman has a patent [127] describing a process for the manufacture of high-IV (>0.75 dL g^{-1}) polyester in the melt phase. It suggests the use of a higher CHDM content comonomer, which enables the higher IV attainable in the melt phase and possible elimination of polyester solid stating.
- Eastman has filed a recent patent application [128] that describes a novel pipe reactor design for either esterification or polycondensation or both. The Eastman patent claims that there are numerous cost- and operation-related simplifications because of the pipe reactor design.
- Eastman patents disclose direct melt to high-IV resin and to the preform process in which the polyester is prepared, formed into useful shaped articles in a single integrated process [127,129–131]. Also described is an apparatus and method for molding polyester articles having a low acetaldehyde content directly from melt formation, using flash tank devolatilization. An acetaldehyde stripping agent is mixed into a polyester melt before devolatilization in a flash tank.

8.6.5. Swollen/Solution Polymerization to UHMW-PET

Using melt and solid state polymerization processes, IV of PET can be raised to 1 to 1.5 dL g^{-1} only with great effort. In the melt processes, further increase in IV is extremely difficult, due to severe thermal degradation and deterioration of key polymer properties. During solid state polymerization, to achieve IV >1 dL g^{-1}, process temperatures required are close to the melting point of PET (due to the restricted chain mobility), thereby increasing the sintering and lump formation tendency in the reactor. Further, a very thin polymer cross section is required. To achieve very high IV PET (UHMW-PET) from low-molecular-weight PET prepolymer (IV ~ 0.4 to 0.6 dL g^{-1}), swollen/solution state polymerization was demonstrated [132–136]. This technique was used successfully for the production of superhigh-tenacity fibers [137,138].

For swollen-state polymerization, solvents are only required to swell PET, but should not dissolve PET, whereas for solution state polymerization, solvents are required to dissolve PET completely. Since swollen-state polymerization (SwSP) is carried out at very high temperatures (>200°C), solvents must have a higher boiling point and a high decomposition temperature. Suitable solvents for swollen-state polymerization are reported [132] as biphenyl–diphenylether mixture, monoethylbiphenyl, diethylbiphenyl, triethylbiphenyl, and hydrogenated terphenyl. Studies revealed that, hydrogenated terphenyl is more suitable especially at a high-temperature reaction: (1) since it does not dissolve polymer even at high temperature, and (2) due to the higher degree of swelling.

It is postulated that swelling increases the surface area (increased mass-transfer reaction rate) and that the mobility of reactive end groups and the frequency of end-group interaction are increased, thereby raising the polymerization rate. Reaction by-products such as glycol and water can be removed by bubbling an inert gas through the reaction medium. The rate of polymerization was also found to be increased by increasing temperature. In the swollen-state polymerization process, particle size (i.e., diffusion length) plays a very important role in determining the rate of IV rise [133,134]. Reaction by-products travel from the inner core to the outer surface and then carried away by an inert gas. If the diffusion distance is reduced, the rate of by-product removal increases, resulting in a rapid IV rise.

During this polymerization process, it was observed that as a result of increased chain mobility, the crystallinity of polymer increases substantially and goes as high as 85%. Formation of a honeycomb-like crystal structure was observed in the electron microscopic analysis, and the final polymer (IV = 2.3 dL g^{-1}) had a melting point as high as 290°C [132]. The high melting temperature may be due to the formation of bigger and more perfect crystals, due to the solvent-assisted crystallization in the solid/swollen state.

On the other hand, IV rise could be restricted during the swollen-state polymerization process, due to the solvent-induced crystallization. To overcome this problem, solution-state polymerization was explored. There were some early attempts [139,140] to increase the molecular weight of PET from 0.62 dL g^{-1} prepolymer in solution, but the IV could not be raised beyond 0.87 dL g^{-1}. This may be due to the somewhat lower polymerization temperatures, around 200 to 220°C. In another attempt by Ma and Agarwal [136], prepolymer of IV = 0.42 dL g^{-1} was dissolved completely in a diphenylether/biphenyl solvent mixture, and further polycondensation was carried out at a high temperature (250°C). The intrinsic viscosity of PET could be increased successfully up to 1.8 dL g^{-1} in 20 h. One of the major advantages of solution-state polymerization process is ease of handling compared to conventional SSP and swollen-state polymerization processes. As compared to SSP and SwSP, higher polymerization temperatures can be employed during solution-state polymerization. This also avoids the lump formation tendency in SSP and SwSP processes, which demands special equipment. Further, solution post-polymerization avoids the restrictions on reactive end-group mobility of the type imposed by crystallization during SSP and SwSP. Solution of high-molecular-weight PET prepared by solution-state polymerization can be spun directly into fibers [141]. Potentially very high modulus and strength can be obtained by such solution spinning of very high-molecular-weight PET [142,143]. A very high degree of crystallization and fibrillar morphology of crystals is also obtained during crystallization from dilute quiescent solutions [141].

8.7. POLY(ETHYLENE TEREPHTHALATE) FOR FLUID PACKAGING APPLICATIONS

PET resin produced by solid state polymerization process has many advantages. Due to the low-temperature operation of solid state polymerization, fewer side reactions and less thermal degradation occur in the process. Therefore, the development of undesirable side by-products, which have an influence on the properties of PET, can be kept within limits that would not affect quality requirements [144]. The side reactions that are being minimized by SSP are the formation of DEG, carboxyl end groups, and vinyl end groups, all of which result in AA formation. Higher DEG content deteriorates mechanical strength and hydrolytic and light stability of resin [145]. Carboxyl end groups reduce the hydrolytic stability of PET, resulting in the polymer degradation during processing, such as injection molding [146]. Vinyl end groups may also polymerize to polyvinyl esters, which are responsible for color formation in PET [144]. The vinyl ester end group of PET can be transesterified under the formation of acetaldehyde, which is extremely important for food-packaging applications, especially for mineral water applications. Also during polymer processing, acetaldehyde gets generated, thereby making the final product undesirable if the AA content is too high.

High-IV PET becomes popular for many food-packaging applications, especially for mineral water, carbonated soft drinks, fruit juices, beer, iced tea, syrups, dairy products, and many more. The largest market segment for PET is soft-drink and water packaging bottles, but the use of other bottle applications has increased rapidly in the last few years. The main drivers for its popularity are glasslike transparency, mechanical properties, reasonably good gas (oxygen, carbon dioxide) barrier properties for retention of flavor and carbonation, limited flavor scalping, ease of handling, and low cost. PET also exhibits a high toughness/weight property ratio, which allows lightweight, large-capacity unbreakable containers and jars. Functional comonomers, additives, or even PET blends are being used to modify PET properties to achieve the performance desired for various packaging applications.

PET containers for beverage applications are manufactured by an injection stretch blow molding (ISBM) process (Fig. 8.11). This process can be divided into two types: a two- or a one-stage procedure. In the two-step process, amorphous performs are produced by a injection molding machine, and then containers are produced by a blow molding machine. An injection molding process is usually carried out at a temperature $15°C$ higher than the end melting temperature of PET resin. The end melting temperature of PET resin varies between 260 and $270°C$, depending on the polymerization technology used. To prevent substantial loss of PET resin IV and also to restrict the undesirable by-product formation, PET resin is dried at 160 to $170°C$ for 5 to 6 h to remove moisture to prevent hydrolytic degradation. The recommended moisture level in dried PET should be less than 50 ppm. A higher moisture content in PET resin gives rise to severe problems, such as higher IV drop, AA generation, poor color, and lack of clarity [147].

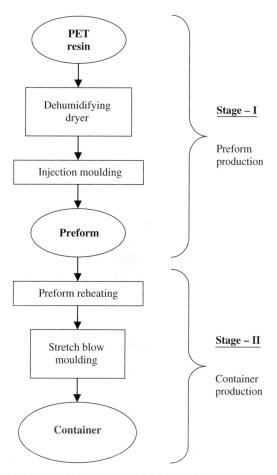

Fig. 8.11. Injection stretch blow molding process.

In the stretch blow molding process, amorphous preform is heated via infrared quartz heaters to achieve temperature, which is around 30°C higher than the glass transition temperature. The selection of blowing temperature is very critical in order to get optimum bottle properties, such as clarity and mechanical strength. A blowing temperature window is selected from differential scanning calorimetry (DSC) data [148] of amorphous preform, as shown in Figure 8.12. If preform blowing temperature is too low, it will impart stress whitening in the bottle, and if it is too high, it will generate crystalline haze.

In the single-step process, amorphous preforms and containers are produced in a single system. The preforms are brought down to blow molding temperature directly. As a result, containers made by this process do not have the same hoop strength as those made by the two-step process. The requirements of several packaging applications are now discussed.

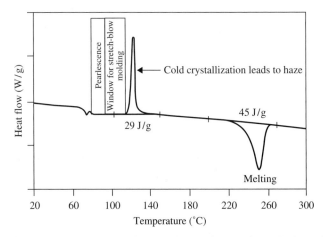

Fig. 8.12. Differential scanning calorimetry graph of amorphous PET preform. (From Fakirov [148] by permission of John Wiley & Sons, Inc.)

8.7.1. Mineral Water

Use of PET for mineral water applications has increased substantially over the past few years. It is the second-largest market segment for PET. PET resin having an intrinsic viscosity of 0.74 ± 0.02 dL g^{-1} is commonly used to produce bottles for mineral water applications. Bottle volume varies in the range 300 mL to 1 L. The lowest acetaldehyde content (to avoid intense off-flavor) and glasslike clarity are two major requirements for bottled water application.

8.7.2. Large-Container Applications

The volume of large containers is usually in the range 5 to 20 L. Preform wall thickness varies in the range 4 to 10 mm. If the rate of crystallization is high, it imparts crystallinity in the preform, making the preform hazy. If such crystalline preforms are blown to produce a bottle, the bottle will lose clarity, develop haze and further reduce the mechanical properties. Due to the higher thickness of preforms, to avoid crystallization, the cooling time for bringing the preform below the glass-transition temperature increases. To obtain fully amorphous preforms, the rate of crystallization of polyester must be very low. Hence, a very low rate of PET crystallization is an essential requirement for the production of large thick-walled transparent containers. To avoid crystallization in the preform during the cooling process, a PET structure needs to be modified to exhibit a very low crystallization rate. This can be achieved by incorporating various comonomers, such as isophthalic acid (IPA), cyclohexanedimethanol (CHDM), neopentyl glycol (NPG), and diethylene glycol (DEG), added at the stage of polymerization [149]. These comonomers are known to retard the crystallization rate of PET (Fig. 8.2) by hindering the chain packing into a crystalline unit cell. Although comonomers are usually incorporated into PET during melt polymerization or melt mixing, it is also possible to carry this out during the SSP process, (e.g., by exposure of solid PET to the vapor of comonomers) [150].

8.7.3. Hot-Fill Applications

Certain beverages, such as iced tea, fruit juices, isotonic drinks, and sausages, have to be filled in containers at higher temperatures to avoid bacterial growth. Filling temperatures depending on the composition, but usually vary between 85 and 95°C, well above the glass-transition temperature of PET (i.e., 78°C). At these higher temperatures, dimensional stability has to be retained since during the hot-filling process, vacuum is generated inside the container. There are several ways to improve the dimensional stability of a container:

1. Heat set process during a blow molding operation
2. Bottle design
3. PET structure modification: high-T_g PET

Heat setting is a process commonly practiced to improve the hot-filling performance of containers. In this process, a blow-molded bottle is held at a mold temperature well above the glass-transition temperature for a few seconds to increase the crystallinity as well as to relax the stresses generated during a biaxial orientation process. Stresses in the bottle will result in dimensional instability under hot conditions. Higher crystallinity in the sidewall portion will provide better thermal stability to withstand the container at a higher filling temperature. PET composition plays an important role in improving the crystallinity during the heat-setting process. The crystallization rate as well as the morphology development are key factors to obtain the thermal stability of beverage container [151,152]. Bottle design is also critical for better dimensional stability during the filling process. Specially designed vacuum panels on the sidewall or base portion of a bottle compensate for the hypotension within the bottle [153].

Higher hot-filling temperature without losing the bottle dimensional stability can be achieved by increasing the glass-transition temperature of PET. A higher glass-transition temperature will improve chain rigidity at the filling temperature and reduce bottle shrinkage due to high vacuum generated during the filling process. Various bulky comonomers can be used to increase the glass-transition temperature of PET. Examples of such comonomers are *tert*-butyl isophthalic acid [154] and naphthalene dicarboxylate [155]. However, it should be noted that a higher glass-transition temperature (chain rigidity) decreases the rate of crystallization of PET. Reduction in the rate of crystallization will have an adverse impact during the solid state polymerization process by increasing the sintering tendency. These comonomers increase the cost of these materials and have not proven to be cost-competitive with heat setting to date.

A polymer blend is also an attractive way to increase the glass-transition temperature for achieving better hot-fill performance. Good literature data are available on PET-PC [156,157], PET-PEN [158–167], and PET-PEI [168–170] blends. However, a polymer blend may not be a commercially viable route because of the higher cost of the second polymer. In addition, the properties of polyester–polyester blends vary with variation in the time and temperature of mixing, making it difficult to produce a uniform product unless completely random copolymer is desired.

In addition to good dimensional stability at higher filling temperatures, the clarity of the bottle has to be retained. The heat set process usually imparts haze to the bottle, due to the increased crystal growth. A bottle produced with high-T_g PET or PET blend will have better clarity, due to the suppression of the crystallization rate as a result of comonomer or addition of a second polymer.

Preform neck crystallization is also a popular industrial process to manufacture containers suitable for hot-filling application. The preform neck is fully amorphous. Thus, since the hot-filling temperature is above the glass-transition temperature of PET, the neck deforms, making the bottle unsuitable for capping. Therefore, before a high-temperature filling process, the preform neck is crystallized to get adequate crystallinity so as to withstand the filling temperature. Due to crystallization, the preform neck becomes opaque, but due to capping, the esthetics of the bottles do not change.

8.7.4. Flavor/CO_2 Retention Through Barrier Properties

Carbonated soft-drink packaging is one of largest markets for PET resin globally. 2-L PET bottles are quite popular for this application; however, the market for small bottles (250 to 500 mL) is also increasing. For carbonated drinks packaged in PET, bottlers are concerned primarily with the rate at which CO_2 escapes from the bottle. If CO_2 diffusion is too fast, the product will go flat on the shelf before it is consumed. CO_2 retention for small bottles is even more critical, due to the higher surface/volume ratio. As a result, the development of high-barrier PET technology at both resin suppliers and bottlers has moved into high gear.

Several barrier technologies are available in the market for improving barrier performance. These technologies can be divided into three major classes: multilayer, coatings (external or internal), and monolayer. However, no clear market winner has emerged among these three barrier technologies.

a. Multilayer In this segment, usually three- or five-layer bottles are produced using coextrusion injection-molded preforms. These bottles contain polyamide as a barrier layer, sandwiched between two PET layers for improved CO_2 retention.

b. Coatings A number of coating technologies have been developed for improving the barrier performance of PET bottles. These coatings can be applied on external as well as internal surfaces of bottles after completion of the blowing process. For example, Sidel's technology (Actis [171], amorphous carbon treatment on internal surface) uses a 150-μm-thick hydrogen rich amorphous carbon layer as a barrier on the internal surface of the bottle. The CO_2 barrier improves sevenfold. Mitsubhishi's DLC (diamond-like carbon) [172] film process involves deposition on the inner surface of the bottle via plasma CVD (chemical vapor deposition). The thickness of the DLC layer is around 0.02 μm. CO_2 barrier performance increases by tenfold or more. SIG Corpoplast has developed a silicon oxide coating technology, Plasmax 12D, for improving the shelf life of CSD bottles. This coating is applied on the inner surface of bottles. Tetrapak has developed a similar technology, Glaskin. The Bestpet process involves coating silicon oxide on the outer surface of PET bottles using physical vapor deposition (PVD) by applying high vacuum under plasma conditions. Epoxy amine is another known coating, with the advantage that it can easily be removed during the washing process and thus allows bottle-to-bottle recycling.

c. Monolayers Wellman has developed a special PET grade called HP867 with a proprietary chemistry for blending with MXD6 up to 4% during molding. It is designed to make clear monolayer CSD containers that meet shelf life requirement for 12- to 16-oz bottles. Several other producers, such as M&G and Mitsubishi, have similar products.

8.7.5. Oxygen Barrier Poly(ethylene terephthalate)

The development of PET bottles for oxygen-sensitive beverages has been on the horizon for many years but is finally gaining momentum. Oxygen scavenging, multilayer, and coating technologies are critical for the introduction of PET in packaging of oxygen-sensitive products. For the shelf life of oxygen-sensitive food products such as fruit juices or beer, the gas barrier properties of PET for oxygen as well as CO_2 are of great importance. The permeation of oxygen may also lead to changes in beverage color, emerging off-flavors, microbial spoilage, and degradation of vitamins [174]. The estimated tolerable oxygen ingress for various beverages is given in Table 8.1.

The use of PET beer bottles is expected to grow by 40% by 2008 according to Schoenwald Consulting in Germany. For beer, the challenge of providing a barrier against carbon dioxide egress and oxygen ingress has been met largely by multilayer PET bottles along with some use of coatings and monolayer bottles.

TABLE 8.1. Estimated Tolerable Oxygen Ingress for Various Beverages

Content	Oxygen (ppm)
Beer	<1–5
Baby foods	1–35
Dried foods (e.g., nuts), snacks, coffee	5–15
Fruit juice and drinks	6–30
Applesauce and fruits	6–30
RTD tea	10–20
Pasta sauces	10–40
Ketchup	10–40

Source: PMMI [173].

However, in Russia and Ukraine, where the largest market for beer in PET containers exists, nonbarrier monolayer PET bottles in 1.5- and 2-L sizes are used due to favorable weather conditions for improved shelf life. In addition, this market requires only a very short shelf life.

a. Multilayer   Among the various technologies available for beer packaging in PET, multilayer is widely accepted compared to coating and modified monolayer bottle technology. The most common multilayer bottle is three-layer PET-MXD6, where an MXD6 layer is sandwiched between two PET layers by a coextrusion technique [175]. These multilayer bottles are used primarily for 500-mL beer bottles. Another multilayer technology contains ethylene vinyl alcohol (EVOH) as a barrier layer [176]. Compared to polyamide, EVOH bottles have better oxygen barrier and transparency, but their major disadvantage is the absorption of humidity and resulting reduction in oxygen barrier properties. Thus, EVOH can be used only in five-layer multilayer bottles, normally in the form of PET/polyamide/PET/EVOH/PET.

Amcor uses BindOx [177] active barrier oxygen scavenging technology. BindOx becomes the middle layer of a three-layer multilayer bottle, and oxygen is actually captured in the BindOx and thus has no chance to reach the product.

Valspar [178] offers a barrier resin under the trade name of Valor. It can be used in a monolayer as well as multilayer structures. Valor offers three different grades, depending on the application. The oxygen scavenger in Valor is moisture activated (i.e., the oxygen scavenger does not activate until the bottle is filled).

b. Coating   The coating technologies are the same as those used to improve the oxygen barrier, as described in Section 8.7.4.

c. Monolayer   Ideally, monolayer bottles are most preferred, due to their processing simplicity and minimal equipment and tooling cost. Blending with a barrier resin, such as polyamide (MXD6 or other), PEN, or oxygen scavenger, or both with PET, would eliminate the use of costly multilayer co-injection machines or coating equipment. Even resin can be modified to make it suitable for packaging of oxygen-sensitive beverages. Several developments have taken place in monolayer PET barrier bottles.

M&G has developed a new ActiTUF technology for monolayer PET bottles combining both active and passive gas barriers. ActiTUF is triggered to react with oxygen only when a bottle is filled with beverage and hence storage of preforms is not an issue [179].

Invista has developed a monolayer passive barrier technology, PolyShield, in which a PET resin is modified for blending with MXD6 polyamide [180]. They have used a proprietary compatibilizer to make bottles crystal clear. This resin is intended to mix with MXD6 use during the molding process. The resin is targeted for beer and juice bottles. Futura Polymers has developed a monolayer bottle from a PET–PTN [poly(trimethylene naphthalate)] [181]. PTN is reportedly based on Corterra PTT technology licensed from Shell Chemical.

Constar International has a monolayer system, Monoxbar [182], which incorporates a proprietary oxygen scavenger for single-serve food and beverage containers for juices, teas, ketchups, and carbonated water. The company has also developed a new monolayer system, DiamondClear, which reportedly offers the same performance for oxygen-sensitive food beverages with the same clarity as standard PET.

Colormatrix offers Amosorb DFC (direct food contact) concentrate. This additive enables oxygen scavenging in a simple monolayer PET package and is added in PET during an injection-molding process.

d. Nanocomposites   Nanocomposite technology is a newer way of improving the barrier properties of PET bottles for both oxygen and carbon dioxide. The improvement in the gas barrier is attributed to the increase in the tortuosity of the diffusive path for permeant molecules. Gas barrier properties with nanocomposites have been shown to improve markedly in a number of polymeric materials [183–187]. Barrier technologies containing nanocomposites are now commercially available and can be used in a monolayer as well as multilayer bottles. Ageis OX is a technology available from Honeywell and contains oxygen-scavenging polyamide as an active barrier and nanocomposites for a passive barrier. Ageis OX is currently used in a three-layer 12-oz PET bottle.

Nanocor–Mitsubishi offers a barrier solution under the trade name M9 [188]. It contains MXD6 polyamide and nanocomposites. Its performance is much better than that of standard PET-MXD6 blend. A three-layer multilayer bottle having M9 as a core layer was found to improve CO_2 and O_2 barriers by 50 and 75%, respectively, compared to PET-MXD6 alone. It is currently being used for 16-oz multilayer PET bottle and is said to have 100-fold lower oxygen transmission rate than PET.

8.7.6. Fast Heating Poly(ethylene terephthtalate)

PET beverage bottles are formed by a process called reheat blow molding (RHB), wherein an injection-molded part called a preform is heated by a bank of heat lamps to a critical temperature and then blown to fill a mold. The time required for the preform to reach the critical temperature is known as the reheat time or reheat rate of the material and varies as a function of the absorption characteristics of the polymer itself as well as of any additives, such as metals, catalysts, toners, dyes, or included foreign matter. The preform heat-up time is a critical limiting step in determining the number of preforms that can be blown into suitable containers over a certain amount of time. Thus, it is desirable to provide polyesters that reheat faster or with less energy.

The heat lamps used in the beverage bottle industry typically are quartz lamps with a broad emission spectrum, from 500 nm to greater than 1500 nm. The emission maximum is typically around 1100 to 1200 nm. PET absorbs poorly in the region between 500 and 1400 nm. Since compounds with absorbance in the range 400 to 700 nm appeared colored to the human eye, compounds that absorb in this range will impart color to the polymer. Thus, to improve the reheat rate of a material, one must increase the absorption of radiation in the region of emission of the heat lamps, preferably in the region of maximum emission.

Several additives are being used in the industry to enhance the reheat rate for molding applications. These include finely divided carbon black, graphite, iron oxides, metal particles, and certain anthaquinone-type dyes [189–197]. Reheat additives can be added at any stage of the polymer preparation process (e.g., esterification or polycondensation). All of these compounds are black- or gray-body absorbers which absorb energy across the entire structure of infrared and visible radiation. As a result, these materials impart a grayness or loss of transparency to the polymer which is in proportion to the amount of material added to the polymer. This effect could be controlled by varying the particle size of the additive, but it could not be eliminated. Selection of reheat additive should be based on (1) high efficiency, (2) excellent thermal stability, (3) chemical resistance, (4) no or minimal effect on the polymerization process (both melt and SSP), (5) smaller particle size, (6) high product clarity and low haze values, and (7) cost-effectiveness.

8.8. CONCLUSIONS

Conventional SSP is a very useful industrial processes for raising the molecular weight of PET from melt polymerized product IV of about 0.6 dL g^{-1} to an IV of about 0.8 dL g^{-1} required for the injection-molding process involved in several applications, including food and beverage packaging. This conventional process has been compared with several new polymerization technologies that span, on one end, the beginning of SSP with prepolymer of lower IV value (0.2 to 0.4 dL g^{-1}), and on the other end, the complete elimination of the SSP process. Several of these technologies have the potential to offer better product attributes and lower plant and operating costs than those of a conventional process. Different fluid packaging applications require different attributes in the PET resin. These are often achieved (1) by the introduction of blends, particulates and comonomers; (2) by employing multilayers and coatings, and (3) by controlling undesirable by-products during the process.

REFERENCES

1. Ravindranath K, Mashelkar RA. Polyethylene terephthalate: II. Enginering analysis. *Chem. Eng. Sci.* 1986;41:2969–2987.
2. Kosuke T, Hiroaki I. Studies on the formation of poly(ethylene terephthalate): 2. Rate of transesterification of dimethyl terephthalate with ethylene glycol. *Polymer.* 1973;14:55–60.
3. Barun D, et al. *Techniques of Polymer Synthesis and Characterization.* Wiley-Interscience, New York, 1972.

4. Vouyiouka SN, Karakatsani EK, Papaspyrides CD. Solid state polymerization. *Prog. Polym. Sci.* 2005;30:10–37.

5. Samant KD, Ng Ka M. Synthesis of prepolymerization stage in polycondensation processes. *AIChE J.* 1999;45:1808–1829.

6. Yang KS, An KH, Choi CN, Jin SR, Kim CY. Solubility and esterification kinetics of terephthalic acid in ethylene glycol: III. The effects of functional groups. *J. Appl. Polym. Sci.* 1996;60:1033–1039.

7. Ravindranath K, Mashelkar RA. Modeling of poly(ethylene terephthalate) reactors: 4. A continuous esterification process. *Polym. Eng. Sci.* 1982;22:610–618.

8. Chegolya AS, Shevchenko VV, Mikhailov GD. *J. Polym. Sci. A.* 1979;17:889–904.

9. Jan FK. Process for the preparation of polyethylene terephthalate (American Enka Corporation). U.S. Patent 3,497,473, 1970.

10. Mellichamp DA Jr. Preparation of glycol terephthalate linear polyester by direct esterification of terephthalic acid (E.I. du Pont de Nemours & Company). U.S. Patent 3,496,146, 1970.

11. Shuya C, Ming-Fa S, Shu-May C. Solid-state polymerization of poly (ethylene terephthalate). *J. Appl. Polym. Sci.* 1983;28:3289–3300.

12. Jabarin SA, Lofgren EA. Solid state polymerization of poly(ethylene terephthalate): kinetic and property parameters. *J. Appl. Polym. Sci.* 1986;32:5315–5335.

13. Ravindranath K, Mashelkar RA. Modeling of poly(ethylene terephthalate) reactors: IX. Solid state polycondensation process. *J. Appl. Polym. Sci.* 1990;39:1325–1345.

14. Ma Y, Agarwal US, Sikkema DJ, Lemstra PJ. Solid-state polymerization of PET: influence of nitrogen sweep and high vacuum. *Polymer.* 2003;44:4085–4096.

15. Duh B. Effect of antimony catalyst on solid-state polycondensation of poly(ethylene terephthalate). *Polymer.* 2002;43:3147–3154.

16. Duh B. Reaction kinetics for solid-state polymerization of poly(ethylene terephthalate). *Polymer.* 2002;43:1748–1761.

17. Zhi-Lian T, Qiu G, Huang N, Claudio S. Solid-state polycondensation of poly(ethylene terephthalate): kinetics and mechanism. *J. Appl. Polym. Sci.* 1995;57:473–485.

18. Heighton HH, Most EE. Process for preparing high viscosity linear condensation polyesters from partially polymerized glycol terephthalates (E.I. du Pont de Nemours & Company). U.S. Patent 3,405,098, 1968.

19. Bamford CH, Wayne RP. Polymerization in the solid phase: a polycondensation reaction. *Polymer.* 1969;10:661–681.

20. Chang TM. Kinetics of thermally induced solid state polycondensation of poly(ethylene terephthalate). *Polym. Eng. Sci.* 1970;10:364–368.

21. Duh B. Method for production of high molecular weight polyester with low catalyst level and low carboxyl content (Goodyear Tire & Rubber Company). U.S. Patent 4,205,157, 1980.

22. Duh B. Method of production of high molecular weight polyester prepared from a prepolymer polyester having an optimal carboxyl content (Goodyear Tire & Rubber Company). U.S. Patent 4,238,593, 1980.

23. Jiongxin Z, Haiyan X, Gao Q, Youwei Z, Nanxun H, Zhilian T. Solid-state polycondensation of poly(ethylene terephthalate) modified with isophthalic acid: kinetics and simulation. *Polymer.* 2005;46:7309–7316.

24. Renwen H, Yan F, Tinzheng H, Geng S. The kinetics of formation of diethylene glycol in preparation of polyethylene terephthalate and its control in reactor design and operation. *Angew. Makromol. Chem.* 1983;119:159–172.

25. Hovenkamp SG, Munting JP. Formation of diethylene glycol as a side reaction during production of polyethylene terephthalate. *J. Polym. Sci. Polym. Chem.* 1970;8:679–682.

26. Besnoin JM, Choi KY. Identification and characterization of reaction byproducts in the polymerization of polyethylene terephthalate. *Polym. Rev.* 1989;29:55–81.

27. Kembowski Z, Torzecki J. The principles of pipe-line design for molten poly(ethylene terephthalate). *Polym. Eng. Sci.* 1982;22:141–146.

28. Schaul JS. Drying and injection moulding pet for beverage bottle preforms. *Polym. Plast. Technol. Eng.* 1981;16:209–230.

29. Rule M. Method to reduce the acetaldehyde content of polymers (Plastic Technologies, Inc.). U.S. Patent 7,163,977, 2007.

30. Wiegner J, Voerckel V, Staeubur H. Compounds useful for reducing acetaldehyde content of polyethylene terephthalate, their use and products resulting therefrom. WIPO Patent WO 2003/082967, 2003.

31. Mark R, Yu Shi. Polyester composition and articles with reduced acetaldehyde content and method using hydrogenation catalyst (Coca-Cola Company). U.S. Patent 7,041,350, 2006.

32. http://www.colormatrix.com/us/news-archive/july-2003-new-aa-scavengers.html.

33. Jennifer M. Consumer demands push growth in additives for active packaging. *Plast. Additives Compound.* 2006;8(5): 30–33.

34. Rule M, Shi Yu. Method to decrease the acetaldehyde content of melt-processed polyesters (Coca-Cola Company). U.S. Patent 6,762,275 B1, 2004.

35. Rule M, Shi, Yu. Polyester with reduced acetaldehyde content and method using hydrogenation catalyst (Coca-Cola Company). WIPO Patent WO 2004/020519, 2004

36. Mahiat B, Waeller J. Acetaldehyde scavenger in polyester articles (Polyone Corporation). WIPO Patent WO 2006/086365, 2006.

37. Prusak P. Use of tocopherol to scavenge acetaldehyde in polyethylene terephthalate containers (Polyone Corporation). WIPO Patent WO 2005/040261, 2005.

38. Franck V, Jean C, Michel V. Thermal degradation of polyethylene terephthalate: study of polymer stabilization. *Polym. Degrad. Stabil.* 1995;49:393–397

39. Rinehart VR. Solid-state polymerization of polyester prepolymer (Goodyear Tire & Rubber Company). U.S. Patent 4,165,420, 1979.

40. Rinehart VR. Solid state polymerization of polyester prepolymers (Goodyear Tire & Rubber Company). U.S. Patent 4,876,326, 1989.

41. Stouffer JM, Blanchard EN, Leffew KW. Production of poly(ethylene terephthalate) (E.I. du Pont de Nemours & Company). U.S. Patent 5,510,454, 1996.

42. Stouffer JM, Blanchard EN, Leffew KW. Production of poly(ethylene terephthalate) (E.I. du Pont de Nemours & Company). U.S. Patent 5,532,333, 1996.

43. Stouffer JM, Blanchard EN, Leffew KW. Production of poly(ethylene terephthalate) (E.I. du Pont de Nemours & Company). U.S. Patent 5,714,262, 1998.

44. Stouffer JM, Blanchard EN, Leffew KW. Production of poly (ethylene terephthalate) (E.I. du Pont de Nemours & Company). U.S. Patent 5,830,982, 1998.

45. Dudley VE. Underwater pelletizer and heat exchanger die plate (Gala Industries, Inc.). U.S. Patent 4,123,207, 1978.

46. Muerb R-K. Device for granulating a thermoplastic, which is extruded from nozzles (Rieter Automatik). U.S. Patent 7,008,203, 2006.

47. S Y-M, S J-Y. Kinetic and property parameters of poly(ethylene naphthalate) synthesized by solid-state polycondensation. *J. Appl. Polym. Sci.* 2001;81:2055–2061.

48. Xu Q., Xu S. Technology and process of continuous solid-state polycondensation for manufacture of high-viscosity polyester chips. *Hecheng Xianwei Gongye.* 1997;20:53–55.

49. Duh B. Process for the production of high molecular weight polyester (Goodyear Tire & Rubber Company). U.S. Patent 4,374,975, 1983.

50. Fakirov S. *Handbook of Thermoplastic Polyesters*, Vol. 1. Wiley-VCH, Hoboken, NJ, 2002, p. 374.

51. Bamford CH, Wayne RP. Polymerization in the solid phase: a polycondensation reaction. *Polymer.* 1969;10:661–681.

52. Duh B. Effects of crystallinity on solid-state polymerization of poly(ethylene terephthalate). *J. Appl. Polym. Sci.* 2006;102:623–632.

53. Medellin-Rodriguez FJ, Lopez-Guillen R, Waldo-Mendoza MA. Solid-state polymerization and bulk crystallization behavior of poly(ethylene terephthalate) (PET). *J. Appl. Polym. Sci.* 2000;75:78–86.

54. http://www.uop.com/objects/UOP%20Sinco%20SSP%20Process.pdf.

55. James M, James AJ. The SSP process of UOP-Sinco. Presented at the 9th World Congress, Polyester 2004, Zurich, Switzerland, Dec. 7–9, 2004.

56. McGehee, JF, Richmond JR, Tegawa K, Nakano Y. Process for the purification of inert gases (UOP Llc). U.S. Patent 6,749,821, 2004.

57. Ferreira C, Naef UG, Borer C. Method for purifying a gas stream (Buhler AG). U.S. Patent 6,548,031, 2003.

58. Ghisolfi G, Giordano D, Boveri G, Al Ghatta, Hussain AK. Process for the purification of inert gases (Sinco Engineering S.p.A.) U.S. Patent 5,612,011, 1997.

59. Rolfe RB. Process for production and use of deactivated gaseous atomic nitrogen for post combustion gas nitric oxide emissions control (Bleiweis; Sol). U.S. Patent 5,547,651, 1996.

60. McGehee JF, Richmond JR, Tegawa K, Nakano Y. Process for the purification of inert gases (UOP Llc). WIPO Patent WO 2005/113117, 2005.

61. Ghisolfi G, Giordano D, Boveri G. Process for the purification of inert gases (Sinco Eng S.p.A). Canadian Patent CA 2143099, 1995.

62. Ghisolfi G, Giordano D, Boveri G. Process for the purification of inert gases (Sinco Engineering S.p.A.) U.S. Patent 5,547,652, 1996.

63. Takumi S, Hashimoto T, Tatsushima M. Method of manufacture of spherical alumina particles (Nikki-Universal Co., Ltd.). U.S. Patent 4,108,971, 1978.

64. Takumi S, Hashimoto T, Akimoto F. High apparent bulk density gamma alumina carrier and method of manufacture of same (Nikki-Universal Co., Ltd.). U.S. Patent 4,301,033, 1981.

65. Stouffer JM, Blanchard EN, Leffew KW. Process for pellet formation from amorphous polyester (E.I. du Pont de Nemours & Company). U.S. Patent 5,540,868, 1996.

66. Leffew KW, Witt AR. Method for increasing solid-state polymerization rate (E.I. du Pont de Nemours & Company). U.S. Patent 6,451,966, 2002.

67. Stouffer JM, Blanchard EN, Leffew KW. Process for forming crystalline polymer pellets (E.I. du Pont de Nemours & Company). U.S. Patent 5,730,913, 1998.

68. McGehee JF, Boveri GR, Sechrist, PA. Process and apparatus for cooling polymer in a reactor (UOP Llc). U.S. Patent 6,703,479, 2004.

69. UOP document UOP 4122–3b 1105PTE0Bu, UOP website.

70. Robert JS. Solid state polymerization (SSP) of low molecular weight poly(ethylene terephthalate) (PET) copolyesters compared to conventional SSP of PET. *J. Appl. Polym. Sci.* 2002;86:230–238.

71. Nadkarni VM, Wadekar SA, Dubey R, Bopardikar AA. An efficient process for the production of polyester (Reliance Industries Ltd., India). WIPO Patent WO 2008/075373, 2008.

72. Nadkarni VM, Ayodhya SR, Wadekar SA, Dubey R, Jadimath SP. Improved process for the production of low molecular weight polyester resin from the low molecular weight crystalline prepolymer (Reliance Industries Ltd.). Indian Patent Application 987/MUM/ 2005.

73. Nadkarni VM, Dubey R, Wadekar SA, Bopardikar AA. A process plant comprising reactor internals coated with low surface energy materials and a process for the production of polyester resin using the same (Reliance Industries Ltd.). WIPO Patent WO 2007/116416, 2007.

74. Pikus I. Radiant heater system for solid phase crystallization and polymerization of polymers (Hosokawa Bepex Corporation). U.S. Patent 5,634,282, 1997.

75. Kimball GJ, Pikus I. Method for thermally processing polyester pellets (Hosokawa Bepex Corporation). U.S. Patent 5,532,335, 1996.

76. Pikus I. Radiant heater system for solid phase crystallization and polymerization of polymers (Hosokawa Bepex Corporation). U.S. Patent 5,497,562, 1996.

77. Herron DJ. Process for solid phase polymerization of polyester (Bepex Corporation). U.S. Patent 4,161,578, 1979.

78. Pikus I. Cooling system for polymer processing (Hosokawa Bepex Corporation). U.S. Patent 6,767,520, 2004.

79. Koehler W. Device for granulating thermoplastic materials (Rieter Automatik GmbH). U.S. Patent 7,296,985, 2007.

80. Cavaglia G. Continuous process for solid phase polymerization of polyesters (K & E S.R.L.). WIPO Patent WO 04/018541, 2004.

81. http://www.uop.com/objects/UOP%20Sinco%20RPET%20Process.pdf.

82. http://www.environmental-center.com/articles/article898/advancedbuhler.pdf.

83. Christel A. Advanced PET bottle-to-bottle recycling. Presented at the 5th Polyester World Congress, Zurich, Switzerland, 2000.

84. Marcus DS, Stephen W, Michael ED, Harry PH. Process for rapid crystallization of polyesters and copolyesters via in-line drafting and flow-induced crystallization (Eastman Kodak Company). U.S. Patent 6,159,406, 2000.

85. Dudley VE. Underwater pelletizer and heat exchanger die plate (Gala Industries, Inc.). U.S. Patent 4,123,207, 1978.

86. http://www.gala-industries.com/downloads/corp_brochures/pelletizer%20FINAL. pdf.

87. Muerb R-K. Device for granulating a thermoplastic, which is extruded from nozzles (Rieter Automatik GmbH). U.S. Patent 7,008,203, 2006.

88. Shima, T, Urasaki T, Oka I. Improved process for polycondensation of high-molecular-weight poly(ethylene terephthalate) in the presence of acid derivatives. *Adv. Chem. Ser.* 1973;128:183–207.

89. Albert JD, Isaac G, James A, Wilson R. Polyester compositions and shaping process. (Imperial Chemical Industries Ltd.). U.S. Patent 3,553,157, 1971.

90. Hiroo I, Shunichi M. Chain extenders for polyesters: I. Addition-type chain extenders reactive with carboxyl end groups of polyesters. *J. Appl. Polym. Sci.* 1985;30:3325–3337.

91. Hiroo I, Shunichi M. Chain extenders for polyesters: IV. Properties of the polyesters chain-extended by 2,2'-bis(2-oxazoline). *J. Appl. Polym. Sci.* 1987;33:3069–3079.

92. Hiroo I, Shunichi M. Chain extenders for polyesters: V. Reactivities of hydroxyl-addition-type chain extender; 2,2'-bis (4*H*-3,1-benzoxazin-4-one). *J. Appl. Polym. Sci.* 1987;34:2609–2617.

93. Hiroo I, Shunichi M. Chain extenders for polyesters: VI. Properties of the polyesters chain-extended by 2,2'-bis(4*H*-3,1-Benzoxazin-4-one). *J. Appl. Polym. Sci.* 1987;34:2769–2776.

94. Bikiaris DN, Karayannidis, GP. Dynamic thermomechanical and tensile properties of chain-extended poly(ethylene terephthalate). *J. Appl. Polym. Sci.* 1998;70:797–803.

95. Bikiaris DN, Karayannidis GP. Calorimetric study of diepoxide chain-extended poly(ethylene terephthalate). *J. Thermal Anal. Calorimetry.* 1998;54:721–729.

96. Akkapeddi MK, Gervasi J. Chain extension of polyethylene terephthalate with poly-acyllactams (Allied-Signal Inc.). U.S. Patent 4,857,603, 1989.

97. Loontjens T. Modular approach for novel nanostructered polycondensates enabled by the unique selectivity of carbonyl biscaprolactam. *J. Polym. Sci. Polym. Chem.* 2003;41:3198–3205.

98. Al Ghatta HAK, Giovannini A, Cobror S. Polyester resins having improved rheological properties (Sinco Engineering, S.p.A.). U.S. Patent 6,447,711, 2002.

99. Khemani. Extruded polyester foams. In: *Polymeric Foams*. American Chemical Society, Washington, DC, 1997, Chap. 5.

100. Torres N, Robin JJ, Boutevin B. Chemical modification of virgin and recycled poly(ethylene terephthalate) by adding of chain extenders during processing. *J. Appl. Polym. Sci.* 2001;79:1816–1824.

101. Maio DL, Coccorullo I, Montesano S, Incarnato L. Chain extension and foaming of recycled PET in extrusion equipment. *Macromol. Symp.* 2005;228:185–199.

102. Yamamoto K, Katagiri M, Tsuyoshi U, Yoshino M, Harashima K. Preparing bicomponent fibres using recycled PET modified with a chain extender. *Kenkyu Hokoku Tokyo-toritsu Sangyo Gijutsu Kenkyusho.* 2004;7:71–74.

103. Awaja F, Daver F, Kosior E. Recycled poly(ethylene terephthalate) chain extension by a reactive extrusion process. *Polym. Eng. Sci.* 2004;44:1579–1587.

104. Daver F, Awaja F, Kosior E, Gupta R, Cser F. Characterization of reactive extruded recycled poly(ethylene terephthalate). Presented at the 62nd Annual Technical Conference, Society of Plastics Engineers, 2004, *Proceedings*, Vol. 1, pp. 981–985.

105. Karayannidis GP, Psalida EA. Chain extension of recycled poly(ethylene terephthalate) with 2,2'-(1,4-phenylene)bis(2-oxazoline). *J. Appl. Polym. Sci.* 2000;77:2206–2211.

106. Incarnato L., Scarfato P, Di Maio L, Acierno D. Structure and rheology of recycled PET modified by reactive extrusion. *Polymer*. 2000;41:6825–6831.

107. Cardi N, Riccardo PO, Giannotta G, Occhiello E, Garbassi F, Messina G. Chain extension of recycled poly(ethylene terephthalate) with 2,2'-bis(2-oxazoline). *J. Appl. Polym. Sci.* 1993;50:1501–1509.

108. Marshall I, Todd A. Thermal degradation of polyethylene terephthalate. *Trans. Faraday Soc.* 1953;49:67–78.

109. Goodings E. Thermal degradation of PET. *Soc. Chem. Ind. (London) Monogr.* 1961;13:211–215.

110. Reimschussel HK. Polyethylene terephthalate formation: mechanistic and kinetics aspects of direct esterification process. *Ind. Eng. Chem. Prod. Res. Dev.* 1980;19:117–125.

111. Kim TY, Jabarin SA. Solid-state polymerization of poly(ethylene terephthalate): III. Thermal stabilities in terms of the vinyl ester end group and acetaldehyde. *J. Appl. Polym. Sci.* 2003;89:228–237.

112. Schumann H-D, Thiele U. Process for direct production of low acetaldehyde packaging material (Zimmer Aktiengesellschaft). U.S. Patent 5,656,221, 1997.

113. Ekart MP, Jernigan MT, Wells CL, Windes LC. Thermal crystallization of a molten polyester polymer in a fluid (Eastman Chemical Company). WIPO Patent WO 97/31968, 1997.

114. Endert E. A new highly economic polyester technology: 2-R single stream PET process (UDHE-INVENTA). 2005.

115. Stibal W, Kaegi W, Ensinger J, Nothhelfer K. Condensation injection molding process for producing bottle preforms of polyethylene terephthalate and/or its copolyesters and resultant preforms (EMS-Inventa AG). U.S. Patent 5,656,719, 1997.

116. Otto B, Schäfer R, Bachmann H, Hölting L, Deiss S. Method for the direct and continuous production of hollow bodies from a polyester melt. German Patent DE 10356298, 2007.

117. Bhatia KK. Continuous polyester process (E.I. du Pont de Nemours & Company). U.S. Patent 5,434,239, 1995.

118. Bhatia KK. Atmospheric pressure polyester process (E.I. du Pont de Nemours & Company). U.S. Patent 5,552,513, 1996.

119. Bhatia KK. Polyesters production process (E.I. du Pont de Nemours & Company). U.S. Patent 5,599,900, 1997.

120. Bhatia KK. Apparatus and process for a polycondensation reaction (E.I. du Pont de Nemours & Company). U.S. Patent 5,677,415, 1997.

121. Bhatia KK. Polyesters production process (E.I. du Pont de Nemours & Company). U.S. Patent 5,688,898, 1997.

122. Bhatia KK. Polyesters production process (E.I. du Pont de Nemours & Company). U.S. Patent 5,849,849, 1998.

123. Bhatia KK. Apparatus and process for a polycondensation reaction (E.I. du Pont de Nemours & Company). U.S. Patent 5,856,423, 1999.

124. Wonders AG, Lavoie GG, Sumner CE Jr. Optimized liquid phase oxidation of *p*-xylene (Eastman Chemical Company). WIPO Patent WO 2006/028769, 2006.

125. Robert L, De Vreede M. Optimized production of aromatic dicarboxylic acids (Eastman Chemical Company). WIPO Patent WO 2006/028809, 2006.

126. Wonders AG, Lin Robert DVM, Partin LR, Strasser WS. Optimized liquid-phase oxidation (Eastman Chemical Company). WIPO Patent WO 2006/028752, 2006.

127. Jerinigan MT, Ekart MP, Bonner RG. High IV melt phase polyester polymer catalyzed with antimony containing compounds (Eastman Chemical Company). European Patent EP 1574540, 2005.

128. Debruin BR. Polyester process using a pipe reactor (Eastman Chemical Company). U.S. Patent Application 20050054814, 2005.

129. Perry ES. Apparatus for producing and processing viscous materials (Eastman Kodak Company). U.S. Patent 3,545,938, 1970.

130. Ekart MP, Jerinigan MT, Cory L, Windes LC. Thermal crystallization of a molten polyester polymer in a fluid (Eastman Kodak Company). U.S. Patent 2005 0154183, 2005.

131. Jerinigan MT, Ekart MP, Bonner RG. High IV melt phase polyester polymer catalyzed with antimony containing compounds (Eastman Kodak Company). U.S. Patent 20050203267, 2005.

132. Susumu T, Yhoichi W, Akira C. Synthesis of ultra-high molecular weight poly(ethylene terephthalate) by swollen-state polymerization. *Polymer.* 1993;34:4974–4977.

133. Susumu T, Futoshi I. Swollen-state polymerization of poly(ethylene terephthalate): kinetic analysis of reaction rate and polymerization conditions. *Polymer.* 1995;36:353–356.

134. Susumu T, Yhoichi W. Swollen-state polymerization of poly(ethylene terephthalate) in fibre form. *Polymer.* 1995;36:4991–4995.

135. Parashar MK, Gupta RP, Jain A, Agarwal US. Reaction rate enhancement during swollen-state polymerization of poly(ethylene terephthalate). *J. Appl. Polym. Sci.* 1998;67:1589–1595.

136. Ma Y, Agarwal US. Solvent assisted post-polymerization of PET. *Polymer.* 2005;46:5447–5455.

137. Susumu T, Shuji C, Katsuya T. Melt viscosity reduction of poly(ethylene terephthalate) by solvent impregnation. *Polymer.* 1996;37:4421–4424.

138. Rogers V. Filaments or films of polyethylene terephthalate (ICI PLC). European Patent EP 0336556, 1989.

139. Farr IE, Auspos L. A manufacture of highly polymeric linear polymethylene terephthalates in solution and precipitation thereof (E.I. du Pont de Nemours & Company). U.S. Patent 2,597,643, 1952.

140. Wan SH, Sang KO, Youg JH. Preparation of high-molecular-weight poly(ethylene terephthalate). *J. Korean Fiber Soc.* 1990;27:624–631.

141. Ma Y, Agarwal US, Yang XN, Zheng X, Loos J, Hendrix MMRM, Asselen OLJ, Lemstra PJ. Nanoscale fibrillar crystals of PET from dilute quiescent solution. *Polymer.* 2005;46:8266–8274.

142. Masayoshi I, Katsumi T, Tetsuo K. Preparation of high-modulus and high-strength fibres from high molecualar weight poly(ethylene terephthalate). *J. Appl. Polym. Sci.* 1990;40:1257–1263.

143. Ziabocki A. Effects of molecular weight on melt spinning and mechanical properties of high-performance poly(ethylene terephthalate) fibers. *Text. Res. J.* 1996;66:705–712.

144. Besnoin JM, Choi KY. Identification and characterization of reaction byproducts in the polymerization of polyethylene terephthalatea. *J. Macromol. Sci. Rev. Macromol. Chem. Phys.* 1989;26:55–81.

145. Bottenbruch L, Binsack R. *Technische Thermoplaste: Polycarbonate, Polyacetate, Polyester, Celluloseester.* Carl Hanser Verlag, Munich, Germany, 1992.

146. Al-AbdulRazzak S, Jabarin SA. Processing characteristics of poly(ethylene terephthale): hydrolytic and thermal degradation. *Polym. Int.* 2002;51:164–173.

147. Schaul JS. Drying and injection moulding PET for beverage bottle preforms. *Polym. Plast. Technol. Eng.* 1981;16:209–230.

148. Fakirov S. *Handbook of Thermoplastic Polyesters*, Vol. 1. Wiley-VCH, Hoboken, NJ, 2002, p. 377.

149. Kanto B. Recent progress in modification of polymers. Lecture abstracts of the 11th Colloquium on Structure and Physical Properties of High Polymers, Japanese High Polymer Society, June 16, 1981, p. 3.

150. Agarwal US, de Wit G, Lemstra PJ. A new solid state process for chemical modification of PET for crystallization rate enhancement. *Polymer.* 2002;43:5709–5712.

151. Appel O, Barriere-Eigenschafter V. PET und PEN Flashen; Gegrankeflaschen mitverbesserter CO_2 Barriere. *Kunststoffe.* 1996;86:650–658.

152. Appel O. Anlagen und anforderungen in PET-kreislaufsystem (Fachtagung Getränke und Lebensmittelever): Packung aus PET in Deutschland—Wohin und Wann. Weiterbildungs und Technologie Forum, Wurzburg, Germany, 1998.

153. Koch M, Jasztat WR. 1995 Heibabfullbare PET Flaschen; Grundlagen der Maschinen und Verfahrenstechnik. *Kunststoffe.* 1995;85:1323–1329.

154. Darwin PRK, Sebastián M-G. Modification of the thermal properties and crystallization behaviour of poly(ethylene terephthalate) by copolymerization. *Polym. Int.* 2003;52:321–336.

155. Karayannidis GP, Papachristos N, Bikiaris DN, Papageorgiou GZ. Synthesis, crystallization and tensile properties of poly(ethylene terephthalate-*co*-2,6-naphthalate)s with low naphthalate units content. *Polymer.* 2003;44:7801–7808.

156. Christian C, Souad M, Mohamed J, Yvan C, Claude R, Frédéric P. Immiscible blends of PC and PET, current knowledge and new results: rheological properties. *Macromol. Mater. Eng.* 2007;292:693–706.

157. Jingshen W, Pu X, Yiu-Wing M. Effect of blending sequence on the morphology and impact toughness of poly(ethylene terephthalate)/polycarbonate blends. *Polym. Eng. Sci.* 2000;40:786–797.

158. Sriram RT, Jabarin SA. Processing characteristics of PET/PEN blends: 3. Injection molding and free blow studies. *Adv. Polym. Technol.* 2003;22:155–167.

159. Yoshitsugu M. Primary structure and physical properties of poly(ethylene terephthalate)/poly(ethylene naphthalate) resin blends. *Polym. Eng. Sci.* 2003;43:169–179.

160. Syang-Peng R. Properties of poly(ethylene terephthalate)/poly(ethylene naphthalate) blends. *Polym. Eng. Sci.* 1999;39:2475–2481.

161. Terry DP, Jabarin SA. Structure and morphology of PET/PEN blends. *Polymer.* 2001;42:8975–8985.

162. Becker O, Simon GP, Rieckmann T, Forsythe JS, Rosu RF, Völker S. Phase separation, physical properties and melt rheology of a range of variously transesterified amorphous poly(ethylene terephthalate)–poly(ethylene naphthalate) blends. *J. Appl. Polym. Sci.* 2002;83:1556–1567.

163. Yu S, Jabarin SA. Glass-transition and melting behavior of poly(ethylene terephthalate)/poly(ethylene 2,6-naphthalate) blends. *J. Appl. Polym. Sci.* 2001;81:11–22.

164. Elinor LB, Syozo M, Taku K, and Shinzo K. Structural development and mechanical properties of polyethylene naphthalate/polyethylene terephthalate blends during uniaxial drawing. *Polymer.* 2001;42:7299–7305.

165. Ophir A, Kenig S, Shai A, Barka'ai Y, Miltz J. Hot-fillable containers containing PET/PEN copolymers and blends. *Polym. Eng. Sci.* 2004;44:1670–1675.

166. Guo M, Zachmann HG. Intermolecular cross-polarization nuclear magnetic resonance studies of the miscibility of poly(ethylene naphthalene dicarboxylate)/poly(ethylene terephthalate) blends. *Polymer.* 1993;34:2503–2507.

167. Masami O, Tadao K. Phase separation and homogenization in poly(ethylene naphthalene-2,6-dicarboxylate)/poly(ethylene terephthalate) blends. *Polymer.* 1997; 38:1357–1361.

168. Jong KL, Woo SC, Yong KK, Kwang HL. Liquid–liquid phase separation and crystallization behavior of poly(ethylene terephthalate)/poly(ether imide) blend. *Polymer.* 2002;43:2827–2833.

169. Adhemar R-F, Adriana de FB. Correlation between thermal properties and conformational changes in poly(ethylene terephthalate)/poly(ether imide) blends. *Polym. Degrad. Stabil.* 2001;73:467–470.

170. Hsin-Lung C, Jenn CH, Chia-Chen C, Rong-Chung W, Der-Ming F, Ming-Jer T. Phase behaviour of amorphous and semicrystalline blends of poly(butylene terephthalate) and poly(ether imide). *Polymer.* 1997;38:2747–2752.

171. http://www.sidel.com/en/products/equipment/actis-custom-barrier-solutions.

172. http://www.mhi.co.jp/tech/pdf/e421/e421042.pdf.

173. http://www.pmmi.org/ms/peconf/m12.pdf.

174. Hertlein J, Bornarova K, Weisser H. Eignung von Kunststoffflaschen fur die Bierabfüllung. *Brauwelt.* 1997;21/22:860–866.

175. Matlack JD, Villanueva JG, Newman BA, Lillwitz LD, Luetkens ML Jr. Molded bottles and method of producing same (Amoco Corporation). U.S. Patent 5,028,462, 1991.

176. *Packaging World*, Jan. 1998, p. 2.

177. Lauren D. Active barrier extending the shelf life of PET. In: *Innovations Technologiques en Plasturgie: Procédés de Transformation Intégré*, Douai, France, Jan. 20, 2005.

178. http://www.valsparglobal.com/packaging/pdf/valORBarrierResins.pdf.

179. Delane R, M & G Presentation on achieving the goal of monolayer resin solutions for beer in PET. Presentation on PET Strategies, Oct. 2003.

180. Karsten F. Polyshield: the new monolayer barrier from Invista. Presented at the Barrier PET Bottle Conference, Brussels, Belgium, Mar. 8–9, 2005.

181. Kulkarni ST, Bisht H, Subbaraman L. Gas barrier PET composition for monolayer bottle and process for producing the same (Futura Polyesters Ltd). European Patent EP 1650260,

182. Monoxbar PET oxygen barrier effectiveness: a comparison of alternatives. Presented at Pack Expo, Nov. 8, 2004.

183. Tie L, Jaewhan C, Ying L, Jerry Q, Peter M. Presentation at Nanocomposites 2001, Chicago, June 25–27, 2001.

184. Williamson DT, Schleinitz HM, Hayes RA. Polyester clay nanocomposites for barrier applications (E.I. du Pont de Nemours & Company). WIPO Patent WO 2006/069131, 2006.

185. http://www.uwstout.edu/rs/2005/article1.pdf

186. Won JC, Hee-Joon K, Kwan HY, Oh Hyeong K, Chang IH. Preparation and barrier property of poly(ethylene terephthalate)/clay nanocomposite using clay-supported catalyst. *J. Appl. Polym. Sci.* 2006;100:4875–4879.

187. Se Hoon K, Sung CK. Synthesis and properties of poly(ethylene terephthalate)/clay nanocomposites by in situ polymerization. *J. Appl. Polym. Sci.* 2007;103:1262–1271.

188. Multilayer containers featuring nano-nylon MXD6 barrier layers with superior performance and clarity. http://www.nanocor.com/tech_papers/NOVAPACK03.pdf.

189. Pengilly BW. High clarity, low haze polyesters having reduced infrared heat-up times (Goodyear Tire & Rubber Company). U.S. Patent 4,408,004, 1983.

190. Pengilly BW. High clarity, low haze polyesters having reduced infrared heat-up times (Goodyear Tire & Rubber Company). U.S. Patent 4,476,272, 1984.

191. Pengilly BW. High clarity, low haze polyesters having reduced infrared heat-up times (Goodyear Tire & Rubber Company). U.S. Patent 4,535,118, 1985.

192. McFarlane FE, Taylor RB. Thermoplastic polyester molding compositions (Eastman Kodak Company). U.S. Patent 4,420,581, 1983.

193. Tindale N. Polyester bottles (Imperial Chemical Industries Plc). U.S. Patent 5,419,936, 1995.

194. Tindale N. Method for the production of polymer bottles (Imperial Chemical Industries Plc). U.S. Patent 5,529,744, 1996.

195. Rule M. Infrared radiation absorbent anthraquinone derivatives in polyester compositions (Eastman Kodak Company). U.S. Patent 4,481,314, 1994.

196. Tung WC, Schirmer JP, Perkins WG, Pazzo NL. Fast heat up polyesters using graphite as an additive (Shell Oil Company) U.S. Patent 6,034,167, 2000.

197. Wu AC, Mcneely GW, Huang X. Infrared absorbing polyester packaging polymer (Hoechst Celanese Corporation). U.S. Patent 5,925,710, 1999.

ABBREVIATIONS AND SYMBOLS

AA	acetaldehyde
ACA or P_1	aminocaproic acid
6-ACA	6-amino-n-caproic acid
Actis	amorphous carbon treatment on an internal surface
12-ADA	12-amino-n-dodecanoic acid
11-AUA	11-amino-n-undecanoic acid
BHET	bishydroxyethyl terephthalate
BPA-PC	poly(bisphenol A carbonate)
CAPEX	capital expenses
CD	cyclic dimer
CHDM	cyclohexanedimethanol
CL	caprolactam
COOH	carboxyl groups
[COOH]	concentration of carboxyl end groups
CSD	carbonated soft drink
CSTR	continuous-stirred-tank reactor
CVD	chemical vapor deposition
DEG	diethylene glycol
DHI	direct high intrinsic viscosity
DLC	diamond-like carbon

Solid State Polymerization, Edited by Constantine D. Papaspyrides and Stamatina N. Vouyiouka
Copyright © 2009 John Wiley & Sons, Inc.

DMT	dimethyl terephthalate
DSC	differential scanning calorimetry
DP	degree of polymerization
E&E	electrical and electronic
EG	ethylene glycol
EVOH	ethylene vinyl alcohol
HMD	hexamethylenediamine
HPPA	high-performance polyamides
HP-SSP	high-pressure solid state polymerization
HT	hydrotalcite
IPA	isophthalic acid
ISBM	injection stretch blow molding
IV	solution intrinsic viscosity
M	monomer
M&EB	material and energy balances
MHET	monohydroxyethyl terephthalate
MW	molecular weight
MXD6	amide compound
NaSIPA	sodium 5-sulfoisopthalic acid
[NH$_2$]	concentration of amine end groups
NPG	neopentyl glycol
NPU	nitrogen purification unit
OH	hydroxyl groups
OPEX	operation expenses
PA	homopolymer PA 66; parallel arrangement
PAs	polyamides
PA1	copolyamide 66 with 1 wt% NaSIPA
PA2	copolyamide 66 with 2 wt% NaSIPA
PA3	copolyamide 66 with 3 wt% NaSIPA
PA 6	polycaproamide; polyamide 6
PA 11	polyundecanamide
PA 12	polydodecanamide
PA 42	poly(tetramethylene oxamide)
PA 46	poly(tetramethylene adipamide); polyamide 46
PA 66	poly(hexamethylene adipamide); polyamide 66
PA 610	poly(hexamethylene sebacamide)
PA 612	poly(hexamethylene dodecanamide)
PBT	poly(butylene terephthalate)
PC	polycarbonate
PCR	post-consumer recycling
PCT	poly(dimethylene cyclohexane terephthalate)
PE	polyethylene
PEs	polyesters

PEI	poly(ethylene isophthalate)
PEN	poly(ethylene naphthalate)
PEPA	2-(2′-pyridyl)ethylphosphonic acid
P-EPQ	tetrakis(2,4-di-*tert*-butylphenyl)[1,1-biphenyl]-4,4′-diylbisphosphonite
PET	poly(ethylene terephthalate)
PFR	plug flow reactor
PMDA	pyromellitic dianhydride
PMMA	poly(methyl methacrylate)
POM	poyoxymethylene
PP	polypropylene
PS	polystyrene
PTA	purified terephthalic acid
PTN	poly(trimethylene naphthalate)
PTT	poly(trimethylene terephthalate)
PVA	poly(vinyl acetate)
PVD	physical vapor deposition
PVOH	poly(vinyl alcohol)
$R^{\bullet}$	radicals
RHB	reheat blow molding
RV	solution relative viscosity
SA	staggered arrangement
SDM	standard deviation of the mean
S-HIP	solid high-IV polycondensation
SHP	sodium hypophospite
SMT	solid–melt transition
SSP	solid state polymerization
SwSP	swollen-state polymerization
TA	terephthalic acid
TGA	thermogravimetric analysis
THF	tetrahydrofuran
TMA	trimellitic anhydride
UHMW-PET	ultrahigh-molecular-weight PET

English Letters

a	specific area for diffusion constant
a_c	cross-sectional area of a solid state polymerization reactor
b	constant
c_{bulk}	concentration of reactive end groups in bulk
c_{byp}	concentration of a generic low-molecular-weight by-product
$c_{\text{byp,gas-phase}}$	concentration of a by-product in the gas phase
$c_{\text{byp,s}}$	concentration of a by-product at the gas–solid interface

$c_{\text{cat}(i)}$	concentration of an i-metal catalyst
c_{DE}	concentration of diester groups
c_{eff}	effective concentration; concentration of reactive end groups in the amorphous domains
c_{E}	concentration of ester groups
c_{Gly}	concentration of glycol
c_{OH}	concentration of hydroxyl end groups
$c_{\text{OH},i}$	concentration of inactive hydroxyl groups
c_{COOH}	concentration of carboxyl end groups
c_P	constant-pressure heat capacity
c_{TA}	concentration of terephthalic acid
c_{W}	concentration of water
C	concentration; condensate molecule; concentration of end groups in an amorphous region
C_i	concentration of inactive end groups; concentration of species i
C_r	concentration of end groups in an amorphous region after IV levels off
C_s	sulfonate group concentration
C_w	water concentration in the polymer phase
d	average distance between neighboring end groups
$\bar{d}$	mean diameter of particles
D_0	initial carboxyl or amine end group excess
D_{byp}	diffusion coefficient of a by-product
D_{EG}	diffusion coefficient of ethylene glycol
D_{Gly}	diffusion coefficient of glycol
D_i	axial or molecular diffusion coefficient of species i
D_{TA}	diffusion coefficient of terephthalic acid
D_{tube}	diffusion coefficient that describes Brownian motion in a tube
D_{W}	diffusion coefficient of water
D_{wp}	diffusivity of water in the polymer phase
E	activation energy
E_a	activation energy of polymerization
$E_{a,1}$	activation energy of a transesterification reaction
$E_{a,2}$	activation energy of an esterification reaction
$E_{a,\text{EG}}$	activation energy of ethylene glycol diffusion
$E_{a,\text{W}}$	activation energy of water diffusion
G	growth rate of crystals
G	gas phase (subscript)
h_0	heat transfer coefficient between the polymer and gas phases
h_i^v	heat of vaporization of species i
IV_u	ultimate IV after a long SSP time
$j_{H/D}$	Colburn factor

$1/J$	deactivation factor
k	Boltzmann constant; apparent or intrinsic rate constant of polymerization
$k_{o,i}$	constant rate of an uncatalyzed i-reaction
k_2	reaction rate constant for second-order kinetics
k_3	reaction rate constant for third-order kinetics
k_{byp}	gas-phase mass-transfer coefficient for the ith by-product
k_{cop}	reaction rate constant for copolyamide SSP
k_{cat}	reaction rate constant for catalyzed SSP
$k_{cat(i)}$	constant rate of an i-metal-catalyzed reaction
k_f	rate constant for forward reaction (polymerization)
$k_{G,i}$	gas-side mass-transfer coefficient
$k_{H^+,i}$	constant rate of a proton-catalyzed i-reaction
k_{hom}	reaction rate constant for homopolymer SSP
k_i	reaction rate constant for reaction i
k_{-i}	reverse constant rate of i-reaction
k_{plot}	rate constant derived from the slopes of a rate expression
$k_{P,i}$	polymer-side mass-transfer coefficient
k_r	reaction kinetic constant; rate constant for a reverse reaction (depolymerization)
K_{eq}	equilibrium constant for polymerization
$K_{G,i}$	overall mass-transfer coefficient in the gas phase
K_i	equilibrium constant for reaction i
l_c	thickness of lamellae
l_r	length of repeat unit of PET
L_0	average of chain-end length
L_t	length of tie chain; chain-end length
$\dot{m}$	mass flow rate of a fluid
M_0	molecular weight of a repeating unit of PET
$\overline{M_c}$	critical entanglement molecular weight
$\overline{M_n}$	number-average molecular weight
$\overline{M_{n_0}}$	initial number-average molecular weight
MW_n	number-average molecular weight
$\overline{M_v}$	viscosity-average molecular weight
n	total number of end groups at the initial stage of SSP; number of carbon in a repeating unit; power of time
n'	total number of end groups after IV levels off
n_1	number of end groups for $R = R_0$ at the initial stage of SSP
n'_1	number of end groups for $R = R_0$ after IV levels off
n_2	number of end groups for $R = L/2$ at the initial stage of SSP
n'_2	number of end groups for $R = L/2$ after IV levels off
n_a	number of end groups available for a reaction
n_g	number of end groups in 1.0 g of PET
$\dot{n}_G$	molar flow rate of gas

n_1	number of lamellae in a polymer molecule
n_t	number of tie chains on a polymer molecule
N	length of a polymer molecule
$N_{P\text{-}G,i}$	diffusion flux of species i between the polymer and gas phases
N_{Re}	Reynolds number
p	ratio of overlapping volume to sphere volume; polymerization conversion
$p_{G,i}$	bulk partial pressure of species i in the vapor phase
p_i^*	partial pressure species i in equilibrium with bulk liquid
$p_{N_2 G}$	partial pressure of nitrogen in the gas phase
$p_{N_2 m}$	log mean of the partial pressures of nitrogen at the sphere surface and in the bulk gas phase
p_t	polymerization conversion
P	system pressure
P	polymer phase (subscript)
P_i	polymer molecule of degree of polymerization i
P_i^{sat}	vapor pressure of species i
Q_{N_2}	nitrogen supply
r	rate of reaction
r_A	chemical reaction rate
r_i	reaction rate of species i
R	pellet radius (modeled as a sphere); radius of the sphere of end-group movement
$\overline{R}$	average radius at the initial stage of SSP
R_0	radius of the sphere of end-group movement when $L > 2R_0$
R_j	reaction rate of reaction j
$\overline{R}_r$	average radius after a long SSP time
s	displacement of polymer chain in its tube
S	length of side of cube occupied by an end group
S/V	surface-to-volume ratio
t	time; reaction time
T	temperature; reaction temperature
T_g	glass-transition temperature
T_m	melting point
$T_p(P)$	polymerization temperature under high pressure
u	velocity
v	overlapping volume between two spheres; molar volume
v_{N_2}	gas flow rate
V	volume that an end group occupies in an amorphous region
W	water
x	ratio of n_1 in n (n_1/n); distance from the center of a polymer particle
x'	ratio of n_1' in n' (n_1'/n')
x_c	mole fraction of crystalline domains

x_i	mole fraction in the polymer phase
$\overline{X_n}$	number-average degree of polymerization
y_i	mole fraction in the gas phase
z	axial position in a solid state polymerization reactor

Greek Letters

β	constant
$\gamma_{P,i}$	activity coefficient of species i in the polymer phase
δ	tube diameter
ε	void fraction of a bed
κ	thermal conductivity
λ	constant
μ	viscosity
μ_0	constant
μ_{tube}	mobility
ρ	mass density; density of PET
ϕ_c	degree of crystallinity
ϕ_v	volume fraction crystallinity
ϕ_w	weight fraction crystallinity
ΔH_j	heat of reaction for reaction j
$\Delta(IV)$	difference between the intrinsic viscosity before and after a solid-phase polymerization reaction
ΔS_j	entropy of reaction for reaction j
$(\Delta W)_r$	reduced weight loss

INDEX

Solid State Polymerization, Edited by Constantine D. Papaspyrides and Stamatina N. Vouyiouka
Copyright © 2009 John Wiley & Sons, Inc.

289